ENVIRONMENT & PLANT RESPONSE

McGRAW-HILL PUBLICATIONS IN THE AGRICULTURAL SCIENCES

Lawrence H. Smith
Department of Agronomy
University of Minnesota
Consulting Editor in the Plant Sciences

E. P. Young
Department of Animal Science
University of Maryland
Consulting Editor in Animal Science

Adriance and Brison   Propagation of Horticultural Plants
Ahlgren   Forage Crops
Anderson   Diseases of Fruit Crops
Brown and Ware   Cotton
Campbell and Lasley   The Science of Animals that Serve Mankind
Carroll, Krider, and Andrews   Swine Production
Christopher   Introductory Horticulture
Crafts and Robbins   Weed Control
Cruess   Commercial Fruit and Vegetable Products
Eckles, Combs, and Macy   Milk and Milk Products
Elliott   Plant Breeding and Cytogenetics
Fernald and Shepard   Applied Entomology
Gardner, Bradford, and Hooker   The Fundamentals of Fruit Production
Hayes, Immer, and Smith   Methods of Plant Breeding
Herrington   Milk and Milk Processing
Jull   Poultry Husbandry
Kipps   The Production of Field Crops
Kohnke   Soil Physics
Kohnke and Bertrand   Soil Conservation
Laurie and Ries   Floriculture
Leach   Insect Transmission of Plant Diseases
Maynard and Loosli   Animal Nutrition
Metcalf, Flint, and Metcalf   Destructive and Useful Insects
Rather and Harrison   Field Crops

*Professor R. A. Brink was Consulting Editor of this series from 1948 until
January 1, 1961.*

ENVIRONMENT & PLANT RESPONSE

MICHAEL TRESHOW

Professor of Biology
Department of Biology
University of Utah

McGRAW-HILL BOOK COMPANY

NEW YORK   ST. LOUIS   SAN FRANCISCO   DUSSELDORF
LONDON   MEXICO   PANAMA   SYDNEY   TORONTO

581.5
Tre

ENVIRONMENT AND PLANT RESPONSE

Library of Congress Catalog Card Number 71-90759

07-065134-5

5 6 7 8 9 0   KPKP   7 9 8 7

PREFACE

The passing season gives way to the new, bringing with it different temperature conditions, a new moisture regime, and a continuing change of the total environment. The growing plant must adjust to each new set of conditions if it is to survive and perpetuate its kind. Not only must the plant adapt to fresh seasonal environmental changes, but growing conditions may, and do, change drastically from week to week and even from day to day. Each new day can present the plant with a set of conditions too harsh for its normal development or too extreme for survival. Even where the impact is not lasting, growth may be temporarily set back or reproduction impaired.

The environment, specifically factors of the physical environment including temperature, light, moisture, soil, and air relations, is responsible for well over half of all the disorders of plants. These noninfectious factors may cause stresses producing anything from mild growth disturbances having negligible effects to severe ailments resulting in critical crop losses or even death.

Disorders caused by stresses of the physical environment can be most elusive. They are little understood and often completely unrecognized. Diagnosing, or determining their cause, requires a knowledge of the cultural, nutritional, and other environmental conditions that favor each species; for what may be best for one type of plant may be harmful or even fatal to another. Rarely in nature do plants grow under fully optimum conditions. It follows that continued improvement of crop production, as well as maximum utilization of the best-adapted ornamental and agronomic plant species, may be obtained only by determining and utilizing, as nearly as possible, the ideal conditions regulating the vigor and productivity of each given species.

Current research is beginning to clarify the role of temperature, water, and light in normal plant development. Much is known about the mechanisms by which metabolic processes operate under normal growing conditions; less is known about the response of plants when air and soil temperatures, moisture, light intensity and duration, and the structure, reaction, and composition of the soil are unsuited to plant development. Much information has been brought to light in recent years regarding the harmful aspects of the pesticides now being used in tremendous quantities to control insect, fungus, and weed pests over vast areas. Much also is known regarding the far greater threat of phytotoxic contaminants in the atmosphere.

Disorders associated with environmental stresses have been described over the years, and their impact on plant processes is revealed in references scattered throughout scientific literature. Despite the importance of recognizing disorders of all kinds, there is no single source to which one can turn for guidance in recognizing the symptoms, diagnosing the ills, or understanding the mechanisms of the disease process. Nor is any comprehensive discourse on the abnormal physiology of nonbiotic disorders available.

This volume attempts to bring together, into a single text and reference, the literature concerning nonbiotic disorders in such a manner that, by carefully observing and studying the symptoms, the reader should be able to diagnose the cause of the various disorders and thereby take the appropriate and necessary steps toward treatment and control. A wide range of symptoms is described together with explanations of steps to be taken in their diagnosis. Many fundamental biological processes, together with the basic physiological and metabolic responses of plants to their physical environment, are explained to help the reader understand the disease process.

The bibliography and selected reference lists at the end of the chapters are intended not only to document the data given but to provide the reader with additional references should he wish to pursue the subject in greater depth.

This book is divided into four main sections. The first is of a general nature and provides background information for better understanding the nature of the environment as a pathogen and for recognizing the symptoms caused by abnormal physiological processes. The second describes normal temperature and light relations and factors associated with climatic stresses. The third treats water relations and discusses factors associated with the soil, and the fourth treats the plant disorders associated with air pollutants and pesticides. Together, they comprise the study of environmental pathology.

The author has written this text to help the student of environmental pathology and biology—especially the physiologist, physiological ecologist, field ecologist, horticulturist, air pollution specialist, and pathologist—to understand the plant in relation to the varied aspects of the environment and to recognize the earliest symptoms of stress caused by an abnormal physiology. It is hoped that the advanced home garden enthusiast will also find this book of interest and value to his horticultural pursuits.

I wish to express my sincere thanks to Donald Packard for his critical review of every chapter; without his efforts this book would not have been possible. I am grateful to Dr. A. Clyde Hill, Dr. Lionel Klickoff, and Dr. I. B. McNulty for help with specific chapters. My thanks also go to Gerald Dean for preparing the photographic illustrations, and to Frances Harner, Suesan Taylor, and Penny Knight for assistance in preparing the manuscript.

MICHAEL TRESHOW

CONTENTS

ENVIRONMENT & PLANT RESPONSE

PART ONE

ENVIRONMENT AND THE PLANT RESPONSE

CHAPTER ONE

THE ENVIRONMENT

No organism is independent of its environment; every living thing constantly influences, and is influenced by, its inorganic and organic surroundings. All components of the universe influence life, and all are part of the environment. Environment includes all the factors and forces prevailing internally and externally on, around, and in the plant. It is light, moisture, temperature, wind, soil, organisms, pollutants, pesticides, and man.

Environment occupies three-dimensional space and extends through time; events of this moment may influence the environment for many years hence. Soil eroded away by today's floods may take millennia to replace. The vegetation destroyed by today's air pollutants may never return. Poisons sprayed in the air and on the land to control insect and disease pests may persist for decades. These are but a few of the environmental hazards which man must recognize and resolve.

The environment is such a complex of factors and interactions of factors that it is impossible to isolate any single component which does not influence another. Yet, to understand environmental structure and operation, this complex must be subdivided or categorized into clearly defined units.

The total environment extends from the microcosm within every cell to the cosmos of the atmosphere and universe. Traditionally, the environment has been classified into biotic components, including all the living things on earth, and abiotic physical components.

Biotic components include the myriads of organisms which inhabit the physical environment, but must also include the internal environment of the organisms themselves. The biotic environment at the most fundamental level or organization begins with the chemical and genetic structure of the individual cell. Genes embedded in the nuclear material of every living cell and comprising a small part of the DNA (deoxyribonucleic acid) molecules of the chromosome, direct the plant's activity by providing the genetic "blueprint" or template for the plant's chemical reactions. The DNA molecules determine the order in which the nucleotides of RNA (ribonucleic acid) are arranged. This RNA pattern is transmitted to the messenger RNA, which ventures from the stable environment of the nucleus carrying the coded pattern of instructions to the ribosomes located along the endoplasmic reticulum in the cytoplasm. There the RNA directs the manner in which amino acids will be aligned and oriented to form the specific enzymes and other protein molecules which will govern the subsequent metabolic activities of the cell. Basically, then, these activities will be determined by the gene. Germination of the seed, the growth and development of the plant, and its ultimate capacity to produce flowers, fruit, and seed are regulated by genes at the subcellular level. Any modification of this internal environment will alter the ultimate function of the plant.

While the internal environment is relatively stable, chromosome and gene mutations (i.e. irreversible alterations of gene composition) occur which redirect the gene activity and can drastically modify plant structure and physiology. Over the short period of human experience, these changes appear slow and subtle. Normal heredity assures that each individual will develop those basic characters handed down by its parents. But in response to a fluctuating external environment, minute chemical changes within the cell cause genotypic alterations which eventually can be distinguished phenotypically in the organism. These continuing changes assure that every individual will differ from every other individual in some slight measure. In this way, some members of a population will be best adapted to a new and ever-changing environment.

Sudden gene and chromosome changes also occur causing the progeny to differ sharply from its parents. The change is often lethal; but, when less extreme, such mutants, or "sports," may provide a new and different individual from which a new horticultural variety may be propagated.

1-1 THE PHYSICAL ENVIRONMENT

The internal cellular environment is constantly influenced and modified by its external surroundings. External physical forces of temperature, precipitation, humidity, wind, light, and soil regulate the rate of cellular reactions and determine if they will proceed normally and optimally. These same factors comprise the physical components of the environment.

Temperature

Temperature is one of the most critical factors of the environment and exerts a profound influence on all physiological activities by controlling the rate of chemical reactions. Every physiological function has temperature limits above and below which it ceases and an optimum temperature at which reactions proceed at a maximum rate. As the temperature deviates from this optimum, the rate of reaction decreases, stopping completely beyond a critical limit. Little or no plant growth is made during the heat of a hot summer day. Rather, growth is largely restricted to the cooler morning and evening hours when conditions are more favorable.

Plant species differ in their tolerance to temperature extremes. Each has its definite requirements for growth and reproduction. There are minimum, maximum, and optimum temperatures for seed germination and every subsequent plant process. If the temperature in any given geographical area is unsuitable for even a single phase of plant development, the species may perish; or if it lives, it may not reproduce or be able to compete successfully with better-adapted biota.

Temperatures play a significant role in determining the distribution of plant species. Green land plants apparently first evolved in the relatively stable, warm climates of tropical regions. Here solar radiation is high throughout the year, and frosts occur only at higher elevations. Plants prospered in this environment and, in the course of evolution, acquired structural and physiological characteristics making it possible for some to survive more demanding environments. Freed of biotic competition, these better-adapted plants gradually extended their range into the colder and drier areas of the world. Extension of the plant's range is a continuing process. It is exemplified by the northward spread of arctic species onto the tundra and the growth of lichens and mosses on freshly exposed rocks laid bare by receding glaciers high in the mountains.

The constant cloud cover characterizing tropical forest areas is lacking in the neighboring deserts, increasing the nightly reradiation of heat from the ground to space and causing greater diurnal temperature variations. Plants extending into this more demanding environment must be adapted to withstand the cooler night temperatures. Still greater modifications are needed for plants to extend toward polar regions, where net heat absorption is minimal. Here, high solar radiation is limited to brief periods during the summer, so that the total heat absorption is more than counterbalanced by the heat loss during the long dark of winter, the light-reflecting snow, and the low angles at which the sun's rays are received. Similar stresses exist at high elevations, where plants absorb little radiation and the plant heat is readily reradiated into the clear night sky. Temperatures at high elevations also remain low for prolonged periods, limiting the potential growth period and imposing an increased stress on vegetation.

The influence of temperature on microclimate and microenvironment is every bit as pronounced as its impact on the general environment. Its influence is reflected in the different kinds of vegetation growing within a small area. Temperature and vegetation types differ considerably from exposure to exposure and place to place over rolling hills and mountain canyons. Plant life on the sunny side of a slope or even a large rock often differs markedly from that on the cooler, shady side owing to the different microenvironment. The cooler soil temperatures prevailing on the north side of a house are often suitable for plants which would become sick or die on the south or west sides. Conversely, other plant species might succeed only under the higher temperature regimes of a southern exposure.

South-facing slopes in the northern hemisphere are generally warmer in the winter than the north-facing slopes, which receive less solar radiation. Temperatures and vegetation are distinctly different. Snows melt first on the warmer slopes, plant growth begins earlier in the spring, and many kinds of plants thrive which would find the colder growing conditions across the canyon unfavorable. On southerly exposures, other species grow normally far north of their usual range. The temperature gradient in canyons is further complicated as cold air flows down the slopes, collecting in depressions and valleys. The different temperature regime here supports an entirely different flora and fauna. Knowledge of air drainage patterns enables the farmer to select the warmer sites for his more tender crops and the camper to pick a comfortable, warm place to sleep.

Color and texture of a soil determines its capacity to absorb heat energy, thus further influencing plant distribution. Dark soils, being warmest, may support entirely different plant populations from lighter soils and may make the difference between success or failure of a commercial planting.

Water

Precipitation, whether in the form of rain, snow, or atmospheric humidity, is closely interrelated with (and even inseparable from) temperature in its influence on plant development. The interaction of water and temperature is clearly illustrated by contrasting the vegetation in desert and tundra areas. Both locations typically receive close to 10 in. of precipitation annually; yet bog and marsh plants such as mosses and sedges cover the wet soil of one, while scattered, thorny shrubs, cacti, and other desert plants are distributed over the dry, barren soil of the other. The difference lies largely in the greater evaporation rate of the hotter desert areas and the frozen condition of soils and subsoils of the tundra during much of the year. The high temperatures of the desert increase both evaporation and transpiration to an extent where water loss is significant. Cold temperatures in the tundra, by significantly reducing the rate of evaporation, permit water to accumulate more effectively.

Water is required for all life processes and often limits plant development. One need go no further than the front lawn to see this. When the grass does not receive sufficient water, its growth slows and ceases long before it starts to look sick and turn brown. Water is required to maintain cell turgidity and to provide a substrate and medium for chemical reactions and for the transport of mineral ions in the plant; also, when transpired from the leaves, water is of some value in cooling and maintaining a plant temperature suitable for metabolic reactions.

Atmospheric precipitation is the main source of water for non-cultivated, terrestrial plants. The total amount of precipitation, its form and availability, and its annual distribution determine the type and distribution of plants found in different locales. The natural vegetation of an area having distinct wet and dry seasons differs greatly from that in areas with the same amount of moisture distributed evenly throughout the year.

Not all the precipitation becomes available to plants as soil moisture. Much of the rain falling on the leaves and branches evaporates without ever reaching the soil. Once on the ground, much of the water is lost by evaporation from the soil. More is lost through runoff, leaching, and transpiration. The effectiveness of the rainfall depends on all these conditions.

Plants respond to the effective moisture available and possess distinct and characteristic structural modifications which characterize the flora of each environment. Adaptations found in desert plants include deep or extensive root systems; marked development of hard, fibrous tissues; and small leaves, which are often gray-green in color and hairy in texture, with a thick cuticle. When plants lacking such modifications are grown under arid conditions, moisture stress may result. The success of growing ill-adapted species depends largely on understanding the degree to which they can tolerate water shortage and their capacity to obtain and conserve water.

Atmospheric moisture is also important to plant growth. Certain desert plants growing in areas where no measurable rainfall has been recorded in decades obtain much of their water from the atmosphere. Water vapor moving into the leaves during moist, foggy mornings is retained during the hot periods of the day. A high humidity helps reduce water loss from transpiration by reducing the vapor pressure deficit and enables many plants to grow in areas of otherwise deficient soil moisture.

Conversely, dry air and wind add greatly to the rate of water loss from both plant and soil. Despite the occasional cooling effect of wind, the net effect is one of desiccation. Evaporation and transpiration may increase to the point where appreciable amounts of water are lost to the atmosphere.

Wind can be physically devastating, and it is the principal agent limiting tree growth in the high mountains. The stunted, twisted trees found at timberline are produced by the action of wind-carried snow and ice particles blasting and killing the exposed buds and limbs. Plants covered by snow or protected by rocks escape this icy blast and survive. Limbs on the semiprotected leeward side of the tree partly escape damage, giving the tree a characteristic one-sided appearance. Wind is also the primary agent in coupling the cold air temperature to the plant, thereby making the plant cold and slowing growth.

Too much water is as harmful as too little. The sick, yellowed, and dying trees sometimes found attempting to survive near a leaky sprinkling system attest the ill effects of water excesses. The solid portion of most soil is fairly constant, but the pore space between these particles varies greatly depending on the amount of water partially or wholly filling the pores. When the air spaces are saturated with water, the plant roots cannot obtain sufficient oxygen for respiration. If it persists, this oxygen deficiency limits respiration, causing a shortage of usable energy. The plant starves, and the leaves turn yellow, wilt, and drop prematurely in much the same way as if the soil were too dry.

Changes in the soil level also modify water content. In one case an underground spring was tapped inadvertently in digging footings for an apartment building. The water table was raised to within 2 ft of the soil line. Birch, maple, and cottonwood trees with roots in the waterlogged area soon declined and ultimately died over a three-year period.

Certain water-loving species are adapted to these wet conditions. Their characteristic thin cuticle and large intercellular spaces permit oxygen to diffuse readily through the leaf and stem so that the root system is able to receive adequate oxygen for respiration even though the soil might be saturated.

Light

Light, another component of the physical environment, also plays a vital role in determining the plant's characteristics, distribution, and survival. Light energy, radiation with wavelengths between 400 and 760 mμ (millimicrons), is required directly to sustain the growth of all green plants. Light intensity, quality, and duration all influence plant development. Too much or too little light is fatal. Extremely high light intensities literally burn up sensitive plants. Low light intensities can be equally harmful to intolerant species unable to obtain the energy needed for survival. The insufficient light received by the lower limbs of forest trees or closely planted orchard trees is directly responsible for their death. Lower limbs inside the orchard die out, and much of the crop must be picked from tall ladders, while the fruit hangs in abundance on limbs at the orchard periphery.

There is a minimum survival light intensity for each species below which more carbohydrates are burned in respiration than produced in photosynthesis. This light requirement varies tremendously among species. Plants evolving in the dense shade of tropical jungles and the floor of coniferous forests endure far less light than those found in the open desert or alpine environments. Ultraviolet light at 14,000 ft may be twice that occurring at sea level and fifty times that falling on the forest floor. Inadequate light limits the available energy and prevents normal development of unadapted species. Should poor light conditions persist, various visible disease expressions of leaf yellowing, drying, and killing will appear.

Marked variation in the duration of sunlight occurs with latitude and from one season of the year to another. Differences in the relative duration of daylight and darkness, that is the *photoperiod*, exert a basic effect on flowering, fruit growth, stem elongation, and virtually every other plant process. Photoperiod plays an important part in controlling reproduction and hence the distribution of all species having critical daylength requirements.

Soil

Soil, a complex physical and biological system providing support, water, nutrients, and oxygen for the plant, is a major component of the physical environment second only to climate in influencing the development and distribution of plants.

The initial characteristics of the soil depend on the parent material from which it was derived, and its final structure and type depend largely on the climate. In areas of high rainfall, much of the plant nutrients are carried below the root zone by the rains, and the soils are relatively sterile. In drier regions, the nutrients are carried toward the soil surface with the rapidly evaporating water. Nutrients may then accumulate in toxic concentrations as mineral salts to form the saline soils of the arid western deserts. The texture of the soil—the relative proportion of various-sized particles—is determined largely by the degree to which the underlying roots have decomposed or degraded it. The distribution of these particles, and the action of plant roots and microorganisms in the dynamic system, are equally important in determining the soil character.

Through centuries of evolution, plants have become adapted to virtually every kind of soil. But each species has its limitations; to place a plant species in a soil to which it is not adapted is imposing a stress situation which may be deleterious or even fatal. A few species evolving in the Great Basin of the Western United States thrive at salt concentrations as high as 3 percent, but this is more than any agronomic species can tolerate. Few cultivated species survive in soils having 0.25 to 0.5 percent salt, and the limitations must be known before crops can be raised successfully on such soils. Other limitations imposed by the soil must also be known before attempting to utilize a soil for agriculture. Soil alkalinity, acidity, drainage, texture, and nutrition can all limit plant development, production, and survival.

The soil environment determines the success of a species and its natural distribution. Local variations in soil conditions may be great. Soil differences sufficient to determine the plant species able to thrive and survive may occur within a single field, yard, or hillside. In one area studied, which appeared to be entirely uniform in slope, exposure, moisture, and temperature relations, the boundary of Douglas fir trees ceased abruptly and gave way to lodgepole pine. Studies of the area's geology revealed an uplifting fault which had exposed an entirely different soil parent material. The fault corresponded exactly to the fir-pine boundary.

Soil provides the reservoir for some fourteen mineral elements essential to normal plant development. Deficiency of any one can inhibit or prevent specific enzymatic reactions and significantly limit plant growth. Nutrient elements are continually used by the plants, and the supply ultimately becomes depleted where the crop is removed year after year and must be replenished. But nature, by the breakdown of plant residues and other organic material, constantly renews the supply.

Excesses of essential elements can also cause plant stresses. The overzealous home gardener applying a surplus of nitrogen to his roses throws them into a vegetative state in which stem growth is excessive but no flowers are produced. High nitrogen concentrations also cause leaf burning and soft fruit of poor quality, and they predispose the plants to infection by bacteria and fungi.

Excessive concentrations of nonessential elements such as aluminum and lithium in the soil environment are also harmful. Excess aluminum in highly acid soils causes leaf yellowing, limited, distorted root growth, and reduced top growth. Lithium toxicity also suppresses growth and further causes leaf yellowing similar to that resulting from certain deficiency diseases. More complex chemicals can be more damaging. The modern weed-, insect-, and fungus-killing pesticides, so indispensable to agriculture, are capable of causing varied undesirable effects if used improperly. Herbicides applied along fence lines and ditches frequently enter the root zone of cultivated plants. Roots absorb the chemicals and transport them to the leaves, causing chlorosis and local killing. Death of the entire plant may follow.

Before the advent of DDT, arsenic was widely used as a foliar spray to control such insect pests as codling moth on apple. Arsenic accumulating in the soil has, even after twenty years, continued to cause necrotic leaf spotting, growth suppression, and killing of sensitive trees planted on the contaminated soil.

Atmosphere

The atmosphere has long been recognized as a vital component of the physical environment. Animal and nongreen plants are dependent on the oxygen provided by photosynthetic reactions occurring only in green plants. The green plants, in turn, are dependent on animals and nongreen plants to continually liberate and renew the supply of carbon dioxide (CO_2) in the environment. Both oxygen and carbon dioxide concentrations influence metabolic activity, and CO_2 is often a limiting factor to plant growth where growing conditions are otherwise optimal.

Oxygen and CO_2 are ordinarily the most important atmospheric constituents, but in many geographic areas smoke, soot, dust, and debris supplement the normal components and comprise a significant hazard. Gaseous and particulate wastes have been contaminating the atmosphere since man built his first fire. As the human population increased and more fuels were burned in the home, in the automobile, and in industry, increased amounts of incompletely burned waste materials have escaped into the atmosphere. Pollutants are no longer confined to the large industrial cities. Contaminants transported in large air masses disperse widely from their points of origin, and plants and animals (including man) are more and more subjected to this polluted environment. Eye irritation and respiratory ills are some of the first harmful effects of these pollutants but not necessarily the most important. Pollutants can weaken or kill leaves of the more sensitive plant species, impair plant growth and limit production. The prematurely early fall senescence of street and garden trees in metropolitan areas often can be attributed to air pollutants. In natural plant communities, vegetation over thousands of once forested acres has been killed by sulfur dioxide and fluorides.

Plants in the *ecosystem,* that is, the total plant and animal communities in their physical habitat, do not all respond to pollutants in the same way. A few may be killed outright, while functions of others are only slightly impeded. Growth and reproduction in some plants may be impaired more than in others, so that the population of sensitive species is reduced while tolerant species thrive and reproduce unimpeded. The overall population of the plant, and subsequently animal, community is altered. In this way a slight modification by one component of the physical environment—an air pollutant—may modify the entire biotic environment.

1-2 THE BIOTIC ENVIRONMENT

The biotic environment includes all living things and the interrelated actions and reactions which individual organisms directly or indirectly impose on each other. It is as difficult to isolate the components and effects of individual biotic factors of the environment as it is to isolate those of the physical factors, since the plants and animals in any community form such a closely knit, interdependent unit.

The green plants, manufacturing food from carbon dioxide and water, comprise the fundamental unit in the ecosystem. All nongreen organisms directly or indirectly depend on green plants for their very existence. The green plants themselves play a substantial part in modifying the environment by shading the underlying area, using water and minerals in the soil, and adding litter as organic matter to the soil. Soil shaded by plants is many degrees cooler than a bare soil surface, which may be many degrees hotter than the air above it. The thick vegetation cover found in forests has a blanketing effect on incoming and outgoing radiation, reducing temperature extremes and providing cooler days and warmer nights than in adjacent clearings. The climate on the forest floor is decidedly cooler, shadier, more humid, and less windy than in the tops of trees, and a far different environment is provided for the organisms which abound there.

Large, dominant green plants modify the physical environment to the point where they profoundly influence growth of other species. The canopy provided by trees also reduces the light falling on the forest floor, preventing the establishment of some species and favoring the growth of shade-loving species which may not survive elsewhere. Everywhere in the forest the climate is modified, so that microclimates and microenvironments are created in which different species may thrive. Out of these microenvironments, they may perish.

In any specific microenvironment, one plant species is always best adapted and able to compete most successfully with other species attempting to become established. Competition, the struggle between organisms for the necessary environmental needs of light, moisture, nutrients, etc., is a constant struggle. Slight changes in either the physical or biotic environments can favor one species over another and alter the plant population. Almost anything needed from the environment may be the object of competition. From seed to senescence, plants are competing for water, light, mineral nutrients, or whatever is in the shortest supply at a given time and place. In the jungle, light is a determining factor; in the grasslands, nutrients are most limiting; in the desert, water is most limiting. Species possessing the most successful means of obtaining light, nutrients, or water will compete most successfully.

Ability to tolerate the parasitic microorganisms in the biotic environment also influences the success or failure of a species. The fungi and bacteria which comprise the parasitic microflora of the environment are unable to derive their nourishment from raw materials. They cannot convert the sun's energy to synthesize the food they need. They are *heterotrophic* and dependent for energy on food materials prepared by the green plants. These heterotrophs may be *parasitic* or *saprophytic* depending on whether they obtain their food from the cells of living or dead plants and animals. The microenvironment of parasites consists of the living cells of the host plant. The cells of the bacteria and the fine, microscopic, thread-like mycelium of the fungi develop in, between, and around these cells. Parasites can be in balance with their host, doing little harm, or they may kill the invaded cells and tissues, causing leaf spots, cankers, blights, rots, decay, and a multitude of other disease conditions. On a larger scale, fungi have destroyed entire populations of plants, leaving only the resistant species. Introduced fungi causing white pine blister rust, Dutch elm disease, and chestnut blight have destroyed their host species over large areas of North America, and other plants have taken their places.

Another group of organisms, the *mycorrhizal* fungi, live within or on the roots of higher plants, helping to provide water, nutrients, and possibly more complex compounds elaborated by the hyphae. In return, the hyphae obtain their nourishment from the roots. These mycorrhiza are essential to the normal development of pines, orchids, and a multitude of other plant species in over a hundred families.

The majority of fungi and bacteria are saprophytic and derive their nourishment from dead plants or other nonliving organic material. They are beneficial to the environment, being essential to sustaining and recycling the soil nutrient supply and atmospheric carbon dioxide. These organisms are the *decomposers*, which break down organic material into simpler compounds and return phosphorus, potash, magnesium, and other essential elements to the soil, from which they can be reabsorbed by the green plants. Without decomposers, organic matter would simply accumulate until the mineral in shortest supply limited further plant growth. In the cooler regions of the world, organic matter often accumulates because low temperatures limit the activity of the decomposers. The situation in tropical regions is reversed, and organic matter is utilized so rapidly that accumulation is almost nonexistent.

Viruses lie in the hinterland between the biotic and physical environments. Although non-living, they cause diseases in much the same manner as fungi and bacteria. The complex virus nucleoproteins develop and reproduce within the plant cell, disrupting normal plant metabolism. The viral nucleic acids redirect the normal cellular nucleic acids and thereby modify the internal environment of the cell. The virus alters plant development in many ways, causing such plant abnormalities as leaf mosaics and yellowing, growth suppression, disturbed reproduction, and reduced yields.

Growth, development, and distribution of all organisms is determined by the sum total of all physical and biological conditions, their extremes, and the interaction between them. Plants live within a limited range of temperature, moisture, light, and soil conditions. Their existence depends on their ability to compete successfully with other plants in the environment, to live in harmony and balance with the microflora, and to tolerate the presence of the parasites in their environment.

From experience and observation, the growth capabilities and environmental tolerance of cultivated plants are known. Each plant species has its own specific environmental requisites for growth and reproduction. As conditions vary from the ideal, the efficiency of metabolic processes decreases, vigor declines, and growth diminishes proportionately. Persistent or extreme variations from the normal physical or biotic environment create a stress condition in which one or more of a host of injuries, diseases, or disorders may appear—e.g. visible leaf spots, cankers, blight, or yellowing.

SELECTED REFERENCES

Billings, W. D., 1964. "Plants and the ecosystem." Wadsworth, Belmont, Calif., 154 pp.

Butler, E. J., and S. G. Jones, 1949. "Plant pathology." Macmillan, London, pp. 158–182.

Daubenmire, R. F., 1959. "Plants and environment," 2d ed. Wiley, New York, 422 pp.

Gates, D. M., 1962. "Energy exchange in the biosphere." Harper, New York, 151 pp.

Gleason, H. A., and Arthur Cronquist, 1964. "The natural geography of plants." Columbia, New York, 420 pp.

Smith, R. L., 1966. "Ecology and field biology." Harper, New York, 686 pp.

CHAPTER TWO

INJURIES, DISEASES, AND DISORDERS

Organisms (or any other agents) which cause a plant to malfunction are known as *pathogens*, from the Greek "pathos" meaning suffering. Contrary to common belief, a pathogen technically does not have to be an organism or virus. A pathogen can be any component of the physical environment, including adverse climate, soil, or air relations. This environmental pathogen can also be referred to by the more specialized term *physiopath*. Physiopaths most commonly cause an injury, but occasionally they may initiate plant diseases.

Abnormalities caused by pathogens are generally called diseases, but the term "disease" should not be restricted to abnormalities caused by biotic pathogens or parasites. The general misconception that pathogens had to be organisms first arose when early mycologists learned that diseases were caused not only by the physical environment but by fungi and bacteria as well. When fungi were discovered to cause diseases, many mycologists and pathologists came to believe that all diseases were caused by such organisms. Abnormalities associated with nonparasitic factors were often regarded as injuries rather than diseases. The more accurate definition of disease is that it involves any chronic deviation from normal regardless of what caused it; similarly, it can be defined as any malfunctioning process caused by a continuous irritation, or as any chronic departure from normal involving the expression of symptoms.

Stresses imposed by either the physical or biotic environments can cause abnormal plant development, often with the expression of visible symptoms; however, visible symptoms may be lacking. A malfunctioning process may arise in the absence of obvious visible effects whenever the plant functions at less than optimum. A disease situation exists whenever plant growth deviates greatly from what is regarded as normal. The amount of growth and production considered normal is largely a matter of opinion, and the retardation might have to be striking before any group of observers could agree that the plants were diseased.

If this inclusive concept of disease is to be accepted, does this mean that every abnormal process or effect must be termed a disease? Certainly not. Malfunctioning is often caused simply by an injury. A distinction should be drawn between an injury and a disease, although sometimes the line of demarcation is vague, and the difference is not always simple or dependent on the causal agent.

The basic difference between injury and disease lies in whether the abnormal condition is acute or chronic. In order to be properly classified as a disease, the abnormality must be chronic, that is, it must involve a continuing irritation. The cause of the effect must persist. When the malfunction is acute and occurs more or less instantaneously, the disturbance is better called an injury. Even though the damage itself remains, if the source of irritation does not persevere, it is still an injury. A few examples of diseases and injuries should help clarify this definition.

A clear example of an acute disturbance causing an injury is the common condition referred to as "tractor blight" caused by tractor wheels or discs, or by other implements tearing away the bark of a plant. In home gardening, a similar malady is sometimes called "lawnmower blight." Here the condition is a well-accepted injury, even though the stripping away of bark may prevent the tree from making normal growth, or even producing normal crops, for several years.

The situation, or at least the definition, may become more complicated if the injury is great enough to cause some of the limbs to die. This limb dieback can develop gradually over the course of a summer or even a period of years. This is an example of a situation where an acute injury initiated a chronic and persisting condition. A similar situation exists when insects or mites feed on leaves and destroy many of the cells but are then brought under control. The insects are gone, but the damage persists. Are these disease situations? No. A disease involves continued exposure to a stress situation. Fungi, bacteria, and viruses remain in the tissue, so there is little question about their causing continuing irritation.

What about physiopaths? What kind of abnormalities are associated with such major environmental factors as temperature? Plant damage can result any time the temperature drops below a critical point even for a rather brief period. By the time the temperature returns to normal, the damage has been done; usually no further damage develops. The plant "suffers" only briefly except for the killed or damaged tissue, and the adverse temperature relations have caused only an injury. High temperatures, on the other hand, can cause either injury or disease. At extremely high temperatures, physical injury occurs, involving leaf burning. At somewhat lower temperatures, the effects are less definitive. The activity of enzymes, the organic molecules which regulate every chemical reaction of plant metabolism and growth, increases with temperature up to a point. Above this point enzyme activity is inhibited, and metabolism is suppressed. In turn, growth is suppressed, and some may say the plant is "sick" even though there are no externally visible symptoms. Low temperatures similarly suppress enzyme activity and cause the plant to function at less than its maximum capacity. This would not ordinarily be considered a disease situation, although if low temperatures continued and caused chronic growth suppression, some question might be raised.

High light intensity may cause the leaves to become yellow or even kill tissues, causing spotting of the leaves and fruit. Damage occurs over a rather brief period, and the condition may best be considered as an injury. Low light intensity, on the other hand, tends to cause a more chronic suppression of growth. Mild yellowing might be the only visible symptom.

Adverse moisture relations can cause acute or chronic injury. Plants growing in a waterlogged soil are subjected to a deficiency of oxygen which persists until the soil dries out. The lack of oxygen limits respiration, which in turn limits normal growth. Without usable energy no new cell constituents can be manufactured to replace those which become used up. Metabolic processes are inhibited, and gradually cells, tissues, and even organs may die. If a waterlogged condition persists, the entire plant eventually succumbs, but even a few days of oxygen deficiency can cause extensive leaf yellowing and defoliation. Even after the soil dries out and oxygen is again available to the roots, growth is abnormal; tissue may continue to die, producing a disease situation.

Inadequate moisture may also last long enough to cause damage which persists even after the water relations have been improved. Usually, though, the plant will resume normal growth once adequate water relations have been restored. In this case only an injury will have occurred.

Adverse soil conditions other than moisture relations may also cause either acute or chronic disturbances. Nutrient deficiencies or excesses, adverse soil alkalinity or salinity, or even unfavorable soil structure or texture may be chronic and result in such disease situations as poor, weak growth, leaf yellowing, and poor fruit production. Most often, however, affected plants resume normal growth as soon as the adverse soil condition is corrected, so it is no wonder that nutrient deficiencies and other poor soil relations are usually regarded as injuries rather than diseases.

Whether air pollutants cause disease or injury, or both, is debatable. If concentrations of certain pollutants—such as fluoride—are sufficiently high, there is no question that the leaf burning which results is an injury. On the other hand, lower concentrations of the same pollutant may cause a disturbed metabolism with no visible expression. Ozone causes acute injury at high concentration and poor growth and production even when no visible injury is apparent. Such pollutants as ethylene, which act systemically as growth regulators, cause persistent growth effects that are unquestionably of a disease nature.

Pesticides may cause either injury or disease, depending on the nature and concentration of the chemical. Direct injury, such as leaf burning, occurs when toxic concentrations of the chemical contact the plant tissue. Chemicals which are taken up by the plant, that is, the systemics, may accumulate in toxic concentrations, causing a chronic disturbance or disease condition that often involves leaf yellowing and twig or limb dieback.

Definitions of a disease should be flexible. A comprehensive term that is essentially synonymous with both injury and disease can be used whenever there is any question as to terminology. The word is "disorder." A disorder can be defined as any plant irritation, whether chronic or acute, whether visible symptoms are present or only metabolic effects are suspected. It is a broad term inclusive of all abnormal conditions whether associated with the biotic or physical environment.

SELECTED REFERENCES

Chester, K. S., 1947. "Nature and prevention of plant diseases," 2d ed. McGraw-Hill, New York, 525 pp.

Duggar, B. M., 1911. Physiological plant pathology. *Phytopathol.* 1:71–78.

Hildebrand, E. M., 1952. Terminology in phytopathology. *Phytopathol.* **42**:283.

Horsfall, J. G., and A. E. Dimond, 1959. "Plant pathology. An advanced treatise," vol. 1. Academic, New York, 674 pp.

——— and ———, 1960. "Plant pathology. An advanced treatise," vol. 2. Academic, New York, 715 pp.

Melchers, L. E., 1915. The grouping and terminology of plant diseases. *Phytopathol.* **5**:299–302.

Sorauer, Paul (trans. F. Dorrance), 1922. "Manual of plant diseases," vol. 1: Non-parasitic disease. Record Press, Wilkes-Barre, Pa.

Stakman, E. C., and J. G. Harrar, 1957. "Principles of plant pathology." Ronald, New York, 557 pp.

Walker, J. C., 1957. "Plant pathology," 2nd ed. McGraw-Hill, New York, 707 pp.

CHAPTER THREE

THE PLANT RESPONSE

Plants will develop normally only as long as each of the many environmental factors remains within a critical range. Whenever temperature, light, moisture, nutrients, or other environmental components exceed this range and become limiting, the plant responds with abnormal growth patterns known as *symptoms*. This abnormal response is the outward sign that a plant is sick. The symptom may be striking and clearly visible (e.g. leaf yellowing or burning) or elusive, being measurable only in poor growth and production.

A plant responds in many different ways to the stress of an abnormal situation. Stress evokes a general type of response, with a concomitant symptom showing many minor variations that depend on the plant, its condition, and the specific pathogen or physiopath. Every plant has its own visible expression indicating an underlying disorder.

The plant response indicates what is wrong and can reveal the cause of the abnormality. When the observer is familiar with the plant response or the symptoms, he can diagnose the disorder and attempt to correct it. The clue in diagnosis lies in becoming familiar with the plant symptoms.

Considering the infinite number of things that can go wrong with the plant, classifying the symptoms may seem impossible. Obviously there are many ways in which to do so. Some pathologists describe up to twenty-five or thirty distinct types. Many of these are really the same fundamental responses even though different tissues are affected. The four groups listed herein are general, yet inclusive, and will satisfy the needs of most. The four basic categories discussed are: *growth response*, *reproductive effects*, *chlorosis*, and *necrosis*.

3-1 GROWTH RESPONSE

Reduced or suppressed growth deviating from what is expected judging by past performance is the basic response of a plant to an adverse or stress situation and the first indication that something is wrong. Growth responses are often slight and difficult to detect. The normal growth rate must be known before any suppression can be recognized. Through years of experience and record keeping, growth expectancies of cultivated plant species, varying with the general growing conditions, have been established. The amount of growth a plant should make also depends a great deal on the opinion of the observer. A poor grower might think 2 in. of new shoot growth normal for a peach tree, while the good grower might not settle for less than 2 ft. Two people rarely agree completely on what constitutes normal growth. "Normal" is an indefinite term encompassing broad limits. Within these limits the plant is usually regarded as normal, even though it may not be very healthy. Beyond these limits, growth is said to be abnormal; the abnormality indicates a disease situation.

Growth is largely influenced by the two fundamental metabolic processes of *photosynthesis* and *respiration*. Photosynthesis is simply the conversion of radiant energy to chemical energy. The chemical energy is stored in various carbohydrates of the cell and released by respiration, which is the conversion of this stored energy to a form which can be used for the plant's many energy-requiring cellular reactions.

Photosynthesis, respiration, and a multitude of other essential plant processes consist of many individual chemical reactions. Plant growth is regulated by these reactions, each controlling a particular step of the total plant metabolism. Thousands of chemical reactions have their own requisites for optimum or maximum activity. The activity or rate of these reactions determines the amount of growth a plant makes. Theoretically then, maximum growth occurs when conditions favor the greatest number of reactions at any one time.

Many factors limit growth, including high temperatures, adverse light relations, inadequate soil moisture or too much water, and nutrient deficiencies or excesses of either essential or nonessential elements. Plant growth can also be limited by many of the chemicals man has cast into the environment. Air pollutants and pesticides are two major growth-limiting components of the environment. When used improperly, pesticides may have long-range and sometimes devastating effects.

Toxic products of various species of insects can also limit growth or stimulate excessive or cancerous growth. The *galls* and *tumors* appearing on leaves or stems are striking growth abnormalities. They are caused by toxins and growth-regulating chemicals introduced into plant cells by insects, particularly certain species of flies and wasps. The chemicals stimulate excessive cell division, cell enlargement, or both. Leaf galls are produced when cells in the mesophyll multiply and enlarge abnormally and become so numerous that blister-like pustules appear. On stems, the cortical as well as the cambial cells proliferate excessively, causing similar overgrowths. Parasitic mites, particularly those in the Eriophyd group, also cause a number of different types of galls. Fungi, bacteria, and viruses induce similar responses. When fungi are responsible, their mycelium or reproductive structures are commonly found intermingled with the gall. Fungus-induced galls include those caused by such organisms as *Exobasidium* on blueberry, *Dibotryon* on stone fruits, *Taphrina*, which causes leaf curl on many different species (Fig. 3.1), and *Physoderma*, which causes alfalfa wart. Bacterial galls are best illustrated by *Agrobacterium tumefaciens*, causing crown gall on many cane, stone, and pome fruit species. By interfering with normal cell division, these galls disrupt the continuity of the food- and water-conducting tissue, thereby starving the plant.

Intumescences are not usually associated with factors in the phys-

FIGURE 3.1 *Leaf curling, puckering, and hypertrophy of peach leaves caused by* Taphrina, *characterizing "leaf curl" symptoms.*

ical environment, but there are exceptions. For instance, ozone can cause local tumors on broccoli leaves by stimulating cell division and enlargement, thus producing blister-like pustules. Prolonged exposure of geranium plants to very high humidity will also induce cell division and tumors. Low temperatures occurring at critical stages of leaf enlargement cause irregularities in the shape and character of the leaf margins and even the shape of the entire leaf. Leaves which are normally oval in shape and smooth-margined may be deeply lobed or variously incised or lacerated. Small local areas over the leaf may be killed so that open, torn areas develop as the leaf enlarges.

A still more striking growth response is the *witches'-broom.* This is a broom-like growth proliferation arising when the normal shoot initiation mechanisms have been thrown out of balance and a dense cluster of new shoots develops from nearly the same point. Witches'-brooms of fir, spruce, and cedar trees are commonly caused by rust fungi, while dwarf mistletoes induce much the same type of malformation on pines and firs. Witches'-brooms are also caused by bacteria and viruses. *Agrobacterium rhizogenes*, causing root growth proliferations, and alfalfa witches'-broom virus are examples.

3-2 REPRODUCTIVE EFFECTS

Crop production depends first on the number of fruit buds set. Basically bud set is a function of the genetic makeup of the plant species, but it is subject to modification by many factors in the external, physical environment. Of these, photoperiod, the relative duration of light and dark, is the most important. Photoperiod controls the formation of the growth regulators and pigments responsible for initiating flower formation; flower buds are initiated only when a suitable photoperiod is provided. Each plant species has its own photoperiod requisites for flower formation and a required dark period, which may range from eight to sixteen hours.

Once formed, flower buds may open and produce fruit the same year or bear fruit the following year after a winter of dormancy. Consequently, growing conditions in each of the two years are important to normal flower and fruit development.

Many plants, particularly fruit trees, must be exposed to a specific number of hours of low temperature before fruits develop normally. Flower buds may have to be exposed to temperatures below 8°C for several hundred hours before they can open. Without this chilling period, buds remain dormant, no fruit sets, and there is no crop. Leaf buds also have a chilling requirement, but it is shorter and more easily fulfilled; if not, however, plant development is abnormal, and growth is weak. Continued failure of leaf buds to emerge causes gradual decline of the tree.

While temperatures below 8°C are necessary for normal bud development, buds are killed outright when temperatures drop below a critical point. Flower buds initiated the year before they open are subject to the vagaries of environment for the better part of a year and often are damaged or killed during the winter or spring months. Bud damage during the winter months is not usually a limiting factor because the buds are hardened and able to withstand the low temperatures. Low temperature damage is far more likely to be critical early in the spring after the blossoms have opened and tissues are most sensitive. Spring frost damage to the open blossoms causes major crop losses in all temperate regions of the world where fruit crops are grown.

After blossom set, reproduction is sometimes limited by poor flower pollination and fertilization. But the plant normally sets far more buds than can mature, and a good crop can be expected even if only a small percentage of the flowers are fertilized. When too much fruit is set, much of the excess drops off. As much as 80 to 90 percent of the blossoms normally drop early in the development of stone fruit trees. Unpollinated flowers and excess fruits continue to drop during periods of stress. Blossom and fruit drop is influenced by temperature, moisture, nutrition, and any other environmental factor causing stress.

Mineral nutrition influences the amount of fruit set, its development, and its subsequent size. High nitrogen levels, for instance, can throw the plant into a lush vegetative state with no flowers forming. Low nitrogen limits fruit size, quality, and yields.

Fruit development can be influenced by various pollutants and pesticides which are capable of limiting flower initiation. Herbicides and blossom-thinning sprays are exceedingly harmful when used carelessly. Chemicals such as the phenoxy compounds, aliphatic acids, and heterocyclic nitrogen derivatives kill blossoms, impair shoot growth, and kill plant tissues when not used wisely.

Soil moisture also influences reproduction, which in turn affects quality and size of the fruit. Moisture availability is one of the critical factors determining fruit size and the capacity of the fruit to "stick" rather than drop during the season.

Fruit is subject to physical damage and biotic diseases throughout its development. Fungi, viruses, and many components of the physical environment can limit fruit growth. Low temperature can damage local areas of cells in the maturing fruit. The dead cells persist, and frost rings and bands appear on the mature fruit. High temperatures cause russeting and scalding, which also persist and mar the fruit quality. Temperature extremes, poor nutrition, and water stress further provide stresses which limit fruit size. Too much water causes soft, succulent tissues that are subject to breakdown and fruit rot.

3-3 CHLOROSIS

Chlorosis, the blanching of the green parts of the plant, is another characteristic response to an adverse environment second only to growth responses in distribution, frequency, and occurrence. In a chlorotic state, the normal green color disappears and leaves become pale green, yellow, or even white.

Chlorophyll is broken down continually in normal usage and must be constantly synthesized if the plant is to stay green. Favorable light, nutrition, water, oxygen, and temperature are all critical to chlorophyll synthesis. When any of these environmental factors is unfavorable, decomposition and depletion of chlorophyll reserves is accelerated, and chlorophyll is broken down faster than it can be synthesized.

Consequently, many stress factors affect chlorophyll levels and may cause chlorosis. Light is required to initiate chlorophyll formation. However, too much light breaks down or oxidizes the chlorophyll back to protochlorophyll, causing chlorosis. This photooxidation is enhanced by high temperatures. Light also provides the energy required for all synthesis reactions, including the production of the chlorophyll molecule. Without light energy synthesis is impossible, and the plant starves.

Chlorosis caused by starvation more frequently results from a lack of nutrients. The chlorophyll molecule itself is made up of magnesium, hydrogen, oxygen, and carbon. A deficiency of any of these elements prevents chlorophyll formation. The additional elements of iron, manganese, zinc, potassium, and sulfur are essential as components of enzymes and coenzymes which catalyze the chlorophyll-synthesizing reactions. A deficiency of any of these elements also limits chlorophyll synthesis, thereby causing chlorosis.

Plants also require an adequate water supply to remain green. Water contributes the hydrogen atom, the most active radical of the chlorophyll molecule. Water further serves to keep the guard cells surrounding the stomata turgid. When water is unavailable, the guard cells lose their turgidity and collapse, and the stomata close. Closing the stomata prevents the exchange of gases between the intercellular spaces and the external environment, limiting intake of carbon dioxide by the leaf, reducing or arresting photosynthesis, and causing the plant to starve from lack of energy.

Even a slight water deficiency is detrimental and can cause chlorosis. Water stress stimulates respiration, accelerating carbohydrate breakdown to a point where the cell's food supply is soon exhausted. This shortage of food, or reserve energy, limits all synthesis reactions. When water is still more deficient, the physical and chemical properties of the cell are altered, so that the cytoplasm becomes dissociated and membrane permeability is damaged. As the breakdown of cytoplasm and membranes proceeds, chlorosis and ultimately death occurs.

Too much soil moisture is equally harmful. When water saturates the air spaces of the soil, oxygen needed by the roots for respiration is limited. If respiration cannot proceed, the energy needed for mineral uptake and synthesis is lacking, and the chlorotic symptoms of starvation soon appear.

Temperature controls the rate of all chemical reactions, including chlorophyll synthesis. When temperatures are below 4° or above 35°C, chlorophyll becomes deficient, and chlorosis appears.

In summary, light, nutrients, water, and temperature relations all indirectly influence chlorophyll synthesis. Since chlorophyll is the primary energy recipient in photosynthesis, a limited supply of chlorophyll reduces photosynthesis and limits available energy. A weak, unhealthy, and spindly plant results from limited photosynthesis, and a chlorotic condition in these plants is the most striking visible symptom of many disorders.

The specific expression and distribution of chlorosis can vary considerably. The mildest form of chlorosis involves simply a slight loss of the green color in local areas of the leaf, a faint mottling, or a rather uniform, pale green discoloration. General yellowing develops as chlorosis becomes more severe, and with extreme chlorosis the entire leaf becomes pale yellow and even bleached.

Chlorosis frequently appears first along the margin of the leaf as the edges and tip take on a pale green color. Where such nutrients as manganese or zinc are deficient, this chlorosis will extend slightly between the large veins in a "feathering" pattern. When iron is deficient, most of the leaf area becomes yellow, with only the larger veins remaining green.

On other occasions, chlorosis is distributed over the entire leaf surface. With this expression, the areas between the veins gradually lose their color, producing a mottled discoloration. Interveinal chlorosis is often caused by high or low temperatures. Viruses can cause similar symptoms, but usually the distribution of chlorosis over the leaf is irregular, giving more of a "splotchy" appearance.

Still another type of chlorosis occurs when the cells underlying the epidermis are killed and the epidermal layer separates from the mesophyll. This symptom is known as *glazing*. Bronzing and silvering are variations of the same plant response. Glazing is caused by low temperatures and certain photochemical air pollutants.

Similar types of chlorosis are caused by thrips, leafhoppers, and mites. These pests destroy the cytoplasm of the leaf cells, causing varied expressions. The lower leaf surface silvering caused by leafhopper feeding resembles symptoms of photochemical pollutants. The obvious difference is the absence of thrips where the pollutants are to blame. Injury from Eriophyd mites (such as the silver mite) gives the leaf a uniform silver or bronze-to-purple discoloration. The two-spot mite is more destructive and causes a russeting or dull brownish-yellow color, which may be limited to local areas or involve the entire leaf.

Regardless of the cause, when chlorophyll is completely lost, other pigment colors become disclosed, so that the leaf turns various shades of yellow, orange, or red depending on the dominant color of the remaining pigments. When chlorophyll breaks down into yellowish droplets, or when the yellow xanthophyll pigments are dominant, the leaf may appear yellow. The leaf appears orange when the carotene pigments are dominant, and reddish when the anthocyanin pigments are abundant and become unmasked. These pigments also provide the normal coloration of the fall season, when temperatures drop and the days become shorter. Then chlorophyll production ceases, and the color of the other pigments is displayed. When all the pigments—xanthophyll, carotene, anthocyanin, and chlorophyll—are destroyed, the leaf becomes bleached, whitish, or brown. By this time the protoplast is destroyed, the cells are dead, and the tissue is necrotic.

3-4 NECROSIS

Necrosis means death or mortification. It may be the death of isolated cells, tissues, or organs, or death of the entire plant. Necrosis occurs whenever cells are killed by any component of the biotic or physical environment. Lethal stresses can be caused by the nonliving components of the environment or by such living agents as fungi, bacteria, and viruses. Any time the stress on the plant becomes too great and the chemical reactions in the cells become irreversible, death ensues.

Plant organs, tissues, and cells, and even the cellular constituents, vary in their sensitivity to stresses. The cell membranes are especially sensitive to high temperature stresses, which readily disrupt the double bonds making up the unsaturated lipid components of the membrane. Damage to the lipid layer permits water to leak through the membrane and causes a disturbed water relation, leading to the death of the cell and necrosis. The protoplast or cytoplasm is also damaged by environmental stresses. High temperatures denature, or break down, the protein components of the protoplast. Stresses also decrease or inactivate cytoplasmic streaming, causing the cell to die from poor distribution and unavailability of chemicals required for synthesis.

Milder stresses which lead first to chlorosis may ultimately kill the cell. Stress first causes an uncontrolled, accelerated burning of the carbohydrates, which soon depletes the reserve food supply. When food and energy reserves are gone, active water and nutrient absorption ceases, and the plant starves. The cell contents dry out (plasmolysis) and die, and the plant becomes visibly necrotic.

The exact color or expression of necrosis ranges from pale yellow or white to dark brown, depending largely on the amount of phenolic degradation materials, mostly tannins and resins, which are formed. The kind of tannins produced depends on the type of cell affected, how fast the cell died, and the conditions which led to death. A rapid death may result in the formation of one group of chemicals, while more gradual death may involve the formation of others. Brownish or reddish phenolic pigments of dying cells are synthesized partly from the degraded living materials, from the phospholipids of the membranes, from the metabolites of the protoplast, or from all of these.

Tissues of the leaves, stems, roots, flowers, or fruit may all become necrotic, depending on the sensitivity of these organs to the particular pathogen or the physiopath involved. The most obvious, common, and characteristic necrotic symptoms appear on the leaves. The margin or edge of the leaf is usually affected first. Since most of the veins end in this region, water shortage and other stresses first arise here. Marginal necrosis is caused by nutrient deficiencies, adverse light relations, drought, heat, frost, pesticides, and air pollutants.

As the stress increases, necrosis characteristically extends from the margin into the intercostal part of the leaf toward the midrib. Ultimately the entire leaf may die. In other cases, though, necrosis appears only in isolated spots. This is particularly true where the disorder is caused by fungal or bacterial pathogens. Fungus spores disseminated through the air land on the leaf surface. When the spores germinate and the fungus mycelium penetrates and infects the leaf, a local area of the tissue is killed. The invading fungus (or bacteria) kills the cells as it advances. The leaf tissue often responds by producing a barrier of cells through which the mycelium cannot penetrate, but cells lying within this perimeter are killed, forming a *leaf spot*. These parasite-induced, often circular leaf spots may be large or small and frequently overlap the veins (Fig. 3.2).

FIGURE 3.2 *Marsonina leaf spot of quaking aspen, characterizing the fungus-related type of leaf spot necrosis.*

Leaf spots can also be minute and irregular in outline, less than 1 mm in diameter, and affecting only a few cells. Such symptoms are referred to as *flecking* or *stippling*. Sometimes these flecks coalesce to produce larger, irregular spots distributed along or between the veins. Flecking symptoms are caused by such physiopaths as ozone, chlorine, or low temperatures. Minute, punctate spots are also produced when insects suck out the plant juices, killing small groups of cells. Similar necrotic spots appear when insects insert their ovipositors to lay eggs, thus killing a local area of cells. When low-temperature injury occurs early in leaf development, the dead areas often drop out, leaving a shot-hole condition similar to that resulting when insects chew out local areas of the leaf.

Necrosis also affects stem tissues. Such injury, known as a *canker*, appears when the cells of the sapwood or inner bark are killed. The discrete, dead areas are sometimes discernible through the bark but become more obvious when the bark is removed (Fig. 3.3). The dead area is then sharply delimited from the healthy white tissue. Cankers may be relatively superficial and restricted to the bark or outer sapwood, or they may penetrate well into the heartwood. Cankers are frequently caused by biotic pathogens, which release toxins or enzymes that digest and kill sensitive host cells in advance of the invading fungus mycelia. *Nectria, Pythium, Ceratostomella, Armillaria,* and *Valsa* are but a few important canker-causing genera of fungi. *Erwinia amylovora,* causing fire blight of many Roseaceous plant species, and *Pseudomonas syringia* are among the most significant bacterial canker-producing pathogens. Insects such as tree hoppers and scale species, which destroy stem tissues, cutting off food and water movement, also are responsible for many types of cankers.

Cankers can be caused by mechanical injuries from tractor implements or lawnmowers, by lightning, and by adverse temperature relations such as sunscald, but more often physiopaths serve only to weaken the tissue to the point where parasites can more readily become established.

Cankers may girdle the smaller shoots and limbs of plants and thereby cut off the water and nutrient supply. The portion of the stem, including the leaves, beyond the canker then dies from starvation and

FIGURE 3.3 *Trunk canker of quaking aspen, showing bark and wood discoloration caused by fungi, also black fruiting structures of the fungus.*

water shortage. Environmental stresses such as heat and poor moisture relations are frequently responsible for this *dieback*, but cankers are also caused by various biotic pathogens, including insects. Scale insects commonly extract plant foods from the stem cells faster than the plant can produce them. Millions of cells are killed by the feeding activity of dense populations of the pests, destroying large branches and trunks of trees.

Cankers often affect root and crown tissues. The soft tissues of young stems and seedlings are quick to respond to a harsh environment. On a hot day, soil temperatures are far higher than air temperatures. The soil conducts this heat to the crown of the young plants, where it causes local plasmolysis and death of the cortical cells, forming cankers at the soil line. The top of the plant topples over, producing a damping-off symptom which looks very much like that caused by soil pathogens. Crown cankers are most commonly caused by fungus and bacterial pathogens.

Fruit and flowers of a plant can be killed and become necrotic at any stage of development. Flower buds are particularly sensitive to low temperatures, and frost is often responsible for serious crop losses. Drought, air pollutants, and other physiological factors can be equally destructive. Local injuries may occur when the fruit is peppered by hail; or damage may be more general, involving large areas of fruit, as in sunscald. Fruit necrosis is often followed by a *fruit rot*, a more advanced stage of tissue breakdown. Once the tissue is damaged, it is readily invaded by fungus or bacterial parasites, which complete the decay process.

Vegetative reproductive organs, including tubers, corms, bulbs, and rhizomes, may also be injured by physiopaths or pathogens. Fruit and vegetable rots are common whenever these organs are stored for any length of time. When storage conditions are unfavorable due to high temperatures, low moisture, or poor ventilation, extensive losses may occur even when crops are stored for only a brief period.

To summarize, within each of the four basic symptom categories of growth response, reproductive effects, chlorosis, and necrosis, there are a number of variations in the precise symptom expression. Many of these symptoms look very much alike, even though the cause differs. The more familiar the observer is with the specific symptoms, the better will he be able to establish their precise cause. Only when the plant response is diagnosed and understood can the disorder be controlled.

CHAPTER FOUR

ELEMENTS OF DIAGNOSIS

Diagnosis is the art of recognizing diseases from their symptoms, a scientific determination of a disorder based on critical scrutiny. Sometimes diagnosis is fairly clear-cut and simple. Such is the case when a parasite can be found and identified, or when a familiar or well-known disease appears which can be recognized from previous experience and the symptoms alone. But disorders are not always easy to identify. When they appear for the first time, or at least for the first time in the experience of the observer, diagnosis is difficult.

To make a diagnosis, the observer must know something about the environmental stresses, pathogens, and parasites common to the area. He should understand the basic procedures of diagnosis, know what background information to seek, and how to interpret it. These basic procedures consist of a series of seven steps delineated in this chapter; they should serve as a logical sequence indicating what to look for in evaluating possible causes of a disorder.

4-1 WHAT DO THE SYMPTOMS LOOK LIKE?

A symptom is any perceptible change in the known structure, appearance, or function of an organism which suggests that it is sick. It includes any external expression characterizing an abnormality, such as *leaf yellowing, necrosis* or *spotting, stem lesions,* or various *malformations* or *misfunctions* as described in the previous chapters.

The first step in diagnosis is the careful, close examination of all the tissues of the sick plant together with other plants in the vicinity to determine precisely what is wrong with them. Evaluation of symptom expression aids in determining the pathogen or physiopath responsible for the disorder. Chlorosis and necrosis patterns are distinct for each type of disorder and provide the clues necessary for diagnosis. For instance, relatively uniform marginal leaf chlorosis or necrosis suggests an abiotic pathogen such as moisture or heat stress, while more irregular patterns imply the presence of a parasite. Recognition of specific symptoms is important, but knowledge of the overall disease picture, or *syndrome,* is even more important in diagnosis since it provides more information and reveals far more than the symptom alone. Syndrome refers to the broader disease expression, including all the symptoms characterizing a particular disease, the different kinds of plants involved, the tissues and organs affected, when symptoms first appeared, and the general distribution of the disorder.

4-2 IS AN ORGANISM OR VIRUS PRESENT?

Diagnosis can be relatively easy when some *sign* of an organism is present. "Sign" refers to the actual body of the fungus or other causal organism. Insects, mites, or their remains might comprise such a sign, and their presence should be sought early in diagnosis. Finding, identifying, and recognizing an insect pathogen as the causal agent is the surest diagnosis. However, the causal agent may have disappeared by the time the symptoms are observed. Other signs of their existence may be evident. Often mites are damaging early in the year and die out from predation or because of an unfavorable environment. In this case their debris or the skeletons of their deceased bodies may remain and can be recognized. Webbing of two-spot mites adheres to the plant long after the mites have disappeared; chlorotic flecking persists throughout the season (Fig. 4.1). When rust mites have been feeding,

FIGURE 4.1 *Chlorotic stipple* of Malva *caused by two-spot mite feeding.*

the rusted leaves provide a striking clue to their earlier inhabitance. Tightly curled leaves would indicate an earlier aphid population.

Chlorotic flecking characterizes thrip as well as mite feeding and may superficially resemble some types of air pollution injury (Fig. 4.2). Similar stippling can also be caused by the ovipositing of several insect species (Fig. 4.3).

When symptoms indicate a fungal origin, the organism, or at least some part of it, may still remain. The presence of a fungus is indicated by microscopic, threadlike filaments which comprise its thallus or plant body. The powdery and downy mildew fungi for instance form a soft, velvety-textured mat over and in the infected areas (Fig. 4.4). Reproductive structures of the fungus may also appear. These minute to microscopic fruiting bodies generally develop after the tissue has been killed and can be found on the lesions or other dead tissues.

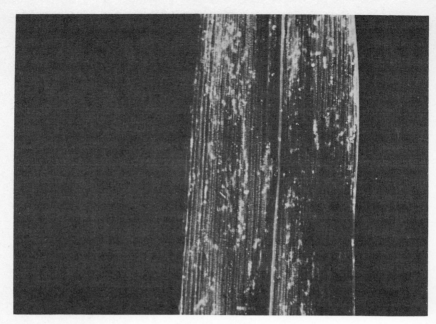

FIGURE 4.2 *Chlorotic stipple of wheat caused by thrip feeding, showing random distribution of lesions over and between veins.*

FIGURE 4.3 *Chlorotic stipple of aspen caused by insect ovipositing.*

FIGURE 4.4 *Downy mildew of alfalfa showing the "cottony" growth of fungus mycelium on the lower leaf surface.*

Even though the tissues surrounding the fruiting bodies are dead, the question still remains as to whether or not the fungus initiated the damage or came in after the tissue had been killed. Fungi often enter dead tissue killed by some other pathogen. When the fungus seems to be growing adjacent to, or on, healthy-looking tissue, it is very likely that a fungus parasite was responsible for killing the tissue.

Parasites may cause disorders even though no organism is evident at the time of diagnosis. Sometimes the parasite has come and gone; at other times its presence is not obvious. A fungus might be inside the host plant tissue but only visible upon microscopic inspection of thin tissue sections. The fungus might have died and "disappeared," or killed the tissue by releasing toxins which moved up through the plant well in advance of the fungus itself. Fungi may be confined to roots, initially killing only that organ of the host plant but subsequently leading to the death of the stem and leaves by causing starvation or water stress.

When the organism does not appear outside the host plant, the pathologist must examine the internal tissues. To do this he tries to grow the fungus or bacteria in culture, examine the thallus and reproductive structures of the parasite which develop, and identify the parasite. Culturing is done by taking a small piece of the diseased tissue and "plating it out." To make such an *isolation*, the surface of the affected plant part is sterilized and a small piece of the sick tissue is placed on nutrient agar in a petri dish. Any fungus which then grows out on the agar presumably arose from within the sick tissue. The fungus was there all along but not visible to the naked eye. By carefully inspecting the distinctive features of this organism in culture, the pathologist can usually identify it.

If the fungus is a suspected parasite, it may be inoculated into healthy plant tissue to see if it will initiate the disease. If the characteristic disease symptoms appear, and the same fungus is reisolated, the organism is considered to have been pathogenic. This *proof of pathogenicity*, where the disease cycle is reproduced, is known as Koch's postulates.

Failure to grow an organism in culture does not give assurance that a fungus or other organism did not cause the disease. Some fungi are obligate parasites which cannot be grown on culture media; sometimes bacterial pathogens are involved which may not grow satisfactorily in the same media used to isolate fungi.

Detection of viruses requires specialized methods, but the types of symptoms resulting from their activity are often distinctive. External symptoms caused by viruses are varied but can be classified mainly in four broad groups.

(1) Most common are the *mosaics*. When irregular mottling and puckering of light and dark green areas cover the leaf, the symptom is usually virus-induced (Fig. 4.5). (2) The cause of *yellows* symptoms, consisting of chlorosis varying from slight blanching to complete bleaching, is more questionable (Figs. 4.6, 4.7). The plant responds to many pathogens in this way. The greatest effect of such yellows-type viruses is in limiting crop production. (3) *Necrotic* diseases are also caused by viruses, but the local lesions and shot-holing of leaves are easily confused with symptoms caused by fungi and bacteria or adverse nutrient relations. (4) Viruses also cause such distorting diseases as vein overgrowths, misshapen leaves, and tumors, which are rarely caused by the physical environment.

4-3 WHAT PART OF THE PLANT IS AFFECTED?

The observer should determine whether the entire plant or only certain parts of the plant are affected. Understanding the response of different plant parts is one of the most valuable aids in determining

FIGURE 4.5 *"Mosaic"-type virus expression on bean.*

FIGURE 4.6 *Yellowing, leaf rolling, and mosaic symptoms characterizing yellow mosaic virus disease of grape.*

the cause of the disturbance. If the entire plant appears to be dying uniformly, the chances are that there is something wrong with the roots or crown rather than with the above-ground organs. The crown is the base of the stem or trunk of a plant located just below the soil

FIGURE 4.7 *"Yellows"-type virus expression on* Malva.

line. When it is damaged, food and water movement is impaired and the above-ground parts suffer. As the water supply to the leaves is cut off, the leaves may turn various shades of yellow or brown, wilt, and finally die.

The entire top is not always affected. When a canker or local root or crown injury is responsible for a top symptom, symptoms will be most striking on the part of the top "fed" by the damaged part of the root system. If one suspects that the root system is damaged in some way, then the roots and crown should be examined. Inspection for tissue discoloration should start just above the soil line. If an area is found where the bark looks a little darker or is a bit sunken, cut into it. Diseased wood will usually be dark brown in color and show up in striking contrast to the moist white tissue of healthy wood. Discoloration is most often caused by a fungus which has invaded the tissue and is feeding on the root cells, but high temperatures and other physiopaths can evoke a similar response. If no discolored tissue is found above ground, and crown or root damage is still suspected to be responsible for the general decline and dieback, the roots should be examined similarly.

The crown and roots must be examined closely to see if the bark is intact or has been "girdled." Often rodents feed on the crown and sever the vascular connection between the roots and stem. Sometimes in tagging trees wires are left that choke the vascular connection as the trunk enlarges. Possibly another root has encircled the crown and choked it off.

The side of the tree trunk which is affected may also provide a clue as to the cause of the disturbance. If damage is consistently on the south or west sides of the trunk, rapidly changing temperatures can be suspected.

When the plant is in a state of general decline, poor soil or an adverse nutrient or water relation should also be considered as a possible cause.

Often only part of the plant appears sick. When a part of the plant shows symptoms, say a limb, one of two things may have occurred. Either the roots which supply the affected limb with water have been damaged, while the rest of the roots remain entirely functional; or a canker or local injury on the affected limb may have cut off the supply of water and nutrients at that point. Limb tissues should be inspected to locate any sunken, dried, or cankered areas which may be causing the disorder. It is also helpful to look closely at the affected limb and its leaves and notice how the leaves may vary from what is considered normal. Leaves on healthy plants should be examined to provide a standard for comparison.

4-4 WHAT IS THE DISTRIBUTION OF AFFECTED PLANTS?

The distribution of sick plants over an area is helpful to diagnosis. If only one plant is sick, then a local disturbance can be suspected. Where symptoms appear on a few adjacent plants in a small area, then a local adverse soil condition or even a parasite might be suspected. Look at the soil and other plants within the affected area. Is there a noticeable difference in the soil, such as color or texture? Maybe the soil in this area was managed differently, consciously or accidentally. Chemical sprays and dusts may have drifted from nearby fields. Under these conditions, symptoms are most likely to appear on a number of plants scattered over large areas of a field or on the entire planting. In residential areas where large acreages or different plantings are not available for inspection, the observer should make comparative observations in as many yards as possible.

If the trouble is limited to a single yard, the disorder may well have been caused by the homeowner. It is then necessary to determine how a particular gardening practice differed from those of the neighbors. If the disorder is more widespread, the cause is likely to be of a general nature, and further study is necessary.

There are a number of reasons why affected plants may be scattered over a large area or show varying degrees of symptom expression. Mainly there is the normal genetic variation among plants, some having greater resistance or tolerance to a particular pathogen than others. In addition, the disorder may have missed some plants in a given field. This would especially be true in a mild to moderate outbreak of some parasitic disease. Where a nonparasitic disorder is involved, the chances are greater that all individuals in a field will be affected at least to some extent.

4-5 IS MORE THAN ONE SPECIES OF PLANT AFFECTED?

Presence of symptoms on more than one kind of plant is one of the best clues that a particular disorder is being caused by some element of the environment. Parasitic diseases are more apt to be restricted to a single species of plant or even a single variety. There are exceptions, of course. Physiopaths such as air pollutants may also hit only a single species. A diagnosis can often be confirmed if the observer knows the relative sensitivity of affected species to various air pollutants and other physiopaths.

Where a disorder is affecting a commercial crop, it is worthwhile to inspect adjacent plantings as well as any weeds in the area. A number of weed species can usually be found near any crop, and if the weeds show similar symptoms, the observer is often assured of an environmental disturbance. Then the local environmental conditions must be evaluated to determine the specific physiopath responsible.

4-6 WHAT ARE THE CHARACTERISTICS OF THE TERRAIN OR LOCATION?

Geographic or edaphic features of the location account directly or indirectly for many disorders and well may be responsible for the disorder in question. Possibly a color difference in the soil may highlight the trouble, but often differences are disclosed only by chemical analysis. There are some other observable clues which may disclose a soil difference. A disease situation is sometimes associated with low areas in a field, where poor drainage and water accumulution modify the microclimate. The humidity in such areas may be higher and the temperature lower. This distinct microenvironment both modifies plant growth and influences development of parasitic organisms.

Differences in the size of plants over a field can also provide a clear sign of soil variation. For instance, plants will be stunted where the topsoil has been removed to level hills. Nutrients in such areas are deficient, and the plants will show the stresses of starvation.

Excessively sandy or clayey soils are also detrimental. If both heavy clay soil and sandy soils occur in the same field, plants growing in the heavier soils may be suffering from suffocation due to lack of aeration, while those in the sandy soil show symptoms of water deficiency because of excessive drainage. This soil texture can readily be determined by rubbing a small amount of soil between one's fingers. Coarse soils feel "gritty," while heavier soils feel smooth or greasy.

The slope of a field influences the development of disorders by determining the light intensity to which the plants are exposed. Light intensity in turn will affect temperature extremes as well as modifying water absorption, soil runoff, evaporation, and plant transpiration. In a similar way, the side of a house on which plants are grown may affect disease development. Certain plants tend to winter-kill on the south or west sides of a house, where daily temperature differences vary greatly in the winter months.

4-7 WHAT IS THE CROP HISTORY?

A disorder may be caused, or at least strongly influenced, by the growing conditions and specific management practices to which the plants have been exposed. Plants do not always respond to stress immediately, and symptoms may not show up for several weeks or months after the actual damage occurred. It is essential to know and evaluate this background before making a diagnosis.

Knowledge of crop rotation, or lack of it, is necessary in diagnosing parasitic diseases and is also useful in diagnosing environmental disturbances. Parasitic organisms may become abundant on one crop and cause extensive damage to subsequently planted, more sensitive crops. Some plant species release toxins which impair growth of other species, and continued cropping of a single species may lead to a buildup of toxic materials.

Where certain pesticides are used, they may accumulate to the point where plant growth and production of more sensitive species grown subsequently can be seriously affected. Consequently, it is vital to know as much as possible concerning the plants raised in previous years and the various treatments they may have received.

Background data on annual crops provides another aid in determining the history of a disease. Even if detailed records are not kept, the grower can usually recall where he got the seed, what pre- or post-emergence herbicides were used and in what concentrations, what pruning and fertilizer practices were followed, and whether any fungicides or insecticides were used and in what concentrations.

Observation can sometimes reveal management practices to the "trained eye." For instance, the amount of nitrogen applied can be estimated from the amount of new growth. Adequacy of pest control practices can be determined from the presence or absence of various insects and diseases. The type of pesticide used can be equally important, since these chemicals in themselves can be highly toxic, with damage often resulting from their indiscriminate or careless use. It is also helpful to know if any chemicals have been applied to adjacent ditch banks, roadways, or fields, since these may also be responsible for the disorder being diagnosed.

The same information should be obtained where the disorder occurs on biennial or perennial species. The commercial grower may have crop records available. The home gardener should try to recall what he did to the plant in question and to nearby trees and shrubs.

Diagnosing disturbances in natural areas such as forests or rangelands may prove more difficult, since management records may be unavailable. In a few instances the use of pesticides may be responsible for plant disturbances. Pesticides of various types are being applied to a greater extent over larger areas of forests and range. Since these can be a factor, they should be considered in diagnosis whenever damage is found to coincide with areas treated. Several hundred acres of aspen in Utah were killed in one instance where 2,4-D had been applied to control a noxious forage species growing in the aspen stands. Similarly, exposure to air pollutants may damage plants. Effects of pollutants may persist long after the "smoke has cleared." This would be especially true where pollutants killed trees or stunted their growth.

Environmental disorders in forest areas are often associated with such climatic factors as temperature and rainfall. Thus, a record of climatic conditions is a vital part of the crop history. Climatic factors should be suspected when injury occurs on unrelated plant species. When this happens, the symptoms on the different kinds of plants usually look much alike. Also, the same kind of plant in nearby locations would be expected to show the same symptoms.

Maximum and minimum temperatures recorded the previous year, along with alternating periods of warm and cold temperatures, probably constitute the most important climatic background information. Winter damage, including limb dieback and needle burning of conifers, is most common when normal cold temperatures resume after a period of relatively mild weather. Damage such as leaf necrosis from high summer temperatures is most common when normal high temperatures follow an unusually cool spring during which a succulent, heat-sensitive foliage has matured.

Precipitation records also may be useful to diagnosis. Exceptionally heavy precipitation may saturate soils for a prolonged period, suffocating the roots and causing oxygen deficiency. Too much precipitation in the form of hail can be destructive through direct physical damage. Too little moisture during either the winter or summer may result in a failure of the plants to absorb enough water for transpiration. Symptoms of leaf rolling and marginal yellowing and burning are commonly associated with this water stress.

When all the above factors are considered and the questions have been answered as completely as possible, the observer should have a general idea of whether an organism or the physical environment caused the disorder being studied; hopefully he should be able to pinpoint the specific environmental factor involved.

PART TWO

CLIMATIC
RELATIONS

CHAPTER FIVE

TEMPERATURE MECHANISMS

P lants live the year around. They may not bear foliage, flowers, or fruit, but they still live to emerge the next year when the environment again becomes suitable for active growth. Often the suitability of the environment, and with it the success or failure of a species, is determined by the maximum or minimum temperatures. While some species of higher plants can survive the hot blasts of the desert winds for hours, or even days, and others may survive prolonged periods of arctic cold, active plant growth is generally confined to a temperature range from about 10 to 40°C. Temperature extremes above and below this impose stresses on the plant's metabolic activities leading to the varied symptoms of leaf chlorosis, fleck, and scorch; needle blights; stem lesions and cankers; fruit scald; and finally complete breakdown. Before discussing these temperature-related disorders, it is important to understand the normal temperature mechanisms and why temperature deviations cause plant malfunctions.

Molecules are constantly in motion. Temperature is a measure of the activity of molecules and the energy involved in their motion. The speed of this motion is proportional to the heat energy to which the molecules are subjected. At low temperatures molecular activity is very slow, and there is not enough energy for metabolic processes to proceed; when temperatures are too high, molecular motion is extremely rapid, and the complex protein molecules come apart. Only when the heat energy level is about 10 to 40°C can chemical reactions proceed normally.

The heat energy possessed by a plant comes to it largely as radiant energy from the surrounding environment. It comes from direct sunlight, from sunlight reflected by clouds or molecules in the atmosphere, and from thermal radiation from the atmosphere, plants, and other warm surfaces. The plant receives additional amounts of energy by conductance and convection from the surrounding air if the air is warmer than the plant.

Far greater quantities of energy originating in the sun are absorbed by the atmosphere and never reach the plant. A substantial amount of infrared radiation is absorbed by water vapor and carbon dioxide in the atmosphere and never reaches earth. The very short ultraviolet light is largely absorbed and blocked out by ozone and oxygen molecules in the upper atmosphere, but considerable ultraviolet of longer wavelengths does reach the surface. Even the small portion of the total radiation reaching the plant might well be lethal if all were absorbed; but plants do not absorb ultraviolet and infrared as efficiently as visible light energy. Some of the absorbed light energy is utilized in photosynthesis, but the remainder can create a substantial heat problem to which the plant must adapt.

The total amount of energy absorbed by the plant depends on the leaf pigmentation or color, the morphological characteristics of the plant, the exposure or orientation of leaves to the sun, and the leaves' capacity to reflect, transmit, or absorb the solar energy.

This absorbed radiation represents an inflow of energy that raises the plant's temperature. If the plant continually absorbed this heat without dissipating any, its temperature would increase steadily until a lethal point was reached. But surplus heat is rapidly and steadily dissipated. Up to 70 to 90 percent of the surplus energy is effectively returned to the environment by reradiation from the plant's surface.

Heat is also transferred to the cooler air by conduction and convection; this convective loss is greatest when an appreciable difference exists between air and leaf temperatures.

Much of the heat energy still retained by the plant is ultimately transferred out of the leaf in water vapor via the mechanism of transpiration. Heat energy is instrumental in converting water in the leaf cells from the liquid to the vapor phase in evaporation. This process utilizes and consumes heat energy from the cells cooling the leaf. Water vapor then passes from the intercellular spaces of the leaf, through the stomata, to the external atmosphere in transpiration. At high temperatures this system of evaporative cooling may account for over half the heat lost by the plant. Slatyer and Bierhuizen (1964) found that a correlation existed between transpiration and leaf temperature; if transpiration was reduced 50 percent, a difference of about $4°C$ between leaf and air temperatures appeared. A complete suppression of transpiration would be expected to give a difference of 8 to $9°C$ between leaf and air temperatures (Fig. 5.1). By reducing leaf temperatures as much as 5 to $10°C$, transpiration provides a satisfactory heat balance, enabling the plant to continue normal physiological activity.

Heat absorbed by a plant during the day normally exceeds the amount lost. As a result, the accumulated heat energy may raise the temperature of the plant, particularly the leaf, above that of the surrounding air. At night the reverse is true; heat is radiated from the plant more rapidly than it is absorbed, causing the plant to be cooler than the surrounding air. Consequently, plant temperatures may be higher or lower than air temperatures and not directly related to them. A passing cloud or a slight breeze are sufficient to cool a sunlit leaf several degrees while not noticeably altering the air temperature (Fig. 5.2). In an exposed field, where radiant energy is highest, leaf temperatures may be 10 to $15°$ above the surrounding air temperature (cited by Leopold, 1964). Leaf temperatures of 40 to $50°$ are not uncommon in still air where air temperatures do not exceed $35°C$ (Gates, 1963). The temperature of bulkier tissues such as fruits and stems may even exceed air temperatures by as much as $30°$.

In summary, the plant temperature is a function of the difference between the gain and loss of heat energy by the plant. This heat balance is determined by the difference in heat absorption by radiation and convection and heat loss by reradiation, conduction, convection, and transpiration.

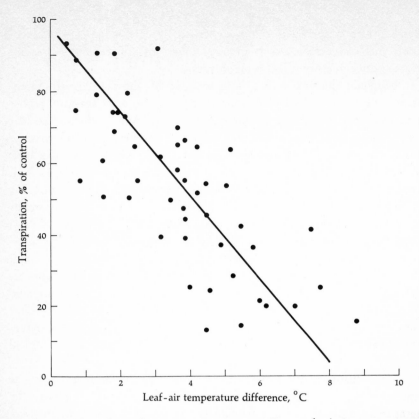

FIGURE 5.1 *The effect of transpiration suppression on leaf-air temperature difference* (after Slatyer and Bierhuizen, 1964).

5-2 THE NORMAL ROLE OF TEMPERATURE

Normal plant development is dependent on a temperature regime suitable for metabolic activity. The processes influenced most strongly by temperature include chemical reactions, gas solubility, mineral absorption, and water uptake.

First, plant development is largely a function of the cellular biochemical reactions. These reactions are controlled by the enzymes; the rate at which the enzymes act and the reactions proceed is a function of temperature. The rate of most chemical reactions doubles with every 10-degree temperature increase up to about 20 to 30°C;* above

---

* The rate of thermochemical reaction depends upon the number of molecules present in the system possessing sufficient energy to react. The heat content is measured by the kinetic energy (KE) of the system and reflects the average KE of the molecules.

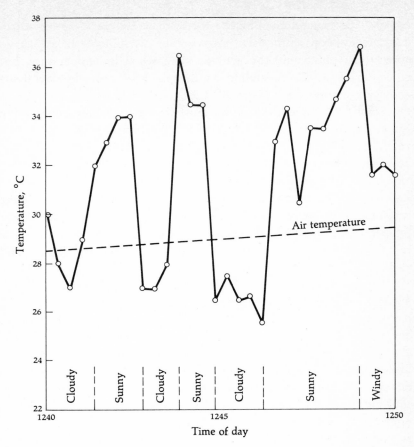

FIGURE 5.2 *Leaf temperatures of* Populus acuminata *as observed on July 22, 1961, in Boulder, Colorado, showing the varying response to sunny, cloudy, and windy conditions* (after Gates, 1963).

this, reactions decrease because the enzymes are gradually denatured or inactivated. The higher the temperature, the more rapid the denaturation. Below about 10°C, the frequency with which the reactants reach the needed energy level is low, and enzyme activity is minimal. Photosynthesis is one of the more important reactions influenced by temperature. Even though it is a photochemical process, it depends on

---

The energy necessary to react, i.e. the activation energy, is a characteristic of the system. When the temperature of a gas system is raised 10°C, the average KE of the molecules increases by approximately 3 percent. The number of molecules with energy equal to or greater than the activation energy increases as much as threefold for the same increase in temperature. Thus the rate of the reaction may even increase two- to threefold for an increase of 10°C.

enzyme activity and must be considered to be indirectly temperature-dependent. Photosynthesis in most plants is negligible below 10°C; above this, photosynthetic activity increases with temperature, but above about 30°C, depending on the plant species, photosynthetic activity again decreases (Leopold, 1964).

Secondly, temperature is the most important factor determining the solubility of gases in the plant cell. Carbon dioxide and oxygen solubility are particularly influenced by temperature (Leopold, 1964). Low temperatures facilitate solubility of these gases, and large amounts can be held in the sap of the plant cells. This suggests that lower temperatures may favor carbon dioxide fixation and increase the carbohydrate reserves, which helps protect plants against still lower temperatures. The high carbon dioxide concentrations present at low temperatures may also increase the acidity of the cell sap slightly, and this could influence the availability of the plant nutrients.

Thirdly, temperatures directly influence the availability and absorption of mineral elements from the soil. In active absorption, the plant absorbs essential mineral ions against a concentration gradient and must actively expend energy to obtain these nutrients. The availability and expenditures of this energy are temperature-dependent. Prolonged low temperatures limit the available energy, and nutrient deficiencies result. Furthermore, the tenacity with which soil particles cling to the mineral ions is regulated by temperature. Nutrients remain tightly bound to the soil at low temperatures, and more energy is required to absorb them.

Fourthly, temperature influences the ability of the roots to absorb water. The viscosity of water doubles as temperatures drop from 25 to 0°C; water is bound tightly to the soil and is absorbed by plants only with extreme difficulty. Low temperatures can significantly limit water uptake—particularly passive absorption, which is strongly influenced by viscosity, even when water is abundant. Optimum absorption of water generally takes place above 30°C, but high temperatures are also conducive to rapid water loss, which may cause moisture stress.

5-3 TEMPERATURE REQUIREMENTS

The optimum temperature for maximum metabolic activity varies markedly from species to species and even among populations and individuals of a single species. Different optimum temperatures also exist for different plant organs and for each successive stage of plant development. The rate of leaf development, for instance, and the size which the leaf ultimately attains are temperature-dependent. Studies with wheat serve to illustrate this. From 10 to 25°C, the increase in leaf size was proportionate to temperature, but a further increase to 30° decreased the leaf area. Still higher temperatures caused leaves to be thicker and smaller but favored stem growth (Friend et al., 1965).

A young stem developing early in the growing season may grow best at one temperature while later growth stages and reproduction require different temperatures (Went, 1953). The lower temperatures of spring are best for young growth of many species, while higher summer temperatures favor flowering, and cooler temperatures are optimum for stem elongation and fruit development. The sequence is exemplified by the tulip: optimum flower initiation occurs at 20°C, subsequent development of flowers is best near 10°C, and later stem elongation is favored by temperatures of 15 to 25°C.

Seed germination, growth, flower formation, and plant hardiness are all influenced by the fluctuations of temperature. Diurnal alternation of relatively high and low temperatures, also the actual day and night temperatures, exert a controlling influence on the growth of certain plants, such as the coast redwood, while night temperatures are most important in regulating the growth of others, such as the digger pine.

A number of plant species including tomato, potato, and pepper are known to develop best when moderate daytime temperatures are augmented by cool nights (Went, 1957). Night temperatures can determine the success or failure of crops in a given area. Tomato plants, requiring cool nights to set fruit, are consistently productive only in regions where night temperatures fall in the optimal fruit-setting range of 15 to 18°C. Low night temperatures also have such beneficial effects as increasing flowering and improving quality and flavor of fruits such as strawberry, apple, and plum. Potatoes, close relatives of the tomato, require night temperatures of 10 to 15°C to form tubers and are productive only in areas where such temperatures are provided. Other species such as African violet and zinnia thrive only when nights are warmer. Still other species are little affected by diurnal variations but prefer relatively cool temperatures below 20°C (Went, 1957).

Reasons why low temperatures are required for normal plant growth are not completely clear, but recent studies have shown that relatively cool temperatures may influence growth by modifying the chemical composition of plants. Beevers (1964) grew rye plants under three temperature regimes and found that plants growing continuously at 12°C had a higher carbohydrate and nitrogen content than similar plants grown in warmer regimes. The relatively high carbohydrate level was considered to result from the slower degradation of carbohydrates caused by decreased respiration at lower temperatures. The limited availability of energy lowered the rate of metabolic processes and growth, although there were no visible symptoms of injury to the plants.

Seasonal variation in temperature may be even more important to normal plant development than daily variation. A prolonged period of low temperature, the rest period, is required for many plants to complete their life cycles. Seeds and buds of most perennials of temperate regions must undergo a period of rest or dormancy before they can develop. Flower and leaf buds of deciduous trees typically require a rest period in order to open in the spring. In general, flower buds require several hundred hours of temperatures below 7°C before they can emerge. Leaf buds of many woody perennials (such as peach, pear, and cherry) require still longer exposure to low temperatures before dormancy is broken. Failure to meet this chilling requirement causes abnormal plant development and twig dieback. When the chilling requirement is barely met, bud development is suppressed and lateral growth sparse. Dormancy which is brought about by an accumulation of growth-inhibiting chemicals in the plant during the growing season is valuable in suppressing growth during periods when prevailing low temperatures might kill the plant.

The relative air and soil temperature, as well as either alone, is important to the normal development and distribution of plants (Daubenmine, 1959). High air temperatures have been reported to stimulate top growth of redwood seedlings regardless of soil temperatures, while low air temperatures, combined with warm soil, stimulate root growth. Roots grown in soil held at 18°C appeared to be healthiest even though they did not make the most growth. Roots developing at 29°C were thinner and darker, and at 8°C were thicker and shorter than normal.

The more temperature is investigated, the clearer it becomes that it exerts a controlling influence over every life process. Diurnal and seasonal temperature variations, relative soil and air temperatures, and temperature extremes are all vital to normal plant development, and all influence plant distribution and survival. Temperature deviations can impose stresses on the plant which lead to abnormalities and a reduced chance for survival.

5-4 MECHANISMS OF TEMPERATURE STRESS

Normal plant functioning depends on proper control of every cellular biochemical reaction. Temperature is among the most critical controls, and every plant species is best adjusted to a particular temperature regime. In any geographic area there are years in which temperature extremes fluctuate beyond the normal range, suppressing growth and causing varied symptoms of malfunction and stress on even the best-adapted species.

As the plant temperature deviates from optimum and approaches either the high or low limits of tolerance, the various enzymes decrease their activity one by one, reducing metabolic activity and suppressing plant development. Even within the range regarded as "normal," low or high temperatures can suppress growth of native and introduced plants in the absence of visible symptoms.

Decreased molecular and metabolic activity are the first responses of plants to low temperatures. This decrease in metabolism is reflected by an inhibition of growth and a general impairment of plant development. When activity diminishes beyond a critical point, the increasing stress structurally damages the cytoplasm, limiting plant development and causing visible symptoms of injury.

More conspicuous symptoms appear when subfreezing temperatures crystallize water, increasing the sugar concentration and osmotically or physically damaging the cells (Levitt, 1951). The relatively high sugar concentration may lead to osmotic damage, but the major damage from subfreezing temperatures is caused by the physical disruption of the cell contents by the movement of water in and out of the cells during freezing and thawing. Ice forming both between and within the cells physically abrades and ruptures the membranes, causing additional injury. This protoplasmic abrasion occurs simultaneously with the deformation of the cytoplasm as the water freezes and thaws and mechanically disrupts the physical structure of membrane molecules. Molecules in the protoplast are further damaged as the protoplast is dehydrated by the formation of extracellular ice. Such internal cellular freezing damage generally produces externally visible injury symptoms.

Subfreezing plant temperatures are more frequent than subfreezing air temperatures and occur far more often than suspected. Even where air temperatures remain above freezing, the plant temperature may drop below freezing because of the heat loss from radiant cooling taking place when the plant tissues reradiate heat energy. Radiant cooling is particularly significant under desert or alpine conditions, where the plant radiates to the cosmic cold of space through the clear transparent atmosphere. If the soil temperature is cooler than that of the leaf, it too accepts radiant energy from the leaf, cooling the leaf still further and causing low-temperature injury.

All parts of the plant are subject to heat loss and injury from radiant cooling, but leaves, having a broad, flat surface, lose heat most effectively. The leaf tips and margins are especially efficient for night cooling, sustain the greatest heat loss, and are thus most readily injured (Geiger, 1957).

The temperature extremes which the plant cell can endure are determined largely by the capacity of the component molecules, especially the proteins and lipids, to maintain their structure. This capacity is to a great extent a function of the sulfhydryl and hydrogen bonding. Hydrogen bonds tend to pull together much more tightly at temperatures below 0°C than at higher temperatures. This pulling shrinks and distorts the protein structure, which in turn reduces—or even destroys irreversibly—their activity and ability to function normally. The type and strength of hydrogen bonding and other types of bonds and hydration of the proteins largely determines the capacity of a cell to withstand high as well as low temperatures.

As the temperature increases, the bonding becomes looser, until at around 30°C bonds tend to "come apart," disrupting many enzymes and denaturing the molecules. If high temperatures persist, bonds in the protein molecule become irreversibly disrupted, and the enzymes become nonfunctional. When an enyzme is inactivated, the metabolic reaction which it regulates is blocked, and the growth processes with which it is associated are suppressed.

Indirect injury may occur at temperatures just below the point of protein denaturation (Langridge, 1963). As the temperature increases, the reaction rates and metabolic activity increase proportionately. As temperatures approach 30°C, the metabolic rate can be very high; the increased activity places a heavy demand for essential elements on the cell, and if these elements cannot be supplied quickly enough, varying degrees of starvation can be expected.

Cellular damage from high temperatures may also result from the formation of toxic substances in certain cells exposed to a localized high temperature. This toxic material may subsequently be translocated to other parts of the plant and cause more widespread injury (Yarwood, 1961).

Regardless of the mechanisms involved, when the temperature exceeds the extremes to which the plant is adapted, growth is impaired and expressions of stress and malfunction appear. This expression is apparent as a plant disorder.

BIBLIOGRAPHY

Beevers, L., and J. P. Cooper, 1964. Influence of temperature on growth and metabolism of rye grass seedlings. II. Variation in metabolites. *Crop Sci.* **4**:143–146.

Daubenmire, R. F., 1959. "Plants and environment," 2d ed. Wiley, New York, 422 pp.

Friend, D. J. C., V. A. Nelson, and J. E. Fisher, 1965. Changes in the leaf area ratio during growth of marquis wheat as affected by temperatures and light intensity. *Can. J. Bot.* **43**:15–20.

Gates, D. M., 1963. Leaf temperature and energy exchange. *Arch. Meteorol. Geophys. Biol.*, **B12**:321–336.

Geiger, R., 1957. "The climate near the ground," trans. by M. N. Stewart. Harvard, Cambridge, Mass., 482 pp.

Langridge, J., 1963. Biochemical aspects of temperature response. *Ann. Rev. Plant Physiol.* **14**:441–462.

Leopold, A. C., 1964. "Plant growth and development." McGraw-Hill, New York, 465 pp.

Levitt, J., 1951. Frost, drought and heat resistance. *Ann. Rev. Plant Physiol.* **12**:195–218.

Slatyer, R. O., and J. F. Bierhuizen, 1964. The influence of several transpiration suppressions on transpiration, photosynthesis, and water use efficiency of cotton leaves. *Aust. J. Biol. Sci.* **17**:131–146.

Went, F. W., 1953. The effect of temperature on plant growth. *Ann. Rev. Plant Physiol.* **4**:347–362.

———, 1957. Climate and agriculture. *Sci. Amer.* **196**:82–94.

Yarwood, C. E., 1961. Translocated heat injury. *Plant Physiol.* **36**:721–726.

SELECTED REFERENCES

Evans, L. T. (ed.), 1962. "Environmental control of plant growth." Academic, New York, 499 pp.

Gates, D. M., 1962. "Energy exchange in the biosphere." Harper, New York, 151 pp.

——— and C. M. Benedict, 1963. Convection from plants in still air. *Amer. J. Bot.* **50**:563–573.

Hudson, J. P., 1957. "Control of plant environment." Academic, New York, 240 pp.

Milthorpe, F. L., 1956. "The growth of leaves." Butterworth, London, 233 pp.

Raschke, K., 1960. Heat transfer between the plant and the environment. *Ann. Rev. Plant Physiol.* **11**:111–126.

Troshin, A. S. (ed.), 1967. "The cell and environmental temperature." Pergamon, Oxford, 462 pp.

Trumbore, R. H., 1966. "The cell, chemistry and function." Mosby, St. Louis, 412 pp.

Went, F. W., 1957. "The experimental control of plant growth." Chronica Botanica, Waltham, Mass., 343 pp.

Whyte, R. O., 1946. "Crop production and environment." Faber, London, 372 pp.

Wolpert, A., 1962. Heat transfer, analysis of factors affecting plant leaf temperatures. Significance of leaf hair. *Plant Physiol.* **37**:113–120.

CHAPTER SIX

DISORDERS ASSOCIATED WITH HIGH TEMPERATURES

Plants grown in the temperate regions of the world are poorly adapted to withstand the stresses imposed by high temperatures and can rarely tolerate temperatures much above 35°C. Normal development may be impaired, and plants may become visibly damaged.

Temperature extremes insufficient to kill the protoplast may still inhibit growth, impair vigor, and suppress production. Growth responses are difficult to detect in the field and often go unrecognized, although production losses from the reduced growth may be significant long before visible symptoms appear.

6-1 MORPHOLOGICAL AND PHYSIOLOGICAL ADAPTATIONS

Plants have evolved morphological and physiological adaptations in order to function within their particular temperature and water regimes. *Mesophytic* plants, which include most of the native cultivated species of temperate climates, and *hydrophytes*, those plants thriving in wet habitats, lack the adaptations enabling them to function normally at high temperatures. While banana plants in Honduras withstand very high temperatures and tropical grasses (corn, *Tripsacum*) do well at 45° or greater, it is mostly the *xerophytes*, or desert plants, which have developed the modifications necessary for withstanding both extremely high temperatures and great aridity. The effectiveness at which plants will function at high temperatures depends on the degree to which the following modifications are developed.

1. *Leaf arrangement and orientation.* Leaf position enables many cultivated species to escape high, lethal temperatures. Leaves oriented vertically to the sun absorb a minimum amount of solar energy and remain several degrees cooler than those exposed more at right angles. If a leaf is turned at a 10° angle away from perpendicular to the incoming light, heat absorption is reduced about 15 percent; when oriented more than about 70°, heat absorption becomes negligible.

2. *Coloration.* Leaf color greatly influences heat absorption. Dark surfaces absorb heat, while light surfaces reflect heat. Whitish or gray-green leaf and stem coloration, characteristic of many desert plants, reflects the sun's rays, limiting heat absorption. Cultivated species having light-colored leaves similarly escape excessive heat absorption.

3. *Cuticle,* a thick waxy layer of cutin covering the leaves of many desert plants, reflects large amounts of the heat and partially insulates underlying cells. Although reflection is often high, transmission is reduced and absorption is higher than for more mesic leaves. Many cultivated plants like onions, beans, peppers, and tomatoes possess a heavy cutin layer partially protecting them from heat damage.

4. *Leaf hairs* have been reported to be of value in shading the living cells (Wolpert, 1962). By absorbing or reflecting much of the incident heat energy, leaf hairs may reduce the temperature of the underlying cells by 1 to 2°C. By improving conductive cooling, leaf hairs act as fins, greatly increasing the leaf surface area and facilitating conduction of heat to the atmosphere.

5. *Transpiration.* Transpiration, the loss of water from plant tissues in a vapor form, is an important mechanism for transferring much of the absorbed heat energy out of the leaf, thereby reducing its

temperature (Gates, 1965). A high rate of transpiration protects plants exposed to the high incidence of light and heat energy. Only through transpiration are they able to transfer the overabundance of heat energy absorbed out of the leaf and in this way to avert lethal temperatures.

6. *Hydration of the protoplast.* The moisture content, or degree of protein hydration in the protoplasm, influences heat tolerance. High temperature tolerance decreases with the degree of hydration. When there are fewer water molecules present, hydrogen is more strongly attached—or, more accurately, attached at more points to the protein molecule. The greater degree of bonding gives the molecule greater stability; the protein molecule is less subject to deformation and can withstand higher temperatures without breaking down or being degraded. The low degree of hydration found in lichens is considered responsible for their capacity to withstand temperatures of 70°C.

Plant species can sometimes acquire heat tolerance by *hardening,* that is, the building of resistance by exposure to gradual increases in temperature. Hardening is determined by the thickening viscosity of the cytoplasm. A gradual increase in cytoplasmic viscosity results in a higher degree of water binding in the cells, which increases the capacity of molecules to adjust to the molecular deformation caused by high temperatures.

Regardless of the plant adaptation, temperatures ultimately reach a point where they cause irreversible chemical alteration, denaturing the protoplast, disrupting the physical structure of membranes, and bringing about cell plasmolysis, along with visible symptoms.

6-2 LEAF DISORDERS

Chlorosis and Variegation

Loss of chlorophyll and the resultant chlorosis form the earliest visible response of leaves to excessively high temperatures. The destruction of the green pigments of the leaf produces a pale green to yellow or even bleached discoloration developing first at the leaf tip and margin, where the degree of radiation and heat stress is greatest. Chlorosis is often followed or accompanied by wilting and necrosis, but if the stress is of brief duration, the symptom may not proceed beyond chlorosis.

Chlorotic mottling and variegation, independent of marginal chlorosis, is another early symptom of heat stress. Variegation is a type of chlorotic mottling showing irregular, intercostal yellow and pale green areas. The symptom is similar to that produced by certain mosaic and yellows-type viruses. The appearance of chlorotic mottling depends on the physiological conditions and maturity of the tissue involved. The degree of injury is usually determined by the age of the leaves and the duration of exposure to high temperatures.

Heat-induced leaf chlorosis was produced in the laboratory by Dr. G. T. A. Benda (1962), who exposed tobacco plants to air temperatures of 35 to 36°C for twenty-four hours. Chlorotic areas developed on actively expanding leaves 2 mm or less in length at the time of treatment. Cells were injured and chlorosis developed at this stage. As the leaf enlarged to full size, the chlorotic areas enlarged proportionately. Apparently the chloroplasts in very young leaves were very sensitive to heat and slow to recover. Cytological abnormalities associated with high temperatures were most recently reported by M. Weintraub (1966), who induced virus-like chlorosis symptoms on tobacco by exposing leaves to 37°C for four to eight days. Inspection of leaf cells under an electron microscope showed that the most striking response to the high temperature was the degeneration of the chloroplast lamellae and swelling of the lipoid granules. Following chloroplast injury, the cytoplasm appeared coagulated. Continuation of stress and cytoplasmic breakdown ultimately ended in degradation of the protoplast, death of the cell, and visible necrosis.

Leaf Scorch

High temperature causes *leaf scorch* directly or indirectly by stimulating excessive evaporation or transpiration. Transpiration is influenced by temperature; as temperatures increase, molecular activity increases, and the water loss by transpiration goes up proportionately. Transpiration may be so rapid at high temperatures that water cannot move up the stem fast enough to keep pace with water transpired from the leaf. If water is not replenished, the protoplasm becomes dehydrated, leaves become flaccid and wilt, chloroplasts and chlorophyll break down, cells die, and chlorosis and necrosis appear. Since water deficiency first occurs in cells nearest the terminus of the vascular system, injury appears first at the tip and margin of the leaf, where many of the veins end.

Leaf scorch affects many different kinds of plants but is particularly common on shade-loving species and on leaves which have developed under cool conditions and are suddenly exposed to the intense heat energy of the summer sun.

A striking example of such temperature response happened in northern California in 1961 (Treshow, unpublished), when on June 16 temperatures suddenly jumped above 38°C in areas which had enjoyed an exceptionally cool spring. Leaf scorch was observed on such varied species as coast redwood, grape, apricot, acacia, lilac, rose, elm, English walnut, plum, box elder, boysenberry, fig, modesto ash, Monterey pine, juniper, euonymous, quince, ginkgo, and hibiscus. Symptoms consisted of leaf tip and margin browning (Figs. 6.1 and 6.2), and occasionally the silvering and glazing of the intercostal areas of sensitive plants (Fig. 6.3). Scorch was particularly intense on leaves exposed at right angles to the sun and on plants growing on south-facing slopes. Grape and apricot fruits were desiccated and shriveled early in their development whenever not protected by the foliage.

S. L. Mielke reported similar necrosis on native broad-leaved species in northern California in 1942. Leaf tissue was killed following a rapid increase in temperature. Vast acreages of oak and other indigenous species turned brown almost overnight despite ample soil moisture.

FIGURE 6.1 *Marginal scorch of oak leaf following sudden exposure of predisposed leaves to temperatures close to 40°C.*

FIGURE 6.2 *Marginal scorch of apricot leaves exposed to high temperatures.*

FIGURE 6.3 *Sunscorch of sensitive, predisposed raspberry leaves exposed to direct sun.*

Horse-chestnut and maple trees are especially sensitive to heat stress and are affected wherever they are grown. Leaves of Norway maples and horse-chestnuts burn severely in Salt Lake City almost every year (Fig. 6.4). Leaf tip and margin burning is most severe on foliage above pavement, where the temperature may be several degrees higher than over lawns.

Tip burn of potato is widespread and exemplifies the type of necrosis occurring on leaves of many herbaceous species (Lutman, 1919). Symptoms first appear as a slight wilting and yellowing of the tips or along the margin of the potato leaflets. Yellowing soon yields to browning, which advances from the tip down the margin. Expression is most severe on middle-aged leaves, since these are in a position to be struck by the sun nearly at right angles to their surface. Leaves are also predisposed and necrosis most severe when a period of cloudy, rainy weather promoting rapid top growth is followed by hot, sunny days.

Another form of necrosis characterizes *onion blight* (Ivanoff, 1938). Here, leaf tip drying and necrosis are accompanied by small, round, white-to-gray spots appearing over the leaf. The disease is

FIGURE 6.4 *Sunscald of Norway maple, possibly associated with both moisture and heat.*

most severe when hot, dry weather follows a wet period. Lutman (1919) reported one year in which 40 percent of the foliage of some plants was killed by mid-August. More sensitive varieties had 90 to 100 percent necrosis when plants were growing on light, sandy soils. The disorder can best be prevented by planting heat-tolerant varieties and providing good soil and moisture.

Sunscald occurs on such leafy vegetable crops as lettuce and cabbage when leaves on the top of the head are exposed to intense heat. The first symptom of sunscald is the water-soaked or blistered appearance of irregularly shaped areas which readily become bleached and parched. The disorder is common in the Salinas Valley of California, where the russeted appearance of affected plants gives rise to the name *russet spotting* of head lettuce. Russet spotting is influenced directly by maximum air temperatures. Lipton (1963) reported that lettuce harvested nine to fourteen days after temperatures exceeded 30°C for two or more consecutive days developed substantially more russeting than lettuce not exposed to these temperatures.

Tip burn of lettuce is a similar temperature-related disorder and one of the most serious diseases of lettuce (Anderson, 1946). Symptoms consist of small, dark brown spots about ¼ in. from the edge of the leaf. The tissue around these spots dies and turns brown. Necrotic spots arise along the petiole and veins of leaves throughout the lettuce head, giving rise to the name of rib rot or rib blight. Tip burn is favored by high humidity as well as high temperature, but there is no proof that the disease is caused directly by high temperature. Anderson (1946) reported that water deficiency or fluctuating soil moisture were even more critical to disease expression than temperature. The difference between maximum air and soil temperature was also important to disease development. Where the disorder is a problem, tip burn is averted largely by planting such relatively resistant lettuce varieties as Great Lakes. Even this variety sometimes develops characteristic spotting symptoms.

Year in and year out, leaf scorching is a chronic condition on both herbaceous and woody plants. It can be avoided only by eliminating such contributing stress conditions as soil compaction and by providing the best growing conditions possible.

Tip burning or necrosis develops on conifer needles at the slightest environmental stress. *Needle blight* is the name often given for necrosis or leaf scorch of pine, spruce, fir, and other needle-leaved, coniferous plant species. Symptoms of plasmolysis, drying out, and burning of the needle from the tip back are basically similar to those described for broad-leaved species. Often a narrow, chlorotic, diffuse transition band appears between the healthy and necrotic tissue. The current year's needles are most sensitive to heat, especially as they are emerging and before the thick, insulating, waxy layer of cutin is formed to partially protect them. The newly developing needles are apt to be burned when they elongate during relatively cool periods and are suddenly confronted with higher temperatures. If preceded by an abnormally cool period, even relatively normal summer temperatures can cause necrosis.

Needles may be affected by high temperatures even though no necrosis or chlorosis appears. Parris (1967) described a *needle curl* condition of shortleaf, longleaf, loblolly, slash, and lodgepole pine which he postulated might be caused by high temperatures in the absence of moisture stress. The same condition is known as the *fused needle* disease in Australia. Needle curl appeared on seedlings growing at temperatures of 37 to 43°C for periods of six to seven hours during one or more days.

6-3 STEM DISORDERS

Stem Lesions

Stem lesions occur on plants of all ages but are most common on seedlings and young plants. Symptoms caused by high temperatures closely resemble stem lesions caused by damping-off fungi in that shrunken, necrotic areas develop on the young seedlings near the soil line. Stem tissue in this area is killed and becomes constricted, and the young plant dies and topples over. Symptoms are distinguished from damping-off in that the temperature-induced necrosis occurs only at the soil line, while fungus-induced necrosis usually extends well below. Also, the earliest lesion symptoms usually appear at the south or southwest side of the stem, are lighter in color, and have definite boundaries. The light color of the dead tissue has given rise to the name *white spot* for the disorder.

Native and cultivated species alike are susceptible. One- and two-year-old plants are affected, and occasionally conifers up to four years old are killed. Losses have been reported on such conifers as Douglas fir, white fir, and white pine. Bates (1924) gave the relative order of heat tolerance of four evergreen seedling species as lodgepole pine, yellow pine, spruce, and Douglas fir. Stem lesions produced artificially on ponderosa pine seedlings exposed to high temperature first appeared dark grayish-green in color, gradually becoming lighter, with tissues shriveling within twenty-four hours.

The extent and severity of heat injury is dependent both on the temperature and the duration of exposure. Cortical cells of white pine killed in thirty minutes at 57°C were killed in less than five minutes if temperatures were over 63°C. Injury to conifer seedlings begins at soil temperatures of about 45°C. This thermal death point may be only a few degrees above the optimum for growth.

Since surface soil temperatures may be higher than air temperature during midday, they are considered to be of greater significance in causing stem lesions and cankers. Even in the shade of lath frames where air temperatures are relatively mild, soil surface temperatures as high as 52°C have been recorded—more than adequate to kill plant tissues.

Heat injury to tender young seedlings of agronomic as well as native species is far more common than the record might indicate, and it contributes substantially to poor stands of crop plants. Injury from high soil temperatures is exemplified by *heat canker of flax* (McKay, 1940; Reddy, 1922) but has also been described on beans, buckwheat, cowpeas, pea, rye, vetch, barley, wheat, maple, and oaks.

Heat cankers are produced when high temperatures, above about 54°C at the soil surface, kill the cortical cells in the outside portion of the young stem so that affected tissues shrink and collapse at or near the soil surface. Plants less than 3 in. collapse. Plants 3 to 5 in. high sometimes survive, but the cortical cells are dead, shrunken, and collapsed, producing a tapered stem and impairing food transport. Since food transport is blocked at this point, carbohydrates accumulate above the canker, causing a slight swelling above the injured area, while the starved root system remains spindly. Plants over 5 in. are rarely affected since their thicker bark provides a more effective insulation.

Damage is most severe when flax stands are thin and the exposed bases of the plants are hit directly by the sun's rays. Thicker seeding will partly avert this hazard, as will planting in rows in a north-south direction. Sowing seed earlier in the season so the plants can reach 5 to 10 in. before temperatures get too high also helps avert injury.

Stem injury and wilting of bean seedlings from heat is fairly common. One incident, occurring near Greeley, Colorado, should serve to illustrate this type of injury (MacMillan, 1923). The first bean planting had been destroyed by late frosts. The replanted seedlings came up normally but wilted just after the first two primary leaves had opened. Closer inspection showed that the stems were shrunken at the ground line and that the upper parts of the plants were leaning over so badly that the leaves touched the ground. The root systems appeared normal. Soil temperatures were measured 1 in. below the ground level and found to average nearly 43°C for a ten-day period in the later part of June when the plants were but three weeks old. By the end of June, soil temperatures climbed to 50°C. Temperatures exceeding 54° at the soil line, sufficient to plasmolyze cortical and cambial cells of the stem, were estimated to have occurred for two to three hours each day.

Modified stems such as potato tubers and onion bulbs are equally sensitive to heat and readily damaged, particularly since the temperature of such tissues may be several degrees higher than the surrounding air. Heat necrosis has been noted in the early potato crop grown in the volcanic ash soils of Idaho. These soils absorb high amounts of heat energy so that soil temperatures are abnormally high. When vascular tissues and the surrounding parenchyma of the tubers are killed, a net necrosis symptom is produced. Affected tissues break down and are readily invaded by pathogenic organisms.

Sunscald

Sunscald, or winter sunscald, refers to the heat-induced cankers which arise on stem tissues during the winter months. Cankers are recognized by the appearance of discolored, dry, sunken tissue on the more exposed limbs and trunks of woody plants. Trees having a relatively thin, noninsulating bark, such as walnut, apple, and peach, are particularly sensitive.

Sunscald is not directly caused by high temperatures but rather by a rapid fluctuation of temperature (Harvey, 1925). This happens most frequently during the winter months, when the deciduous trees are bare of their protective foliage. The cambial temperature of unshaded peach tree limbs has been reported to reach as high at 30°C while the air temperature remained below 0°C. A temperature rise of 10°C, requiring but a few minutes, is primarily responsible for the scalding. The sensitive cambial cells are unable to adjust rapidly enough to accommodate fluctuations, and the rapid change can be lethal even when the maximum temperature is well below that ordinarily tolerated.

Death of the cambial cells often causes the bark to separate from the wood. Some splitting and cracking may then develop, but more often only a canker appears. Such cankers and cracks are most significant in providing access for parasitic fungi and bacteria.

6-4 FRUIT DISORDERS

Fruits may be injured by high temperatures alone, and heat-induced fruit scalding is observed almost every year in some areas of the country. Harvey (1923) found that a striking difference in surface temperature existed between the sunny and shaded sides of several kinds of fruit. Strawberries had temperatures of 34.4°C on the south side—some 8° higher than on the north side of the fruit and 12° above the air temperature. Comparable differences were also reported on cucumber and gooseberry. Apple fruits were as much as 12.5° hotter on the sunny side of the fruit than the shaded and 16.5° higher than the surrounding air (Brooks and Fisher, 1926). Temperature of the green parts of the apple was much higher than that of the red, and it was postulated that the red coloring might prevent overheating.

When temperatures of the half-grown apples exceed about 37°C, tissue just beneath the skin is killed and becomes "mushy." When the outer cells of the fruit are killed, the plant often repairs the damage by forming cork tissues beneath the wound. This results in a new but russeted skin. The dead outer tissues separate and eventually slough off as the fruit grows. The corkiness and sloughing may protect the fruit but mars its quality and may render it unmarketable.

High temperatures, even in the absence of immediate symptoms, can produce *water core,* a disease characterized by a watery appearance of the flesh near the core of the fruit (Brooks and Fisher, 1926; Harley, 1939). The high temperature disrupts the normal enzyme activity and membrane permeability, leading to excessive amounts of water accumulating in the intercellular areas. Water core is most common in states (such as Virginia) where rainfall near the time of harvest is normally ample, or in orchards in semiarid regions which are irrigated when temperatures are relatively high. Overmature apples are most susceptible, so the disease can be prevented by picking at the proper time. Delicious, Stayman winesap, and Duchess varieties are among the more commonly affected.

Exposed fruits of the tomato may also be "burned" by heat. Tissues develop a blistered, water-soaked appearance, giving rise to bleached or yellowish slightly depressed areas (Harvey, 1924).

Kelsey spot, or *heat spot,* of plums provides another example of a disorder caused by high temperature stress (Proebsting, 1937). The disorder is recognized by the appearance of shallow surface depressions of a slightly darker color than the normal skin. Necrosis of the underlying flesh may extend to the pit. Kelsey spot occurs when temperatures above 40°C persist for several hours. European varieties are far more tolerant of high temperature than the Japanese varieties and are safer to grow in hotter regions.

6-5 SEED GERMINATION

Seed germination is markedly influenced by temperature (Goodin et al., 1966). High soil temperatures can inhibit seed development or even kill seeds, preventing germination. Germination was found to decrease steadily as the temperature increased above 20°C and to cease completely at 45°C. Soil temperatures in the Imperial Valley of California frequently exceed this and may reach 70°C at the time of fall planting. At a ½ in. depth, dry seedbeds during September rarely dropped below 40°C, although marked contrast in temperature measurements occurred depending on slope, exposure, and moisture content. Where such extreme temperatures prevail even at night, a satisfactory stand would be difficult to establish, and subsequent production would be seriously limited.

BIBLIOGRAPHY

Anderson, E. M., 1946. Tipburn of lettuce. *Cornell Univ. Agr. Exp. Sta. Bull.* 829, 14 pp.

Bates, G. G., and J. Roesner, Jr., 1924. Related resistance of tree seedlings to excessive heat. *U.S. Dept. Agr. Bull.* 1263, pp. 1–16.

Benda, G. T. A., 1962. Heat-induced variegation, a model disease. *Phytopathol.* **52**:1307–1308.

Brooks, C., and D. F. Fisher, 1926. Some high-temperature effects in apples: Contrasts in two sides of an apple. *J. Agr. Res.* **32**:1–16.

———— and ————, 1926. Water-core of apples. *J. Agr. Res.* **32**:223–260.

Gates, D. M., 1965. Heat transfer in plants. *Sci. Amer.* **213**:76–84.

Goodin, J. R., R. M. Hoover, and G. F. Worker, Jr., 1966. High temperature effects on sugar beet germination in Imperial Valley. *Calif. Agr.* **20**:14–15.

Harley, C. P., 1939. Some associated factors in the development of water core. *Proc. Amer. Soc. Hort. Sci.* **36**:435–439.

Harvey, R. B., 1923. Condition for heat canker and sunscald in plants. *Minn. Hort.* **51**:333–334.

————, 1924. Sunscald of tomatoes. *Minn. Stud. Plant Sci., Stud. Biol. Sci.* **5**:229–234.

————, 1925. Conditions for heat canker and sunscald in plants. *J. Forest.* **23**:392–394.

Ivanoff, S. S., 1938. Onion "blight." *Tex. Agr. Exp. Sta. 51st Annu. Rep.,* pp. 260–261.

Lipton, W. J., 1963. Influence of maximum air temperatures during growth on the occurrence of russet spotting in head lettuce. *Proc. Amer. Soc. Hort. Sci.* **83**:590–595.

Lutman, B. F., 1919. Tip burn of potato and other plants. *Vermont Agr. Exp. Sta. Bull.* 214.

MacMillan, H. G., 1923. The cause of sunscald of beans. *Phytopathol.* **13**:376–380.

McKay, R., 1940. Heat canker of flax. *J. Dept. Agr. Eire* **37**:383–386.

Mielke, S. L., and J. W. Kimmey, 1942. Heat injury to leaves of California black oak and some other broadleaves. *Plant Dis. Rep.* **26**:116–119.

Parris, G. K., 1967. "Needle Curl" of loblolly pine. *Plant Dis. Rep.* **51**:805–806.

Proebsting, F. L., 1937. "Kelsey Spot" of plums in California. *Proc. Amer. Soc. Hort. Sci.* **34**:272–274.

Reddy, C. S., and W. E. Brantzel, 1922. Investigations of heat canker of flax. *U.S. Dept. Agr. Exp. Sta. Bull.* 1120, 18 pp.

Weintraub, M., and V. T. John, 1966. Cytological abnormalities induced by high temperatures in *Nicotiana glutinosa. Phytopathol.* **56**:705–709.

Wolpert, A., 1962. Heat transfer, analysis of factors affecting plant leaf temperature. Significance of leaf hair. *Plant Physiol.* **37**:113–120.

SELECTED REFERENCES

Beadle, N. C. W., 1940. Soil temperatures during forest fires and their effect on the survival of vegetation. *J. Ecol.* **28**:180–192.

Beevers, L., and J. P. Cooper, 1964. Influence of temperature on growth and metabolism of rye grass seedlings. II. Variations in metabolites. *Crop. Sci.* **4**:143–146.

Blackman, G. E., 1956. Influence of light and temperature on leaf growth, in F. Milthorpe (ed.), "The growth of leaves," pp. 151–169. Butterworth, London.

Boyce, J. S., Jr., 1959. Brown spot needle blight on Eastern White pine. *Plant Dis. Rep.* **43**:420.

Calvert, A., 1964. Effects of air temperature on growth of young tomato plants in natural light conditions. *J. Hort. Sci.* **39**:194–211.

Dippenaar, B. J., 1937. Cause and control of heat spot of plums. *Farming in S. Africa* **12**:83–85.

Griffin, H. D., 1965. Maple dieback in Ontario. *Forest Chron.* **41**:295–300.

Hellmers, H., 1962. Temperature effects upon optimal tree growth, in T. T. Kozlowski (ed.), "Tree growth." pp. 275–287. Ronald Press, New York.

Marsden, David H., 1952. Some common non-infectious disease of shade and ornamental trees in New England. *Proc. 28th Int. Shade Tree Conf.,* pp. 73–79.

Yarwood, C. E., 1961. Translocated heat injury. *Plant Physiol.* **36**:721–726.

CHAPTER SEVEN

DISORDERS ASSOCIATED WITH LOW TEMPERATURES

Leaves, limbs, buds, flowers, and fruits of many ornamental, cultivated, and even native plants are readily injured by low temperatures. Annual crops such as bean, potatoes, flax, corn and cotton; herbaceous perennials such as spinach; shrubs such as roses, euonymus, and camelia; tree fruits such as apple, cherry, and peach; and even timber species such as pine, oak, and maples may be affected in any given year. Seedlings of such comparatively hardy crops as wheat and alfalfa are killed in northern areas during some years and must be replanted. Less striking but more significant damage occurs when low temperatures inhibit growth and limit production. The extent of injury is determined by how low, how fast, and for how long the temperature drops.

The sensitivity of plants to low temperatures, and the amount and severity of injury, depends to a great extent on the physiological condition, or predisposition, of the plant. Mineral nutrition, especially the nitrogen level, has a particularly strong influence on low-temperature tolerance. Tissues too high in nitrogen are "soft," with relatively large and thin-walled cells. Plants with moderate to low nitrogen content are "harder" and more tolerant of temperature extremes. High sodium and calcium also favor damage. Conversely, for reasons not understood, high potassium levels provide some degree of protection from low temperatures. The carbohydrate food reserves are also critical to predisposition; injury is usually most severe when photosynthate reserves are low, possibly due to the lower osmotic value and decreased water-binding capacity.

Predisposition is further influenced by soil moisture. Plants growing in wet, waterlogged, and saturated soils are generally far more sensitive to frost injury than those in drier soils.

Age further influences sensitivity (Gardner, 1944). The softer, more succulent tissues of young plants are normally far more sensitive to frost than older tissues, but the response varies among species. Striking differences in winter hardiness of some juniper species were reported where limbs bearing the scale-like, adult leaves were killed and those bearing the acicular, juvenile leaves emerged undamaged. Reasons for such anomalies are not known.

Damage caused by low temperatures is commonly referred to as "frost injury," a term which implies that the temperature dropped below freezing and that ice formed in the affected tissues. But since temperatures above freezing can be injurious, the term *low-temperature injury* is more accurate and preferred.

Low-temperature injury occurs when a plant gives off more heat than it absorbs. Loss of heat may lower the plant temperature critically in either of two ways: (1) Loss of heat by conduction, and (2) loss of heat by radiation. Heat loss from conduction, occurring when the surrounding air is colder than the plant itself, takes place when a mass of cold air moves through an area, dropping both air and plant temperatures often well below freezing.

Radiation occurs as heat passes into the atmosphere from the warmer surfaces of the plant. Plants continually radiate heat; at night plants may pour more heat into the atmosphere than they can absorb and hence become colder than the surrounding atmosphere. Radiation is most rapid when the sky is clear and the atmosphere quiet, with no wind to stir the air layers and no clouds to blanket the earth. Under such conditions of minimal heat absorption, leaf temperatures may drop 3 to 6° below the surrounding air temperature.

Radiation is carried on primarily by the outer leaves. Inner leaves are covered and protected by neighboring leaves, so they radiate less heat and are less likely to cool down enough to be injured. On a larger scale, trees can serve as a screen to protect nearby smaller plants by reducing radiation.

Radiation frost damage is generally far more variable in severity over an area than damage from conductive frost. Injury is spotty from one part of a local area, as in an orchard, to another and even varies from tree to tree and limb to limb.

Both radiation and conduction frosts cause a drop in plant temperatures to a point where leaves, stem, flower, or fruit may be damaged. Symptoms range from growth suppression to the complete killing of the plants. Typical leaf symptoms include mild chlorotic flecking, general chlorosis and bronzing (Fig. 7.1), laceration, curling, puckering, crinkling, necrosis, and even defoliation. Stem injury includes cracking, splitting, various cankers, and dieback. Flowers and fruit responses include blossom killing, fruit drop, and various fruit markings and abnormalities.

FIGURE 7.1 *Blanching of broccoli leaves attributed to low-temperature injury.*

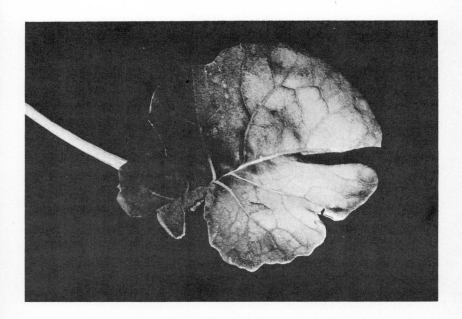

Leaf Chlorosis and Frost Banding

Scarcely a winter or spring passes in which low temperatures do not cause leaf chlorosis on at least a few plant species. Conifer species particularly are affected. Every winter, conifers throughout the world assume a yellow-green hue resembling nitrogen deficiency. Perry and Baldwin (1966) showed that this chlorosis was caused by a disruption of the photosynthetic apparatus and a breakdown of chloroplasts caused by the winter cold. The amount of chloroplast breakdown varies among species; in sensitive plants, breakdown begins with the first hard frost and is most severe in foliage exposed to direct sunlight. Chlorophyll is released to the cytoplasm and often converted to the yellow pigment chlorophyllin. Chlorosis persists through the winter and disappears in the spring following a week or more of warm weather when normal chloroplast structure and leaf color is restored.

Chlorosis of cultivated plants may be general over the leaf or may appear as distinct bleached bands across the blade of young plants. *Frost banding*, as this is called, has been reported on sugarcane, wheat, oats, and barley (Richards, 1934; also see Fig. 7.2). The bands of chlorotic tissue developing at the soil line were attributed primarily to radiation of heat from the plant to the soil cooling the soils and base of the plants several degrees below freezing and preventing chlorophyll formation.

FIGURE 7.2 *Low-temperature injury on one-month-old barley leaves.*

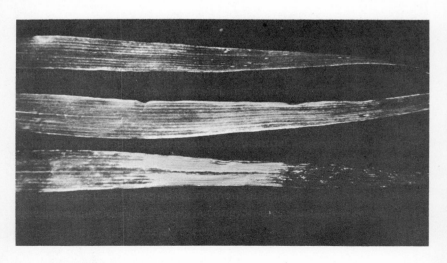

The emerging leaves exhibited three types of injury: the most conspicuous consisted of spots or bands of yellow, chlorotic tissue varying in size from a mere speck to ¾ in. wide. Individual cells superficially appeared alive and active, but chlorophyll development was permanently inhibited. A second type of lesion differed only in the extent of injury. All the chlorophyll in the bands was destroyed and bleached. A third type of injury, causing the greatest damage, occurred when vascular tissues were killed at the soil line, disrupting water and nutrient movement. Often tissue was killed outright, so that the plant collapsed. This type of injury primarily affected plants at the very crest of ridges between furrows or on the windward side of such exposed areas where temperatures were lowest.

Chlorosis of a "splotching" or mosaic type was reported on young sugarbeet plants by C. W. Bennett (1963). Symptoms were similar to those caused by such destructive yellows-type viruses as sugarbeet yellows; but the yellowing was attributed to low temperatures during early stages of plant development. Light yellow mottle developed on the first pair of true leaves when they were less than half grown. Chlorosis continued to appear on the emerging leaves but was less apparent on successively developing leaves. This type of chlorosis occurred over a wide geographical area, and over 50 percent of the plants in some California fields were affected. There was no indication, however, that ultimate root yields or sugar content at harvest were affected.

Leaf Necrosis, Albinism, and Malformations

Temperatures near the freezing point may cause chlorosis, while still lower temperatures kill the cells outright, causing necrosis. Necrosis symptoms range from a faint marginal band of necrotic tissue to the bleaching, browning, or blackening of tissues of the entire leaf or shoot. Every plant species is subject to such injury. No crop, ornamental, or even native species is consistently spared from low-temperature injury.

The response of aspen to late frosts provides one well-documented example of low-temperature injury (Marr, 1947). Aspen are widespread throughout North America and well adapted to moderately cold temperatures throughout most of the year. But aspen grow in areas where May and June frosts occur with persistent regularity, and leaves often emerge before the threat of frosts has passed. The tender young leaves and new shoots of trees in early emerging clones are especially vulnerable. The trees usually survive such frosts, but the deformed leaves with their "scorched" brown or blackened tips and margins, and the frequent presence of dead twigs, or shoot blights, bear witness to the earlier damage.

Late spring and summer frosts similarly cause varying degrees of damage to crops almost annually in many agricultural areas. Injury to such field crops as sugarbeets, potato, and corn is striking. Low-temperature injury on corn is characterized by necrotic leaf tips; less striking symptoms accompanying this necrosis include a yellow flecking along and between the veins contiguous with the dead areas. The amount of injury is determined by the age of the plants, the soil moisture conditions, and the temperature extremes (Treshow, unpublished).

Chlorotic flecking is not restricted to corn. Late spring frosts are of distressing frequency in Utah and have provided ample opportunities to observe plant responses. Pigweed, raspberry, pea, pear, potato, bean, and alfalfa are regularly affected. Flecking may be the sole symptom or may be accompanied by more general chlorosis and necrosis (Fig. 7.3).

Cold injury on alfalfa leaves is expressed by pale brown to white dead areas developing along the leaf margins or in interveinal areas of tissues along the veins. Bleached striations develop, extending from near the midrib to the margin of the leaflet blade.

FIGURE 7.3 *Flecking symptom together with marginal necrosis, characterizing low-temperature injury on alfalfa.*

Low-temperature injury is more serious when it occurs in the winter and affects the crown and root tissues (Jones, 1928). On alfalfa, phloem parenchyma and xylem are damaged first. Dead cells often form a sheath surrounding the vascular cylinder; the cells exterior to it die and slough off, leaving a roughened surface much like that produced by mechanical injury. Parenchymatous cells beneath the injured phloem begin to divide, forming a new cork layer. More severe injury shortens the life of the plant and provides a "court" of entry easily penetrated by fungus and bacterial parasites.

Spring frosts also cause various types and degrees of injury including cupping, crinkling, twisting, and curling to leaves of stone fruit and apple trees. Distortion apparently results from the death of groups or clusters of cells before the leaf is fully expanded. When individual cells or groups of cells are killed early in leaf development while leaves are still in the bud or just emerging from it, the surrounding undamaged cells mature and enlarge normally, destroying the continuity of the tissue. The differential cell enlargement causes various malformations as the leaf enlarges. Circular, elliptical, or irregular holes are produced when dead tissue drops out. Leaves of such stone fruit trees as apricot, cherry, peach, and prune are commonly affected. Shot-holing is often restricted to leaves of the lower branches, where leaf temperatures are lowest and injury most severe.

Low temperatures early in leaf development cause various degrees of lacination, incising, or slashing. This is especially common on peach leaves (Gigante, 1946). When cells are killed at later stages of maturity when they are more fully enlarged, marginal leaf necrosis or tip bleaching may develop (Fig. 7.4). Leaves then tend to curl and never reach normal size.

Leafy vegetable crops are damaged by low temperatures whenever early frosts occur before harvest. Immediately after thawing the frozen areas appear water-soaked, and the tissues become flaccid, pithy, or spongy and lose their flavor. In tender leaf tissues such as cabbage, Italian broccoli, kale, and mustard, the epidermis is often loosened in localized spots along the veins so that the tissue appears blistered. Low temperatures may cause the epidermis to separate from the mesophyll, producing a glazing and blistered appearance (Fig. 7.5).

If plants are suitably preconditioned or hardened by slow cooling, they can withstand remarkably low temperature extremes. During hardening, the permeability of the outer plasma membrane increases; this favors the rapid resorption of water from the melting intercellular ice crystals. Also, due to the water binding, the free water content of

FIGURE 7.4 *Low-temperature injury on common lilac showing characteristic marginal necrosis.*

the cell decreases so that the amount of soluble sugar and protein increases, lowering the freezing point of the protoplast.

The rate of temperature drop is often more significant than the temperature extremes in causing injury. If plants can be cooled to about $-30°C$ without injury, then there is almost no limit to how much more cold they can tolerate. Pine leaves have been cooled to $-90°C$, cabbage leaves to $-15°C$, and mulberry leaves to $-210°C$ without being killed; stems of Siberian elm have been cooled to $-196°C$ without injury. Northern forest species of black currant and birch, hardened gradually, were able to withstand $-253°C$, and apricot $-60°C$, rather than their usual limit of $-5°C$. Such extremes were possible only following gradual hardening, which facilitated a high sugar accumulation (Troshin, 1967).

Needle Blight of Conifers

Needle-leaved conifers grow in some of the coldest regions of the world and are normally well adapted to low temperature extremes; but if the trees are not hardened, or temperature fluctuations are too great, needles of even the hardiest conifer trees will be killed. Seemingly overnight, needle tips, or even the entire needle, turn reddish or brown as the tissues become plasmolyzed, desiccated, and killed. The environmental conditions responsible for this necrosis may occur over a broad geographic area, be confined to a relatively narrow elevation zone, or even be confined to an area containing but a few trees. Only the tops of trees are affected at the lower edge of the damage zone, while only the lower parts of the trees are affected at the top of the zone. Occurrence of sharply defined bands, or belts, of damaged trees has led to the name of *red belt* injury for this type of damage (Hubert, 1918 and 1930).

Red belt appears most frequently when a sudden temperature drop during the winter months follows an unseasonably warm spell. Henson (1923) explained the occurrence of red belt as follows: At a time when valley bottoms are filled with cool air, warm, dry chinook winds flow over the mountains or passes of the northern Rocky Mountains into the valleys and spread over the cool air without mixing. The warm air strikes against the sunlit slopes, which are unprotected by local cool air drainages. Rapid warming and drying occur along this belt. At night, local air drainage would add to and deepen the pool of cold air in the border zone between the two air masses; the

FIGURE 7.5 *Radiation frost injury characterized by bronzing of half the leaf.*

next day it would be shallower, again subjecting trees in the border zone to alternating warm and cool temperatures. It was believed that red belt was caused by the alternate subjugation of the trees to chilling and warm dry air while available water was limited. Red belt has also been attributed to a sudden temperature rise alone when water was unavailable.

J. A. Baranyay (1965, unpublished) attributed the disease to an alternation of warm, chinook winds with normal, low temperatures. In one instance, temperatures in December rose suddenly from -30 to $2°C$ and stayed above freezing for five days. Necrosis first appeared in February, and affected leaves dropped by June. Buds survived, and trees looked almost normal by July. Where the disease occurred in two successive years, however, affected trees were killed. He found that red belt was severe eight years out of eleven in the Bow Valley area of Alberta; about 800 acres of mostly lodgepole pine were killed in 1964 alone. Red belt was also extensive in the northern United States following the winter of 1947–1948, the warmest and the fourth driest winter on record (Stoeckeler, 1949; Voigt and Voigt, 1948).

A similar incident was reported in New York (Curry and Church, 1952). Damage was frequently most severe on the south and west sides of trees, suggesting that bright sunshine raised the surface temperature and stimulated excessive transpiration under conditions where water replenishment from the frozen trunk was impossible. Entire stands of timber in the Adirondack mountains of New York looked as if they had been scorched by fire. Nearly 50 percent of all the conifers, including red spruce, hemlock, white pine, and balsam fir received some permanent injury. White spruce, white cedar, black spruce, and red pine showed relatively little damage.

The severity of needle damage varies with the degree and duration of low-temperature exposure. When injury is mildest, only the previous year's needles become necrotic. When stress is more severe, older needles and ultimately the youngest shoots are damaged.

The amount of permanent damage from red belt depends on the severity of necrosis. Red and Scotch pine in the Great Lakes states recovered completely when one-fourth or less of their foliage was injured (Stoeckeler, 1949). Recovery was still satisfactory even where one-half of the needles were damaged; but where close to three-fourths or more of the needles were necrotic, recovery was poor, and trees usually died.

In the Black Hills of South Dakota, winter damage to ponderosa pine is particularly common when chinook winds rapidly raise the temperature from freezing to as high as 4 to 15°C within a few hours. The effect of the sudden temperature increase is essentially one of drying and is discussed in more detail in chapters dealing with moisture relations.

Damage is not limited to trees in northern states or regions. The winter of 1950–1951 was more severe than normal in Haiti and brought needle reddening and burning to extensive acreages of Haitian pine (Pedersen, 1953). Minimum temperatures in January are usually above 3°C, but in January, 1951 temperatures dropped suddenly to −4°C. Seedlings were killed and saplings severely injured. Temperatures were lowest in flat or low-lying areas of poor air drainage, where nearly 90 percent of the trees were killed.

Needles of the coast redwood can be killed even by relatively mild temperatures. In one incident, severe needle burning and killing of seedlings was attributed to an eight-day period in December during which the temperature remained between −5 and −1°C. The relatively low relative humidity accompanying this cold snap was held partly to blame. Failure of the seedlings to survive such temperatures is believed to play an important role in limiting the northern distribution of redwoods.

The extent of injury also depends on the time of year low temperatures occur (Day and Peace, 1937). Trees highly resistant to cold during winter can be severely damaged by April or May frosts. Under controlled laboratory conditions, 50 percent of the three- to four-year-old Douglas fir trees exposed to −11°C in March were killed. In May a temperature of only −3°C killed 50 percent of the exposed trees, indicating the gradual increase in sensitivity as buds develop and shoots elongate.

Studies of an April freeze in New Mexico illustrate the influence of elevation on damage and distribution of native vegetation. Phillips (1947) reported damage to subalpine species in New Mexico forests above the 9,000-ft level. Trees from 9,000 to 9,500 ft were breaking dormancy in mid-April when the temperature dropped to −10°C. Foliage was killed on 90 percent of the Douglas fir trees under 10 ft high. However, new growth came out and trees recovered completely within eight days. Ponderosa pine saplings were killed outright and older plants burned critically. The overall sensitivity of trees to late frosts was given in order of increasing tolerance as limber pine, white fir, ponderosa pine, Douglas fir, and alligator juniper. Limber pine at lower elevations, where leaves had come out earliest, was most severely damaged. The frequent restriction of this species to exposed rocky crags at the higher elevations may be due to the delayed dormancy of trees in such areas; by the time leaves emerge, danger of needle-killing frosts has passed.

Various studies have shown that seeds of the same species, but collected from different geographic areas, differ in their sensitivity to frost (Minckler, 1951). Seeds from trees growing in areas of more severe winters consistently produce more cold-tolerant seedlings, which can be regarded as distinct ecotypes.

7-2 STEM DISORDERS

Frost Cracks

Frost cracks, that is the radial splitting of tree trunks caused by a rapid decrease in temperature, are widespread in distribution and surprisingly common in northern hardwood forests. One study in an oak-hickory association (Fergus, 1956) revealed frost cracks in 13 percent of all the oaks. Their frequency in old trees was greatest, with cracks present in 94 percent of all trees over 10 in. diameter at breast height (dbh). Cracks were anywhere from 6 in. to 21 ft long.

Frost cracks develop when tree trunks or limbs lose their heat too rapidly (Fig. 7.6). The outer layers of bark and wood cool most rapidly and are subjected to appreciable tangential tension, causing marked shrinkage and cracking following a sudden temperature drop. The cracks close with the warming days of spring, but the damage has been done, and the scars remain. Affected timber is of poor quality, and cullage is great.

Cracking is most serious when the bark is separated from the wood; but if the bark is tacked down and reunited soon enough, the wound will knit and growth be resumed. Separation of the bark and wood leads to a characteristic blister-like cavity in the affected annual rings, giving rise to a bark condition called *blister shake* (Tryon and True, 1952). Blister shake is the name given to injury in the basal part of the trunks of yellow poplar; it is characterized by the development and separation of cavities between the annual rings in early years of growth.

FIGURE 7.6 *Severe splitting of sweet cherry trunks. Symptom is far more pronounced than typical.*

Stem Necrosis, Dieback, and Shoot Blight

Winter injury is more serious when the crown and upper root system is damaged. Above-ground symptoms then consist of reduced growth, poor foliage color, and discolored bark. These are general symptoms of a weak plant and are caused by many stresses.

Cold injury can damage wood of fruit trees to the extent that their survival is imperiled. One report from Georgia (Anonymous, 1965) indicated that survival of numerous peach orchards was threatened by low winter temperatures. Some growers were said to be losing as much as half their crop each year because of winter dieback of fruiting limbs. While trees cannot be protected from injury, winter-damaged trunks and stems can sometimes be revived by bridge grafting, heavy pruning, and fertilization. When fruit trees are damaged, blossom thinning is helpful in preventing overbearing and easing the strain on the tree.

Cane fruits are also subject to cold injury. Winter dieback is a particular problem when poorly adapted varieties are grown in colder regions.

Such relatively herbaceous plants as alfalfa suffer stem and crown injury from low temperatures (Jones, 1928), which causes a breakdown of the phloem parenchyma and xylem followed by a decay of the crown and tap root. Such injury shortens the life of the plant and provides a convenient entry court for parasitic bacteria and fungi.

Temperatures slightly above freezing can also kill plants. Many commercial corn strains are reported killed by a succession of cold nights in which temperatures range from 1 to 7°C. Most strains, however, can withstand −1 to 0° and a few even as low as −2°C.

Plants introduced to new areas by nurserymen and home garden enthusiasts, who are constantly attempting to improve the landscape by extending the northern range of various ornamental species, are frequently damaged. Nurserymen all too readily introduce otherwise desirable plants which, at best, can be expected to tolerate the expected low temperature extremes only every other year.

Even native plants are damaged by low temperatures and winter drying. Desiccation is particularly serious when air temperatures are mild and the soil is frozen. When a snow cover exists, it may reduce water loss and provide a significant measure of protection. Limbs of young pine, spruce, juniper, and fir trees above the snow line may be killed during periods of low temperature stress, while the tissues below the snow remain uninjured.

Damage to pine shoots is caused by the same factors described under leaf injury. Periods of mild, unseasonable weather followed by a temperature drop to freezing and then a quick rise in temperature are especially destructive. The longer the period of warm temperature, the greater the amount of injury. Killing of Scotch pine shoots was simulated by placing young plants at 18 to 21°C prior to exposure to freezing temperatures (Studhalter, 1942). One cold night followed by a warm spell is all that is needed to kill the leaders and terminal bud. The loss is especially great if the buds are wet when frozen.

A single late freeze can be devastating even to native, indigenous species. Beal (1962) believed that a late May freeze, where on two consecutive nights temperatures reached +1 and −1°C, was responsible for killing about 3,000,000 ft of white oak in Bland County,

Virginia. Mature oaks growing in the valleys and hollows, and occasional hickory and red oak, were also killed. In a similar incident in Pennsylvania, temperatures of −5 and −6°C on June 10 and 11, 1941, killed practically all the new growth of scrub oak (Clarke, 1946). Smaller oak trees were killed to the ground.

Aspen are frequently damaged by late spring frosts. Earliest clones to break dormancy are vulnerable to damage for the longest period and are most severely injured. When freezing temperatures follow a few days of temperatures over 10°C, the youngest shoots turn black and become brittle.

Shoot blight develops rapidly in response to a sudden temperature drop when the new young shoots of deciduous or evergreen plants are emerging in the spring. Stem tissues are highly succulent and sensitive to every kind of stress. When an unseasonably cold air mass moves through the area, or when conditions are favorable for radiation frosts, the low temperatures kill the young tissues of woody trees and shrubs, leaving only a shriveled black and blighted dead shoot. New growth generally arises from adventitious buds at the base of the newly killed shoots within a few weeks of the injury, but the shoots and leaves are severely dwarfed and aborted. When food reserves are inadequate, two-year-old wood may also be killed, and a more general dieback symptom appears.

Temperatures of 2 to 3°C can kill the tender succulent new growth of such conifer trees as Douglas fir and blue spruce. Winter injury is a serious problem on American arborvitae, killing tips of branches on the southwest side of the plant (White and Weiser, 1964). Low temperatures alone were ruled out as the cause when studies showed that plants could withstand −125°F. Rapid temperature fluctuations and the rate of temperature drop may be more important in producing injury. A rapid temperature drop of 10° (from +2 to −8°C) recorded on the southwest side of trees was sufficient to kill the tissues. Leaf temperatures can drop 10°C in only one minute on the southwest side of an arborvitae when the sun moves behind an object. Injury was reproduced artificially using cold-acclimated plants.

Low temperatures even in the absence of a rapid drop can cause stem injury, including cankers (Barnard and Ward, 1965). Stem cankers were produced artificially by wrapping crushed dry ice in canvas and placing it around tree trunks early in May when the cambium was active. Earlier treatment, before the "sap" began to flow, was not harmful.

Low temperatures contribute greatly to the distribution of plant species (Shreve, 1914). Ponderosa, Jeffrey, and sugar pines grow in essentially the same kinds of habitat, but new seedlings of the latter are most sensitive to low temperatures and are killed and eliminated in areas where temperatures drop to about $-7°C$ in the late spring.

Low summer temperatures, while not causing visible injury, can significantly limit the growth rate of oak. Kozlowski (1962) made daily measurements of increment growth during the early summer and found that growth was closely related to minimum temperatures.

If plants fail to become dormant in the fall, or an early winter occurs, they fail to become suitably hardened and are predisposed to winter injury. During late, warm autumns, trees continue to grow actively as long as temperatures remain above freezing during the day. With the onset of low winter temperatures, the sensitive, unconditioned cambium and vascular tissues are killed.

Effects of early winters, with the abrupt invasion of cold air in late autumn, have been thoroughly studied in the Pacific Northwest. Sudden, early temperature drops in 1924, 1935, and again in 1955 killed or damaged thousands of acres of fruit trees, ornamentals, and even native species (Duffield, 1956; Daubenmire, 1956). In 1955, wood of apple and cherry trees was killed over thousands of acres, and trees over still more extensive areas were damaged to an extent where pathogenic fungi became established in the weakened tissue. Cankers, heart rot, and limb breakage followed, and affected trees have been going out of production ever since.

Daubenmire (1956) reported that fifty-two species or varieties of woody plants were damaged or killed in the Pullman, Washington, area in 1955. Injured patches of bark appeared to die out at varying times over a period of several months. Browning of the damaged wood was due in part to a condensation of storage carbohydrates which apparently was transformed into gums and tannins. An unusually heavy needle drop of conifers also developed, and the total damage was appraised at 66 million dollars.

Top Dying of Conifers

The sudden drying, browning, and death of the tops of otherwise thrifty pines is too readily blamed on insects or porcupines. More often, such dieback is caused by winter injury when sudden temperature changes kill the cambium.

One incident, involving top dying of exposed Coulter pine and big-cone spruce in southern California, typifies this type of winter injury (Wagener, 1949). In March, 1944, needles in the tops of exposed trees began to fade, and by late summer dying tops numbered in the hundreds. Injury was most pronounced on trees growing on ridge tops and other fully exposed sites. Scattered, more exposed, trees were completely killed. Top injury appeared on trees up to 100 ft tall.

Temperature records in this instance revealed an extended growing season caused by an unseasonably wet November and December. January began cool, but between January 12 and 23 mild temperatures prevailed, reaching highs over 15°C. On January 24, the temperature dropped suddenly to −5°C. The abnormal drop and sudden stress were beyond the tolerance of the sensitive Coulter pine. Vascular tissue was killed, and dieback developed.

Similar dieback on hundreds of small groups of Douglas fir in northwest Oregon was described by T. W. Childs in 1961. The red-topped "flags" of affected trees in a twenty-five- to thirty-year-old stand stood out prominently several months after an early winter freeze. Close scrutiny showed that irregular callus layers formed around the trunk of dead trees nearest the snow line 8 to 12 ft above the ground. Frost rings were plainly visible behind the lines of initial callus formation. Adjacent to the most severely damaged groups of trees, trunks of many of the living trees were irregularly swollen, and bark was cracked. The variation in the degree of damage among trees was ascribed to the different rate at which the trees had become fully dormant before the freeze hit. Trees dormant earliest survived the sudden low temperatures.

Low-temperature Breakdown

Potatoes and many other storage organs are damaged whenever temperatures approach the freezing point for the tissue (Wright and Diehl, 1927). Breakdown symptoms on potato consist of discolored brownish blotches in the flesh and brownish-black discoloration of the stem which give rise to the name *mahogany browning*. Irregular patches of vascular, cortical, and pith tissue are damaged. The disorder is most serious when affected tubers first become dark or discolored during or after cooking.

Sweet potato is injured when stored for ten days or more at temperatures below 13°C; the physiological changes occurring at these temperatures cause a brownish discoloration affecting local scattered areas in proximity to the vascular elements in the central part of the sweet potato (Lauritzen, 1931).

Surface or skin pitting occurs in many fruits and vegetables exposed to temperatures slightly above freezing for a week or longer. Cucumbers, peppers, and snap beans develop pitting symptoms at 0.5°C in ten to twenty days. Pitting has also been reported to be important on citrus fruits, summer squash, and several other fruits and vegetables exposed to low temperatures (Ramsey et al., 1938).

Vegetables such as celery and onion may be subjected to freezing before harvest or during transit. Petioles of celery become flabby and water-soaked and the leaves papery. Elliptical, isolated, sunken lesions may appear on the petioles. Affected tissues soon turn brown, so they are obvious at harvest and are culled.

7-3 FLOWER AND FRUIT DISORDERS

A specific set of temperature and environmental conditions is required for the formation and maturation of flower buds and blossoms. Once the buds are formed, they must still withstand the stress of low temperatures throughout the following winter and spring. When winter ends and the buds begin to open in the spring, they continue to be subject to frost injury. Spring frost damage occurs almost annually in many fruit-growing regions of the world. The frost-blackened pistils of the flower and the dead buds resulting from winter freezes are easily recognized. But the damage is not always this clearly defined. More frequently, less striking injury occurs which is difficult to recognize. The pistil may be damaged even though there are no externally visible symptoms; a discolored, frost-killed, or frost-damaged ovule can be just as effective in arresting fruit development. Ovule damage can be detected only by cutting through the young fruit and may consist only of slight browning of the young embryo within. Many of the injured fruits are subsequently shed, while others, less severely injured, may continue to grow but be undersized and irregular in shape. Damaged fruits often abort and drop.

Injury to ovules prevents normal development of the ovary. Even when only a few of the ovules are damaged, the associated parts of the carpels may fail to develop, thus giving rise to malformed and sometimes pitted fruit (Simons and Lott, 1963). Russeting, the formation of a corky layer of epidermal cells when localized injury to the surface cells stimulates cork formation just beneath this layer, is a particularly common response to low temperature. Symptoms consist of rough, brownish areas on the skin in a band extending around the base or middle of a fruit. Underlying growth is retarded, causing a slight constriction around the fruit.

The exact temperature required to kill blossoms of different kinds of plants is difficult to establish. Actually, there are so many interacting and predisposing factors to consider, as well as the stage of flower or fruit maturity and the duration of the low temperatures, that it is impossible to provide a precise lethal threshold temperature for flowers of even a single species.

In a general way, flowers of most stone and pome fruit trees are killed when air temperatures remain below -2 to $-3°C$ for more than a few hours. The female strobili of conifers are reported to be killed at these same temperatures (Campbell, 1955). The critical lethal temperature for unopened buds is substantially less than that for open blossoms and may be as low as 20 to $30°C$ below freezing. Sensitivity fluctuates directly with the temperature changes during the winter months. Freezing tests showed that the killing point of peach flower buds in the winter may be as much as $8°$ higher after a warm period and may fall 2 to $3°C$ after a cold spell (Chaplin, 1948; Mowry, 1964).

Low temperatures also markedly influence reproduction of long-season, hot-weather crops such as rice and red kidney beans (Wierenga and Hagan, 1966). Rice required temperatures above $7°C$ to initiate flowering, and higher temperatures are needed for optimal grain set. For subsequent optimal production, irrigation water must be above $18°C$ (Anonymous, 1964). Red kidney beans were grown at a constant soil temperature at $25°C$ and subjected to $10°C$ temperatures at various stages of growth. A single cold treatment early in the cycle (three-leaf stage) decreased the shoot and root yield 17 percent. Lowering the soil temperature at the nine-leaf stage did not affect vegetative growth, but bean yields were 15 percent less than when plants were grown at soil temperatures of $25°C$. Intermittent soil temperature reductions lowered yields as much at 34 percent.

Pollination, even though by wind, will also proceed normally only at temperatures above freezing. The reason for reduction is obvious for such insect-pollinated species as stone fruits, since bees normally fly only at temperatures above $18°C$, but pollination of wind-pollinated species such as rye and wheat is also impaired by low temperatures.

The more that temperature relations are studied, the clearer it becomes that every stage of plant development and reproduction is served best by a specific critical temperature, and any deviation from this will cause a stress reflected in poor growth, reproductive failure, and sometimes visible symptoms.

BIBLIOGRAPHY

Anonymous, 1964. Cold water and air can cut rice yields, may improve variety. *Crops and Soils* **16**:17.

Anonymous, 1965. Dying-tree problem. *Amer. Fruit Grower,* Western ed., June, pp. 34–37.

Barnard, J. E., and W. W. Ward, 1965. Low temperatures and bole canker of sugar maple. *Forest. Sci.* **11**:59–65.

Beal, J. A., 1962. Frost killed oak. *J. Forest.* **24**:949–950.

Bennett, C. W., 1963. Noninfectious yellow splotching of leaves of young sugarbeet plants in western United States. *Plant Dis. Rep.* **47**:63–64.

Campbell, T. E., 1955. Freeze damages shortleaf pine flowers. *J. Forest.* **53**:452.

Chaplin, C. E., 1948. Some artificial freezing tests of peach fruit buds. *Proc. Amer. Soc. Hort. Sci.* **52**:121–129.

Childs, T. W., 1961. Delayed killing of young-growth Douglas fir by sudden cold. *J. Forest.* **59**:352–453.

Clarke, W. S., Jr., 1946. Effect of low temperatures on the vegetation of the Barrens in central Pennsylvania. *Ecology* **27**:188–189.

Curry, J. R., and T. W. Church, Jr., 1952. Observations on winter drying of conifers in the Adirondacks. *J. Forest.* **50**:114–116.

Daubenmire, R., 1956. Climate as a determinant of vegetation distribution in eastern Washington and northern Idaho. *Ecol. Monogr.* **26**:131–154.

Day, W. R., and T. R. Peace, 1937. The influence of certain accessory factors on frost injury to forest trees. II. Temperature conditions before freezing. *Forestry* **11**:13–29.

Duffield, J. W., 1956. Damage to western Washington forests from the Nov. 1955 cold wave. *Pac. Northwest Forest and Range Exp. Sta. Res. Note* 129.

Fergus, C. L., 1956. Frost cracks on oak. *Phytopathol.* **46**:297.

Gardner, V. R., 1944. Winter hardiness in juvenile and adult forms of certain conifers. *Bot. Gaz.* **105**:408–410 (*Forest. Abstr.* **6**:112).

Gigante, R., 1946. Laciniation of peach leaves caused by cold. *Boll. Staz. Pat. Beg. Rono.* **20**:125–136.

Henson, W. R., 1923. Chinook winds and red belt injury to lodgepole pine in the Rocky Mountain parks area of Canada. *Div. Forest Biol. Sci. Serv. Dept. Agr.* **28**:62–64.

Hubert, E. E., 1918. A report on the red belt injury of forest trees occurring in the vicinity of Helena, Montana. *Mont. State Forest Bien. Rep.* **5**:33–38.

————, 1930. Forest-tree diseases caused by meteorological conditions. *U.S. Mon. Weather Rev.* **58**:455–459.

Jones, F. R., 1928. Winter injury of alfalfa. *J. Agr. Res.* **37**:189–211.

Kozlowski, T. T., 1962. Daily radial growth of oak in relation to maximum and minimum temperature. *Bot. Gaz.* **124**:9–17.

Lauritzen, J. I., 1931. Some effects of chilling temperatures on sweet potatoes. *J. Agr. Res.* **42**:617–627.

Marr, J. W., 1947. Frost injury to aspen in Colorado. *Bull. Ecol. Soc. Amer.* **28**(4):60.

Minckler, L. S., 1951. Southern pines from different geographic sources show different responses to low temperatures. *J. Forest.*

Mowry, J. B., 1964. Seasonal variation in cold hardiness of flower buds on 91 peach varieties. *Proc. Amer. Soc. Hort. Sci.* **85**:118–127.

Pedersen, Arthur, 1953. Frost damage in the pine forest. *Carib. Forest.* **14**:93–96.

Perry, T. O., and G. W. Baldwin, 1966. Winter breakdown of the photosynthetic apparatus of evergreen species. *Forest Sci.* **12**:298–300.

Phillips, F. J., 1947. Effect of a late spring frost in the southwest. *Forest. Irrig.* **13**:484–492.

Ramsey, G., J. S. Wiant, and G. K. Link, 1938. Market diseases of fruits and vegetables. *U.S.D.A. Misc. Publ.* 292, 74 pp.

Richards, B. L., 1934. Frost-banding of cereal seedlings. *Utah Acad. Sci., Arts, Letters* **11**:2.

Shreve, F., 1914. The role of winter temperature in determining the distribution of plants. *Amer. J. Bot.* **1**:194–202.

Simons, R. K., and R. V. Loss, 1963. The morphological and anatomical development of apples injured by late spring frosts. *Proc. Amer. Soc. Hort. Sci.* **83**:88–100.

Stoeckeler, J. H., 1949. Winter injury and recovery of conifers in the upper midwest. *Lake States Forest. Exp. Sta.* no. 18.

Studhalter, R. A., 1942. Apparatus for the production of artificial frost injury in the branches of living trees. *Science* **96**:165.

Troshin, A. S. (ed.), 1967. "The cell and environmental temperature." Pergamon Press, Oxford, 462 pp.

Tryon, E. H., and R. P. True, 1952. Blister shake of yellow poplar. *Bull. W. Va. Agr. Exp. Sta.* 350T.

Voigt, G., and K. Voigt, 1947–1948. Causes of injury to conifers during the winter of 1947–1948 in Wisconsin. *Dept. Soils, Univ. of Wis.*, pp. 241–243.

Wagener, W. W., 1949. Top dying of conifers from sudden cold. *J. Forest.* **47**:49–53.

White, W. C., and C. J. Weiser, 1964. The relation of tissue desiccation, extreme cold and rapid temperature fluctuations to winter injury of American arborvitae. *Proc. Amer. Soc. Hort. Sci.* **85**:554–563.

Wierenga, P. J., and R. M. Hagan, 1966. Effects of cold irrigation water on soil temperature and crop growth. *Calif. Agr.* **20**:14–16.

Wright, R. C., and H. C. Diehl, 1927. Freezing injury to potatoes. *U.S.D.A. Tech. Bull.* 668.

SELECTED REFERENCES

Das, T. M., and A. C. Hildebrand, and A. J. Riker, 1966. Cine-photomicrography of low temperature effects on cytoplasmic streaming nucleolar activity and mitosis in single tobacco cells in microculture. *Amer. J. Bot.* **53**:252–259.

Daubenmire, R., 1957. Injury to plants from rapidly dropping temperature in Washington and northern Idaho. *J. Forest.* **55**:581–585.

Davis, J. R., and H. English, 1965. A canker condition of peach seedlings induced by chilling. *Phytopathol.* **55**:581–585.

Day, W. R., 1928. Frost as a cause of disease in trees. *Quart. J. Forest.* **22**:179–191.

—— and T. R. Peace, 1934. The experimental production and the diagnosis of frost injury on forest trees. *Oxford Forest. Memo.* **16**, 60 pp.

Egeberg, R., Jr., 1963. Inherent variations in the response of aspen to frost damage. *Ecology* **44**:153–156.

Eguchi, T., 1963. Effect of low temperature on flower and seed formation in Japanese radish and Chinese cabbage. *Proc. Amer. Soc. Hort. Sci.* **82**:322–331.

Faris, J. A., 1926. Cold chlorosis of sugar cane. *Phytopathol.* **16**:885–891.

Fryer, L. D., 1947. Damage to *Pinus radiata* by climatic agents. *Aust. Forest.* **11**:57–64.

Groves, A. B., 1946. Weather injuries to fruits and fruit trees. *Va. Agr. Exp. Sta. Bull.* **290**:1–39.

Hackett, W. P., and H. T. Hartman, 1964. Inflorescence formation in olive as influenced by low temperature, photoperiod and leaf area. *Bot. Gaz.* **123**:65–72.

Levitt, J., 1956. "The hardiness of plants." Academic, New York, 278 pp.

McGuire, J. J., and H. L. Flint, 1962. Effects of temperature and light on frost hardiness of conifers. *Proc. Amer. Soc. Hort. Sci.* **80**:630–635.

Parker, J., 1960. Survival of woody plants at extremely low temperatures. *Nature* **187**:1133.

——, 1963. Cold resistance in woody plants. *Bot. Rev.* **29**:123–201.

Sakai, A., 1960. Survival of the twig of woody plants at −196°C. *Nature* **185**:393–394.

Sellshop, J. P. F., and S. C. Salmon, 1928. The influence of chilling above the freezing point on certain crop plants. *J. Agr. Res.* **37**:315–338.

Tryon, E. H., and R. P. True, 1962. Radical increment reduction in American beech caused by a freeze. *Phytopathol.* **52**:1221.

—— and ——, 1964. Relative susceptibility of Appalachian hardwood species to spring frosts occurring after bud break. *W. Va. Agr. Exp. Sta. Bull.* **503**.

Went, F. W., 1953. The effect of temperature on plant growth. *Ann. Rev. Plant Physiol.* **4**:347–362.

CHAPTER EIGHT

LIGHT MECHANISMS

Light provides the fundamental source of energy which directly or indirectly drives the life processes of nearly every living organism. Without the radiant energy of sunlight to synthesize the plants' energy-rich sugars, life—except for a few chemosynthetic bacteria— would disappear.

8-1 WHAT IS LIGHT?

Light is a tiny, narrow band of visible radiant energy within the electromagnetic spectrum comprised of wavelengths from 390 to 760 mμ*

---

* Radiation most often is measured in foot-candles, the light intensity 1 ft from a standard candle, and in the metric system in luxes, the light intensity 1 m from a standard candle. 1 ft-c (foot-candle) is the same as 10.764 luxes. The amount of total energy is a preferred expression and can be measured in gram calories per square centimeter per minute, 1 of which is equal to about 6,700 ft-c on a clear day.

(Fig. 8.1). The shorter, high-energy wavelengths below the visible range are known as *ultraviolet,* while the longer wavelengths on the far side of the visible range are the *infrared.*

Radiation is characterized by its wave motion and wavelengths, but in its interaction with matter, such as a leaf surface, it acts as though it were composed of small packets of energy called quanta or *photons.* Each photon contains energy equal to Planck's universal constant of action (1.58×10^{-34} cal sec) times the velocity of light (3×10^{10} cm/sec) divided by the wavelength. Consequently, the energy in a single photon is inversely proportional to its wavelength (the shorter wavelengths possessing the greatest energy).

High intensity of radiation in the far ultraviolet and shorter wavelengths is normally negligible, since it is absorbed by ozone in the atmosphere, but at higher elevations, where far less radiation has been absorbed, ultraviolet radiation may be sufficiently intense to damage plants; the large amounts of energy transferred to the molecules in the leaf cells alter chemical bonds and disrupt their structure.

Infrared radiation, characterized by longer wavelengths, contains enough heat energy to stimulate chemical reactions. This heat energy

FIGURE 8.1 *The electromagnetic spectrum most significant to biological reactions, showing major regions of energy absorption by water vapor, ozone, and carbon dioxide.*

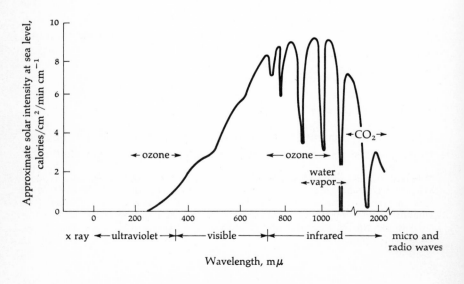

is absorbed by the water in the cells, sometimes raising the temperature to lethal levels sufficient to denature enzymes.

Fortunately, the most harmful radiation is absorbed by the atmosphere and never reaches the plant. Some wavelengths are absorbed by oxygen; more are absorbed by ozone. Ozone, produced by the action of high-energy radiation on oxygen, occurs in a distinct layer in the stratosphere and absorbs most of the ultraviolet radiation below 290 mμ and much of the infrared in the range from 760 to 1,200 mμ (Fig. 8.1). Infrared radiation beyond 2,000 mμ is absorbed largely by water vapor, carbon dioxide, and ozone. Plants reflect or transmit most of the remaining radiant energy, and only a small amount remains to be absorbed. The visible light energy, largely between the extremes of ultraviolet and infrared, is utilized for normal development. But before light energy can be utilized, it first must be absorbed by photoreceptive materials known as pigments. Chlorophyll, located in the chloroplasts of plant cells, is the dominant light-absorbing pigment in green plants; it is the one essential to converting photon energy to chemical energy.

8-2 CHLOROPHYLL AND PHOTOSYNTHESIS

Most plants are structurally organized to receive the greatest amount of light possible. The large surface/volume ratio of the leaf, coupled with the large intercellular surface area and lamellar chloroplast structure, all facilitate maximum light absorption. Chlorophyll is formed when its precursor, called protochlorophyll, is exposed to light. Plants require light first for forming chlorophyll from protochlorophyll and secondly for developing and maintaining orderly lamellar layers of grana in the chloroplast. Seedlings kept in the dark rarely develop any chlorophyll; rather, they become yellowish as the xanthophyll and carotene pigments develop.

Chloroplasts, the intracellular organelles in which the chlorophyll of higher plants occurs, also undergo changes with light exposure. During the first few minutes of irradiation, proplastids in dark-grown seedlings are converted to chloroplasts. While the shape of chloroplasts varies somewhat among species, they are characteristically disc-shaped to ovoid, about 1 to 2μ in length, and enclosed by a double membrane. The membrane encloses a substance or stroma which is interlaced by double-membraned lamellae. Before being exposed to light these lamellar structures are arranged randomly, but upon exposure to light the lamellae become oriented into flat layers arranged like stacks of coins and known as *grana*. Chlorophyll molecules are localized on these layers, which provide for a maximum surface area and light absorption.

The initial light reactions of photosynthesis, photolysis of water and photosynthetic phosphorylation, take place on the grana. Later stages, in which carbon dioxide is incorporated into the carbohydrates, take place in the stroma.

Photosynthesis, by absorbing light energy, converting it into a form of chemical energy which the living cell can use, and storing this energy for use as needed, provides the basic mechanism for synthesis of carbohydrates, fats, and protein. The process of photosynthesis enables plants to take two inorganic substances, carbon dioxide and water, and "reshuffle" their atoms to make the organic compounds on which all life depends. The details of photosynthesis are treated extensively in numerous texts and will be treated here only briefly, but a few fundamental mechanisms should be clarified.

Photosynthesis consists of light reactions and dark reactions. Light reactions are of two interrelated kinds. The first, known as *cyclic photosynthetic phosphorylation,* utilizes light energy to synthesize adenosine triphosphate (ATP), the energy-carrying molecule common to all life. The energy to produce the high-energy bond of ATP is obtained when an electron is transferred from a donor at one energy level to an electron acceptor in another. Chlorophyll is the key molecule in this energy transfer. Light striking the chlorophyll molecule excites one of the electrons in chlorophyll to an energy level sufficiently high to eject it from the molecule. Having lost this high-energy electron, the chlorophyll molecule is now able to accept another. The "lost" electron is captured by specific nucleotides which extract the energy from the electron. Void of its energy, the electron is then returned to the chlorophyll molecule. The energy of the electron is transferred in a phosphorylating enzyme system to the high-energy bond of ATP. The energy is then available for the dark reactions of photosynthesis or any other process requiring energy.

Light is secondly used in *noncyclic photophosphorylation,* in which light energy raises the energy of electrons from water. Light energy splits the water molecule into H^+ and OH^- ions. Electrons are removed from the OH^- ions, leaving (OH) radicals which subsequently combine to form water molecules and oxygen gas, which is evolved. The electrons are passed via cytochromes to chlorophyll molecules that have been activated by light to eject an electron and become electron acceptors. The energy released is used by a phosphorylating enzyme system to form ATP.

The dark reaction, or "synthesis" stage, of photosynthesis incorporates carbon dioxide into organic materials, the first of which is a five-carbon sugar called ribulose diphosphate. The newly formed six-carbon chemical immediately splits to give two molecules of phosphoglyceric acid, which are reduced to give two molecules of the three-carbon sugar, triose. These condense to form fructose, glucose, and other sugars, which can subsequently be combined either to produce disaccharides, starch, fats, and other storage materials, to go into cell wall metabolism, or to be immediately utilized in respiration.

Once produced, some 75 percent of the total photosynthate is incorporated in polysaccharides; much of this is used as building material for the cell wall. Another 15 to 20 percent of the photosynthate is consumed in respiratory activity. The remaining photosynthate serves as substrate for carbohydrate, fat, and protein metabolism. Consequently, all chemical synthesis, energy, and plant growth depend on adequate photosynthesis to provide the needed photosynthate.

The rate of photosynthesis and the amount of photosynthate subsequently produced depend on a number of environmental factors, including light, suitable temperature, and adequate carbon dioxide, water, and nutrients. A deficiency of any one of these can limit growth, reduce vigor, and predispose plants to biotic and abiotic diseases.

Light quality, intensity, and duration are all vital to normal plant development, but the intensity of light is the most critical variable influencing photosynthesis. Photosynthetic rates are inhibited whenever light intensities exceed an optimal range or fall below it. The light intensity must be great enough to enable photosynthesis and food production to proceed faster than the rate at which carbon compounds are oxidized in respiration. A compensation point, at which carbon dioxide is neither absorbed nor evolved, is reached when the light energy absorbed by the plant for photosynthesis is equal to the amount of energy burned in respiration. Below this, the photosynthate produced is inadequate for survival. The compensation point varies tremendously among species and especially between shade and sun plants. Optimal light may range from full sunlight for alpine plants to less than 1 percent of full sun for mosses and red algae. For seedlings of shade-loving forest trees growing under otherwise optimal conditions, the compensation point may be as little as 1 to 2 percent of full sunlight or as much as 30 percent.

Light requirements are extremely variable and difficult to establish due to shading effects of the plants themselves; while optimal intensity might be established for individual leaves, requirements of the whole plants are more complex. The outer canopy of leaves might be receiving ample light while at the same time inadequate light limited photosynthesis by the inner, shaded leaves. Under far brighter conditions, the inner leaves might be receiving optimal light while at the same time the light received by the outside leaves was so intense that it decomposed the chlorophyll and suppressed photosynthesis.

Shade or lack of shade also affects photosynthesis by influencing the leaf temperature. Since photosynthesis consists of enzymatic as well as light-driven reactions, it is temperature-dependent and limited by low temperatures. Although the temperature coefficient (the Q_{10}) of the light step in photosynthesis is close to 1.0, the dark steps have a Q_{10} of 2.0 or more, and temperatures below about 20°C are often limiting.

In the field, next to light itself, low carbon dioxide concentrations most frequently limit photosynthesis, The optimum amount of CO_2 for photosynthesis is much higher than the normal ambient concentrations. Carbon dioxide can be particularly limiting in dense stands of vegetation where the plants deplete the supply faster than it can be replenished. Localized depletion of CO_2 can also be significant where air is layered over the leaf surface in still air. Air layering is particularly limiting to photosynthesis around large, flat leaves; a slight breeze, carrying with it a fresh supply of carbon dioxide, greatly increases the rate of photosynthesis.

Carbon dioxide enters the leaves of higher plants mainly through the stomata, and the extent to which the stomates are open has a major influence on the rate of photosynthesis. When stomata are closed, entry of CO_2 is prevented and photosynthesis ceases; even when they are partly closed, photosynthesis is substantially reduced. Water is indirectly one of the most significant factors affecting the size of the stomatal aperture. An inadequate water supply limits guard cell turgidity, causing the cells to collapse and thus closing the stomata; CO_2 flow into the leaf is reduced, limiting photosynthesis.

The amount of water actually needed for photosynthesis is exceedingly small when compared with the amount required to maintain the plant—well under 0.1 percent of the total amount used.

Indirect effects of water shortage on the stomatal aperture are generally thought to become obvious long before water directly limits photosynthesis. Water balance and availability in the plant depend on the amount of water available in the soil and the rate of transpiration. While light does not affect the soil moisture, it does influence the rate of transpiration since it is absorbed by the leaf, raising its temperature and increasing transpiration and water loss.

While some light is necessary for plants, too much can be detrimental. Intense light not only promotes rapid transpiration but breaks down the chlorophyll molecules by photooxidation. Photosynthesis can be limited by water availability through its influence over other life processes—particularly the degree of hydration of the protoplasm. Removal of water from the protoplasm—that is, dehydration—disrupts colloidal structures and suppresses photosynthesis by impairing enzymatic efficiency. Dehydration further limits photosynthesis by physically damaging the molecular structure of essential organelles, ultimately causing death of the cell.

The age of the individual leaves also influences photosynthetic activity. This is best demonstrated with such evergreen species as palms and conifers, whose photosynthetic ability reaches its peak as the leaves mature but declines after that. A decrease in photosynthesis with age has also been reported in annual species.

The rate of photosynthesis is further influenced by too much oxygen. Surprisingly, the normal 21 percent concentration of oxygen in the atmosphere has been demonstrated to inhibit photosynthesis. An ample oxygen supply favors a rapid respiratory rate, allowing respiration to compete more successfully than photosynthesis for intermediate chemicals needed for both processes. When the oxygen concentrations are low, photosynthesis competes more successfully for these intermediates and is accelerated. Another reason why low oxygen levels favor photosynthesis is the probability that one of the early molecules of chlorophyll formation is effectively quenched by oxygen.

8-3 PHOTOMORPHOGENESIS

The quality, intensity, and duration of light all exert marked formative effects on plant growth, development, differentiation, and reproduction. This control, independent of photosynthesis, is known as *photomorphogenesis*.

In order for light to exert any morphogenic effect, it must first be absorbed by the photoreceptive pigments. Pigments differentially absorb various wavelengths of light and reflect others, giving the tissue a characteristic color. The green color of plant foliage is due to an abundance of chlorophylls, while the orange and red of other tissues are imparted by the dense concentrations of carotene and anthocyanin pigments. Other pigments normally present in plant cells are usually masked by the more dominant pigment.

Phytochrome and Photoperiod

One such inconspicuous pigment, the pale blue chromoprotein *phytochrome*, is apparently the major photoreceptive pigment concerned with photomorphogenesis. Phytochrome, which is present in the cytoplasm of all green plants, is responsible for regulating such diverse processes as seed germination, root development, shoot growth, tuber and bulb formation, dormancy, flowering, and fruit coloring.

Botanists long have speculated about the presence of pigments and hormones regulating the above processes; in the last few years, with the isolation of phytochrome, definite plant activities have been attributed to it. The phytochrome mechanism is not yet completely understood, but a few generalities can be provided.

The phytochrome pigment has two forms, each acting in its own way, which readily convert from one to the other. One form has an absorption peak and is active at red light of 660 mμ. This is designated P_r for red. The other form of phytochrome is designated P_{fr} for the "far red" and has an absorption peak at 730 mμ.

Exposure of plants to daylight converts the pigment to the P_{fr} form. In the presence of mostly infrared irradiance, phytochrome slowly reverts back to the P_r form. In sunlight, where both red and far red wavelengths are present, the conversion of P_r to P_{fr} predominates. Generally effective quantities of P_{fr} are produced with light exposure of a few minutes.

The P_{fr} form of phytochrome is biologically active and responsible for the basic responses of plants to daylight and darkness. The P_r form lacks this activity, so that no influence is exerted in the dark. The rate of phytochrome conversion in the dark is sufficiently slow to provide a *"photoperiod"* timing mechanism which synchronizes various phases of plant development with the relative duration of daylight and darkness.

The more fundamental question is how this timing mechanism works. It is probable that phytochrome is in some way tied to certain growth regulators or auxins, and that changes in the concentration and activity of growth substances have something to do with morphogenic responses, but it is difficult to tie these to light reactions. Some explanations of the mechanisms which regulate this interaction have been offered, but few facts have been established.

FLOWERING   Photoperiod research has been particularly effective in determining the light relationships regulating flowering. For the majority of plants, the length of night determines whether the plant will produce flowers or remain indefinitely in a vegetative state.

When enough phytochrome is in the P_{fr} form, a series of chemical reactions presumably is initiated which measures the length of darkness and initiates reactions which cannot proceed without it. According to Mohr (1962), P_{fr} acts as an active enzyme which initiates flowering and other morphogenic responses. Phytochrome is probably not directly responsible for promoting flowering but presumably exerts its influence by some mechanism involving the hypothesized flowering hormone *florigen*. The presence of such a hormone was postulated in the 1800s, but its existence has yet to be established. Possibly a single flowering hormone exists, but more likely several interacting chemicals are involved. One current theory is that some process or chemical may normally be present in the plant which inhibits flower formation. If this is the case, phytochrome might catalyze some process neutralizing this inhibitor and allowing flowering to proceed.

It has been suggested that florigen reduces the activity of flowering-inhibiting auxins to a level which permits the initiating of the flowering response. Flowering can sometimes be stimulated by spraying plants with an antiauxin chemical.

Another theory postulates that there may be two steps involved in flowering, the first mediated by gibberellin and the second by a flowering factor, *anthesine*. Together they may constitute the flowering hormone known as florigen. In long-day plants there is enough anthesine but not enough gibberellins; in short-day plants gibberellin is high and anthesine is low. This could account for the promotion of flowering by applications of gibberellins to short-day plants. In some way phytochrome interacts with these chemicals to regulate their activity.

Plants can be divided into three groups depending on the day-length required to induce flowering: *short-day, long-day,* and *day-neutral.* Short-day plans will flower if the light exposure is less than a critical length of about fourteen hours and the dark period is continuous, while long-day plants require a daylength in excess of about fourteen hours; often fifteen or sixteen hours of daylight are necessary. Day-neutral plants are insensitive to daylength and will bloom regardless of the photoperiod.

The photoperiodic regulation of flowering has numerous practical and economic implications, but only a few have been utilized. Plants grown for their foliage have been retained in a vegetation state by providing a photoperiod unsuitable for blooming; also, photoperiods of species grown for their flowers have been adjusted so that plants bloom more consistently at a desired time of year.

SEED GERMINATION Seeds of many plants that require light to germinate do so poorly or not at all when buried beneath a layer of soil. A single exposure to light may be sufficient to induce germination of some seeds, while several daily exposures may be required by others. Peppergrass seed, for instance, does not germinate at all in the dark, while darkness reduces germination of lettuce seed as much as 70 percent. Seed germination of still other species like June grass, American elm, and many cucurbits is inhibited by the presence of light (Went, 1956).

The influence of light on germination may be closely tied to temperature (Dale, 1965). *Betula pubescens* seeds have no light requirement when chilled, but required light when not chilled. Long days resulted in a higher percentage of germination than short days at 15°C; but at 20°C good germination occurred under both short and long days.

NATURAL DISTRIBUTION OF PLANTS Plant populations tend to adjust genetically by the process of natural selection to whatever environmental conditions prevail. Species failing to adjust perish. The duration of light and dark periods is among the most stable factors to which plants have adapted (Daubenmire, 1959). Since daylengths are constant with latitude for any given season, photoperiod provides a reliable index of the time of year and plays an important role in determining the natural distribution of plants. The influence of photoperiod provides a natural mechanism for confining plants to their range and restricting a species to areas where temperature and light requirements would be most suitable. Photoperiodic regulation of flowering, dormancy, and other life processes provides plants with a consistent means of making vegetative and reproductive growth at times of the year most suitable for their perpetuation and continued survival.

Daylight at the equator is about twelve hours throughout the year, but as one proceeds toward the poles, the daylength becomes progressively shorter in winter and longer in the summer. Since the flowering process is regulated by daylength, short-day plants could not reproduce toward the equator, and long-day plants could reproduce near the poles only during periods of spring or fall.

Even among species possessing satisfactory means of vegetative reproduction, which could offset a lack of sexual reproduction, the daylength may not be suitable for the proper coordination between seasonal changes and for accumulation of adequate food reserves.

STEM GROWTH Vegetative growth is similarly largely controlled by the duration of light. Photoperiod, coupled with temperature, regulates the time when leaf and flower buds emerge and shoots begin to grow in the spring, also the time when growth should cease and plants become dormant in the fall. Following the emergence of a shoot, its rate and duration of elongation is controlled by photoperiod. Extending the daylength in fall will prolong the duration of growth in most cases, so long as a suitable temperature is provided. Wareing (1948) has kept yellow poplar growing continuously for eighteen months by providing long daylengths. Red maple seedlings make persistent and continuous growth at longer photoperiods, while stem elongation ceases after two to four weeks when an eight-hour daylength is provided (Downs and Borthwick, 1956). Artificially lengthening the day in Florida has permitted northern conifers to grow over a 220-day season so that twice the growth can be made over that in native areas (Watt and McGregor, 1963).

Plants grown in areas where the photoperiod is unfavorably long may continue to grow late into the fall. Failure to become dormant soon enough may delay hardening enough for severe winter injury to develop.

Wareing (1951), studying Scots pine seedlings, found that trees receiving a twenty-hour daylength grew taller than plants provided with shorter daylengths. After the first year, the number of nodes is predetermined by the number of initials produced the preceding year, but the extension of internodes appears to be directly dependent on daylength.

Wareing (1953) also showed that daylengths determined the time when dormancy of European beech was broken. Shoot growth began in the spring when daylengths exceeded twelve hours. Even at a very low light intensity, buds emerged from dormancy when exposed to daylengths longer than twelve hours but remained dormant when daylight was less than twelve hours.

While the amount and duration of tree growth is often related to daylengths, exceptions are fairly common. Red maple, for instance, makes the greatest growth with sixteen hours of daylight, while most pines make the greatest growth when less than twelve hours of daylight are provided. This differential preference determines the ability of a species to compete in its environment by regulating the time of year when a plant makes most of its growth and has the greatest demand for water and nutrients.

The optimal photoperiod for shoot growth varies among species and ecotypes of a single species, especially when plants have evolved in widely separated geographical areas of their range (Decker, 1954). Northern and southern races of Scots pine and alder grew poorly in ten-hour photoperiods, but when continuous light was provided, more closely approximating the natural long, summer days, plants from northern regions grew far more than those of southern regions. Similarly, seedlings from seed sources at a northern latitude of 66° did well when provided long photoperiods but poorly under short-day conditions. Seedlings from seeds collected at more southerly latitudes where natural daylengths were more uniform grew well under both long- or short-day conditions.

Height of annuals and perennials also is often regulated by photoperiod (Went, 1956). Cosmos plants which received a twelve-hour light period were considerably taller than those grown under longer daylengths. Under short-day conditions the stem of some plants is so short that the leaves form a rosette. Radish, spinach, lettuce, dill, and many other long-day plants are affected this way. Cereal plants commonly make their best vegetative growth under short-day conditions. Strawberries, which require short photoperiods for blossoming, develop runners more extensively under long photoperiods.

There has been considerable speculation regarding the mechanisms by which photoperiod affects stem elongation. The normal inhibition of stem elongation caused by light is largely prevented if gibberellins are applied to the plant. Gibberellins are presumed to be links in the intermediary causal chain whereby the phytochrome and high-energy reactions affect stem elongation. Visible radiation may inhibit stem elongation by decreasing cell wall plasticity, and gibberellic acid may act by preventing the radiation-induced decrease of plasticity.

It is also entirely possible that cell elongation and phytochrome act independently and that the assumed change of gibberellic acid level is not part of the causal chain between phytochrome and the inhibition of cell elongation.

Another possible mechanism explaining stem elongation is that growth is regulated by indole acetic acid (IAA). Ultraviolet light may inactivate IAA and thereby retard growth. Determinations of auxin content before and after ultraviolet light irradiation have shown that such irradiation reduces auxin levels. Since IAA does not absorb wavelengths in the visible portion of the spectrum, although this is the portion that has a destructive effect on IAA, there must be a photoreceptor capable of absorbing visible wavelengths of light and then causing photodegradation of IAA. Two pigments, carotene and riboflavin, whose absorption spectra closely resemble the action spectra for destruction of IAA, are proposed to be involved in IAA destruction and subsequent regulation of shoot growth.

LEAF DEVELOPMENT   Daylength markedly influences leaf development (Milthorpe, 1956). Leaf size of long-day plants is often greatest when plants are exposed to short days; the reverse also may be true. For instance, strawberry, a short-day plant, developed larger leaves under long days. However, no such effects could be found in cereal plants. In *Phleum pratense*, a long-day plant, sixteen-hour days caused the leaves to be significantly larger than twelve- or fourteen-hour days.

Leaf retention is also regulated by photoperiod. Leaves of deciduous trees remained attached later in the season when daylengths were long than when daylengths were short. Under long-day conditions some deciduous plants in species of willow, lilac, and rose keep their leaves almost all winter and become essentially evergreen (Downs and Borthwick, 1956). Studies with lettuce seedlings revealed a slower leaf production and smaller leaves under nine-hour days than sixteen-hour days. The response of leaves to light seems to vary with every species, and too few species have been studied to generalize.

Photoperiod also influences leaf shape (Milthorpe, 1956). Two varieties of *Sesamum orientale* produced only entire-margined and simple leaves when exposed to ten- and eleven-hour photoperiods, although serrate leaves were produced at twelve-hour days. At fifteen- and sixteen-hour photoperiods, compound leaves were produced as well.

8-4 LIGHT INTENSITY

STEM GROWTH   Light intensity, as well as duration, affects growth. The capacity to compete for light determines the success or failure of both cultivated and native species; inadequate light limits development of many species. Light intensity may be limited by shade from larger plants and structures and in urban areas may be substantially reduced by air pollutants. It is conceivable that in some instances the shading effect of particulate materials and gaseous wastes in the atmosphere may be as detrimental as the toxic effect of the pollutants.

Stem elongation is retarded as much by excessive radiant energy as by too little; a high light intensity inhibits cell elongation and limits growth of most higher plants. Despite such inhibition, overall maximum growth, as determined by shoot elongation and leaf expansion, is made by most sun-loving plants in full sunlight of 10,000 fc or more. But this is only because sufficient light of 2,000 to 3,000 fc is then available to leaves in the inner, shaded parts of the plants. In full sunlight stems are thicker, with well-developed xylem and supporting tissues and internodes relatively shorter than when grown in the shade.

One significant study of light requirements was conducted with ponderosa pine. Trees grown under various shade conditions for periods up to ten years grew best in full sunlight; trees receiving 50 percent light were only slightly shorter but had grown only half as much in diameter. White spruce made its best growth under partial light conditions. Hardwoods generally do best at low light intensities. American and winged elm, sycamore, river birch, red maple, and alder all made maximum height growth at 20 to 50 percent relative illumination. Bark thickness has also been reported to be partly a function of light. Bark of eastern white pine is substantially thinner, smoother, and more sensitive to sunburning when the cells develop in the shade (Huberman, 1943).

LEAF GROWTH   Light intensity influences not only stem growth but the development and ultimate expansion of leaves as well. High light intensities encourage growth of smaller but denser and heavier leaves, while shading produces much larger, thinner leaves with thinner epidermis, less palisade, more intercellular space and spongy parenchyma, and more numerous stomata. When shading reduces the light intensity to 2,000 fc, the ultimate leaf area may be increased 15 to 55 percent while the plant weight is cut nearly in half (Milthorpe, 1956).

At high light intensities, individual cells of the leaf blade are usually smaller than in subdued light or shade. This in turn results in smaller but thicker leaf blades, denser but smaller stomata, smaller vein islets, and more conducting and mechanical tissue. The cuticle and cell walls are generally thicker, intercellular spaces smaller, and blades of harder texture. These morphological modifications help make the plant more resistant to temperature and drought stress and infection by fungus and bacterial parasites.

Compared with plants grown in the shade, individuals developing in full sun are also characterized by more weakly developed spongy parenchyma and a better-developed palisade layer. In fact, in full sun palisade tissue frequently develops on both sides of the blade.

Many plants develop different anatomical leaf modifications depending on whether they develop in the sun or shade. Typically, shade leaves are thinner and have a larger surface per unit weight, a thinner epidermis, less palisade tissue, more intercellular space and spongy parenchyma, and less supportive and conductive tissue but more stomata than leaves maturing in the sun.

The typical shade leaf seems well adjusted to carry on photosynthesis as long as it is protected from the detrimental effects of too much light. Their morphology enables them to absorb maximum amounts of light and survive at low light intensities.

Light quality may also influence leaf expansion. The effect of different wavelengths on size and shape of expanded leaves varies considerably among species. Cells in the under surface of tomato leaves expand irregularly and incompletely in red light, so that the mature leaves are curled. Expansion in blue light is normal. However, cucumber and pea leaves expand normally even when only red light is provided.

Leaf color is markedly influenced by light quality and duration.

A prolonged daily exposure to light can prevent chlorophyll formation in the leaf, thereby revealing more of the yellow caratenoid pigments and causing chlorosis. Even under normal light conditions, chloroplasts are fewer and smaller than when leaves develop in the shade. In tomato, for example, chlorophyll development is inhibited when plants receive more than eighteen hours of daylight during a twenty-four-hour period. Under continuous light no chlorophyll is formed, and the leaf is completely blanched. Chlorophyll is not the only pigment influenced by light. Anthocyanin has long been known to be formed only in the presence of light and may be excessive in unshaded plants grown for their green foliage color. Photo responses of mustard seedlings include not only the synthesis of anthocyanin but inhibition of hypocotyl lengthening, hair formation along the hypocotyl, enlargement of cotyledons, increase of the negative geotropic reactivity, opening of the plumule hook, and formation of tracheary elements in the vascular bundles of the hypocotyl (Butler, 1963).

Virtually every plant process is directly or indirectly influenced by light. Light of the correct quality, quantity, and duration is essential to normal plant development, and adverse relations may be responsible for abnormal development and disease.

BIBLIOGRAPHY

Butler, R. O., 1963. The effect of light intensity on stem and leaf growth on broad bean seedlings. *J. Exp. Bot.* **14**:142–152.

Dale, J. E., 1965. Leaf growth in *Phasealus vulgaris*. 2. Temperature effects and the light factor. *Ann. Bot.* **29**:293–308.

Daubenmire, R. F., 1959. "Plants and environment," 2d ed. Wiley, New York, 422 pp.

Decker, J. P., 1954. The effect of light intensity on photosynthetic rate in Scotch pine. *Plant Physiol.* **29**:293–308.

Downs, R. J., and H. A. Borthwick, 1956. Effects of photoperiod on the growth of trees. *Bot. Gaz.* **117**:310–326.

Huberman, M. A., 1943. Sunscald of eastern white pine, *Pinus strobus* L. *Ecology* **24**:456–471.

Milthorpe, F. L., 1956. "The growth of leaves." Butterworth, London, 223 pp.

Mohr, H., 1962. Primary effects of light on growth. *Ann. Rev. Plant Physiol.* **13**:465–488.

Wareing, P., 1951. Growth studies in woody species. III. Further photoperiodic effect in *Pinus sylvestris*. *Physiol. Plant.* **4**:41–56.

————, 1953. Growth studies in woody species. V. Photoperiodism in dormant buds of *Fagus sylvatica* L. *Physiol. Plant.* **6**:692–706.

Watt, R. F., and W. H. D. McGregor, 1963. Growth of four northern conifers under long and natural photoperiods in Florida and Wisconsin. *Forest Sci.* **9**:115–128.

Went, F. W., 1956. The role of environment in plant growth. *Amer. Sci.* **44**:378–398.

SELECTED REFERENCES

Allard, H. A., 1932. Length of day in relation to the natural and artificial distribution of plants. *Ecology* **13**:221–234.

Billings, W. D., 1957. Physiological ecology. *Ann. Rev. Plant Physiol.* **8**:375–392.

Bohning, R. H., and C. A. Burnside, 1956. The effect of light intensity on rate of apparent photosynthesis in leaves of sun and shade plants. *Amer. J. Bot.* **43**:557–561.

Giese, A. C. (ed.), 1964. "Photophysiology." Academic, New York, 377 pp.

Hillman, W. S., 1962. "The physiology of flowering." Holt, New York, 164 pp.

Racker, Efrain, 1965. "Mechanisms in bioenergetics." Academic, New York, 259 pp.

Reifsnyder, W. E., and H. W. Lull, 1965. Radiant energy in relation to forests. *U.S.D.A. Tech. Bull.* 1344, 111 pp.

Strothman, R. O., 1967. The influence of light and moisture on the growth of red pine seedlings in Minnesota. *Forest Sci.* **13**:182–191.

Thut, H. F., and W. E. Loomis, 1944. Relation of light to growth of plants. *Plant Physiol.* **19**:117–130.

Van der Veen, R., and C. Meijer, 1959. "Light and plant growth." Macmillan, New York.

Wareing, P., 1948. Photoperiodism in woody species. *Forestry* **22**:211–221.

Wassink, E. C., and J. A. J. Stolwijk, 1956. Effects of light quality on plant growth. *Ann. Rev. Plant Physiol.* **7**:373–400.

CHAPTER NINE

LIGHT STRESS
AND RADIATION

Disorders caused by adverse light relations are subtle and insidious, often involving only impaired growth and reduced vigor. When the stress is more acute, leaves gradually lose their color, turning first pale green then yellow; limbs may die back a little each year, with the plants becoming increasingly weaker and more sensitive to other stresses.

9-1 LIGHT INTENSITY

The effect of light, either too much or too little, is greatest on photosynthesis. Too high a light intensity impairs photosynthesis by rapidly photooxidizing chlorophyll, so that the remaining available supply is inadequate to absorb sufficient light energy. As the old chlorophyll is destroyed in the most exposed upper palisade cells, leaf color fades.

Even though chlorophyll is continually being replenished, its break-down by intense illumination is so fast that renewal fails to keep pace, causing leaves to develop a pale green or yellowish cast. High light intensity may also inhibit chlorophyll synthesis, further limiting the chlorophyll content. Leaves are slightly protected from chlorophyll loss by the migration of chloroplasts to the center of the cell, but this defense mechanism is limited, and unshaded, southerly oriented leaves may become distinctly yellow. High light intensities may also impair photosynthesis directly by the oxidizing of one or more of the enzymes participating in photosynthesis.

Plants vary greatly in their response to light, and optimum light requirements are unknown for most species; in a particular study with Scotch pine, however, photosynthetic activity was found to be inhibited as the light intensity reached 9,300 fc (Gerhold, 1939). Studies by Dr. A. C. Hill (unpublished) in Utah, on the other hand, showed that carbon dioxide uptake of alfalfa increased up to full sunlight of close to 12,000 fc. Shirley (1945) found that maximum dry weight increase of red pine seedlings occurred at 98 percent of full sunlight, although maximum height growth was reached at half this light intensity.

During periods of excessively high light intensity, the leaf color of conifers like Scotch pine and of many broad-leaved species begins to fade. When the incident light intensity decreases, color returns to normal. Chlorosis due to high light intensity most seriously affects shade-loving plants unaccustomed to intense light. Their cells are large and thin-walled and their leaf blades usually large and thin, consequently sensitive to both moisture and light stress. Sun plants are most affected by high light intensity when predisposed by a period of cloudy, overcast weather during which more succulent "soft", intolerant tissues developed. Germinating seedlings initially growing in the shade may be especially sensitive to excessive light once the protective canopy is removed. Continued exposure to intense light may kill the protoplasts of leaf, stem, or fruit cells and cause brown-ing of localized areas of tissues. Controlled studies with ultraviolet light showed that plant growth was also inhibited by high light intensities, but once light intensity was reduced, normal growth was resumed and there was no permanent damage to the plants (Shirley, 1929).

Insufficient light limits the radiant energy available for photo-synthesis, causing food reserves to be depleted faster than they can be stored. For example, the photosynthetic rate of Scotch pine was shown to decrease in direct proportion to decreasing light intensity from 6,400 to 1,800 fc. Below 1,800 fc, the photosynthesis rate dropped still more steeply (Gerhold, 1959).

Strothman (1967) found that removal of competition for light invariably produced a substantial growth response of red pine seedlings. Unshaded seedlings produced more needle fascicles per seedling, longer needles, larger terminal buds, thicker stems, and larger roots than shaded plants; also, unshaded plants made over twice as much growth in height. Poorer root growth in the shade plants was attributed by Shiroya et al. (1962) to the limited photosynthate translocated to the roots. The effect of low light intensities in reducing root growth helps explain the adaptive success of certain plants in the forest understory. Only root systems of plants adapted to low light intensities can compete successfully for water or nutrients; unadapted species are soon crowded out.

Light as a pathogen is usually a coupled disease reaction. Rarely does light act alone. *Onion blast* and *tomato sunscald,* for instance, result from the inability of plants to adjust to high temperatures, bright sunlight, and low relative humidity after a period of cloudy, wet weather. Under cooler conditions, with subdued light, the plant produces abnormally soft top growth. The water requirement under such conditions is not great, and adequate water can be obtained by the limited root system produced. When high light and temperature conditions resume, the tissues exposed to direct sunlight are subjected to unaccustomed stresses; the limited root system is unable to absorb sufficient water to replace that lost during rapid transpiration, and desiccation, chlorosis, and leaf scorch follow.

The interaction of light with temperature, as well as the effect of low light intensity on growth, is shown in the results of a study where young aspen cuttings were exposed to various light and temperature conditions (Farmer, 1963). As light intensities were reduced from 1,700 to 500 fc, plant height diminished correspondingly when temperatures of 24°C during the day and 22°C at night were provided. Yet there was no significant height difference when plants were grown at only slightly lower temperatures of 21°C during the day and 19°C at night.

A few disorders have been attributed to light alone. *Sunscald* of bean pods is the best-documented example (MacMillan, 1923). The disorder was reported in Colorado, where, due to the high elevation, ultraviolet light radiation is typically great and the relative humidity sufficiently low that little light is filtered out. The earliest visible symptoms of sunscald were minute reddish or brownish spots appearing on the outer valves of the most exposed pods. These spots gradually lengthened and formed short streaks paralleling the suture. In a day or two the spots increased in size to 3 to 4 mm and developed into slightly sunken, water-soaked lesions which often coalesced, producing larger, irregular lesions.

Leaf symptoms consisting of deep brown lesions occurred at the higher light intensities when groups of epidermal cells were killed. Cells were stimulated to produce excessive chlorophyll, causing them to assume an abnormally deep green color; more intense ultraviolet radiation killed the cells, producing an overall mottling or mosaic symptom. This was followed by the development of large islands of dead tissues on the stem and petiole.

To learn the role of temperature in lesion development, Dr. MacMillan exposed plants to temperatures up to 55°C for half an hour. No symptoms developed. But when plants were exposed to artificial ultraviolet light (λ 2,300) for half an hour at 25°C, lesions indistinguishable from those occurring naturally appeared. Under microscopic observation the epidermal cells were seen to retain their form but to be filled with a brown, pigmented substance of an undetermined chemical nature. Cells containing chlorophyll took on deep green color. Epidermal cells gradually lost water, dried out, and died. Other cells remained green or chlorotic. As a result, many shades of green and brown coloration developed, giving leaves a mottled appearance. Scald was prevented when a window glass was used to protect the plants by blocking out the ultraviolet light.

Similar sunburning has been recognized on soybeans, cowpeas, and lima beans (Gibson, 1921). Foliage of these species showed minute, brick-red spots on the upper surface of the leaves. The lesions were usually confined between the veins, but in more severe cases reddening extended over the veins or followed along them longitudinally. The ultimate size of the spots depended largely on the extent to which the lesions coalesced. When the spots reached a diameter of about 4 mm, the center became necrotic and brownish in color and often cracked open, providing an open wound readily subject to invasion by parasitic fungi. Identical lesions were simulated artificially by concentrating light rays with a magnifying lens.

Kreitlow and Kilpatrick (1967) described a light-related leaf spot of alfalfa, red clover, and other forage legumes. The foliar lesions were round, sunken, usually under 1 mm in diameter, and scattered over the upper leaf surface. On the stems and petioles, lesions were generally linear and ranged in color from brownish to reddish-brown or black. The older leaves were most severely spotted; affected leaves became chlorotic, shriveled, and dropped. Affected stems often collapsed and died. Spotting was greatest on plants raised in growth chambers under fluorescent lights at the highest light intensity, but it also occurred in the field.

9-2 DURATION OF LIGHT

Photoperiod profoundly influences plant development, and deviations from normal cause abnormal growth and reproduction. Studies with Scotch pine showed that short daylengths caused premature fading and discoloration of needles (Stutzman, 1954). If long days were extended into the fall and winter, the needles retained their deep green color into December even though freezing temperatures occurred on a number of occasions. Such a change in photoperiod is not usually directly harmful but is indirectly significant in keeping plants in an actively growing state sensitive to low-temperature injury. The short days of fall are needed both to arrest stem elongation and to allow plants to become hardened before winter. When tissues fail to become hardened, winter injury is likely. This was demonstrated by Dr. Paul Kramer at the Duke University campus in 1937. He showed that shoots on a hedge of *Abelia grandiflora* near some electric lights were killed back badly while the branches between the lights remained healthy. The hedge was trimmed to a uniform height on September 25; by October 20, when the first light frost occurred, plants on each side of the lights, for a distance of about 5 yd, were distinctly different in appearance than those farther away. Limbs bore numerous new soft, succulent shoots with pale green leaves which failed to become hardened. Leaves farther from the lights were bronzed and dark green. The difference was so pronounced that it was evident from 200 yd away. The succulent shoots were killed up to 3 yd from the lights, where the intensity was greatest. Black locust seedlings exposed to varying photoperiods reacted the same way. Where a long photoperiod was maintained, the plants were killed by freezing temperatures in November. Proximity to electric lights, causing growth to continue late into the autumn, might be expected to be deleterious and contribute to winter killing of almost any tree or shrub normally dormant during the winter months.

One of the most subtly remarkable disorders for which photoperiod was directly responsible was brought to light in 1947 by Dr. Kenneth Post of Cornell University. Chrysanthemum flower disorders described as bull heads, long necks, uneven flowering, late flowering, and flower failure were all disclosed to be related to photoperiod.

To understand the disorder, it is necessary to understand a little about chrysanthemum culture. Chrysanthemums are brought into bloom commercially by covering them with black cloth to shorten the daylength artificially and ensure earliness. Later varieties are prevented from blooming by providing supplementary light to lengthen the days. Flower bud formation in chrysanthemums is initiated when the days are shorter than about $14\frac{1}{2}$ hours. The date at which this occurs depends on the amount of cloudy weather during a critical period. Clouds in the morning or evening reduce daylength as much as an hour. As few as two short days will cause the crown bud to form on early varieties, but the buds stop developing if a 14- to $14\frac{1}{2}$-hour-long day is resumed. Daylengths less than $13\frac{1}{2}$ hours are needed for normal development. If a prolonged period of $13\frac{1}{2}$- to $14\frac{1}{2}$-hour days then follows, the plants continue to bud, but the buds do not develop normally. Even a slight increase in daylength of only five to thirty minutes after budding starts causes crown buds to develop abnormally and form a long neck with bract-like leaves on the stem. If the daylength is nearly short enough for normal bud development, a few of the upper buds produce flowers while the lower buds abort. A dark, cloudy day or two interrupting the normal long days can simulate short days sufficiently to induce plants to form buds prematurely. When the longer days resume, the flower buds develop abnormally, often abort, and the crop may be ruined.

Continuous light may also be directly harmful to plants, but the response is interrelated with temperature (Withrow and Withrow, 1949). At 26 to 30°C, symptoms of exposure to continuous light involve a sharply defined chlorosis appearing first on basal areas and veins of actively expanding leaflets. Studies with tomatoes showed that when temperatures were dropped to 23°C, symptoms also included small, dark brown, necrotic lesions. At 17 and 14°C, symptoms consisted of a faint mottling of actively growing leaves followed by a gradual but uneven loss of chlorophyll, producing a mottle-leaf expression that closely resembled the mosaic symptoms caused by viruses. At all temperatures, the leaves were small, distorted, and chlorotic. However, plants recovered when again exposed to light periods of sixteen hours. Fluctuating temperatures could offset this injury even when light was continuous.

Chlorosis from too long a daylength, even though not continuous, can occur when the light intensity is very low (Hillman, 1956). Tomato plants subjected to twenty-four hours of irradiance at 4 and 20 fc became chlorotic. Chlorosis was pronounced at 20°C, but at 12°C photoperiod had relatively little effect on chlorophyll concentrations.

Plants in natural field situations are not likely to be exposed to continuous light, but house plants or plants growing on porches or in foyers may be continually exposed to a light environment. Offices and homes are often subjected to these conditions; the same problem can also arise around farms or other areas subjected to continual light.

9-3 IONIZING RADIATION

Since World War II a new, invisible, ionizing irradiation has been added to the environment in constantly increasing amounts. Radiation from nuclear explosions, radioactive wastes from nuclear reactors, X-ray machines, and particle accelerators may all influence the well-being of plants.

Some opportunity to study the effects of ionizing radiation occurred in the South Pacific atolls, where fallout from atomic tests was pronounced; another arose near Atlanta, Georgia, where an unshielded nuclear reactor was operating. More controlled studies concerning the influence of ionizing radiation have been conducted on plants and ecosystems at Brookhaven Laboratories on Long Island by Dr. George M. Woodwell (1962, 1963). Sources of gamma radiation from cesium 137 and cobalt 60 were suspended in a tower in such a way that they could be raised or lowered from safe distances. Pines, being especially sensitive, were killed by six months' exposure to 20 to 30 r per day. Deleterious effects were produced by exposures to only 1 or 2 r per day. Death of the terminal shoots and growing points was one of the more obvious symptoms of injury.

Cellular effects of gamma radiation are numerous and diverse and include damage to nucleic acids, chromosomes, mitochondria, and cell membranes. The site of principal damage, though, is the cell nucleus.

Within the nucleus, radiation damage is closely correlated with the amount of energy absorbed by the chromosomes. Plants having many small chromosomes are generally more resistant to ionizing radiation than plants with a few large chromosomes. Plants with polyploidy possess additional tolerance to the effect of radiation. In all cases the radiation exposure necessary to cause direct effects is far greater than exposures from naturally occurring sources or present levels from fallout. The potential threat from nuclear reactors used for generating electric power may have to be considered in the future (Eisenbud, 1964).

BIBLIOGRAPHY

Eisenbud, Merril, 1964. "Environmental radiation." McGraw-Hill, New York, 430 pp.

Farmer, R. E., Jr., 1963. Effect of light intensity on growth of *Populus tremuloides* cuttings under two temperatures regimes. *Ecology* **44**:409–411.

Gerhold, H. D., 1959. Seasonal discoloration of Scotch pine in relation to microclimate factors. *Forest Sci.* **5**:333–343.

Gibson, F., 1921. Sunburn and aphid injury of soybeans and cowpeas. *Univ. Ariz. Agr. Exp. Sta. Tech. Bull.* no. 2.

Hillman, W. S., 1956. Injury of tomato plants by continuous light and unfavorable photoperiodic cycles. *Amer. J. Bot.* **43**:89–96.

Kramer, P. J., 1937. Photoperiodic stimulation of growth by artificial light as a cause of winter killing. *Plant Physiol.* **12**:881–883.

Kreitlow, K. W., and R. A. Kilpatrick, 1967. A physiogenic leaf and stem spot of some forage legumes. *Plant Dis. Rep.* **51**:619–622.

MacMillan, G. H., 1923. Cause of sunscald of beans. *Phytopathol.* **13**:376–380.

Post, K., 1947. Chrysanthemum troubles of 1947. Mimeographed report, Dept. of Floriculture and Ornamental Horticulture, Cornell Univ., Ithaca, N.Y.

Shirley, H. L., 1929. The influence of light intensity and light quality upon the growth of plants. *Amer. J. Bot.* **16**:354–390.

———, 1945. Reproduction of upland conifers in the Lake States as affected by root competition and light. *Amer. Midl. Nat.* **33**:537–612.

Shiroya, T., F. R. Listes, V. Slankis, G. Krotkov, and C. D. Nelson, 1962. Translocation of products of photosynthesis to roots of pine seedlings. *Can. J. Bot.* **40**:1125–1135.

Strothman, R. O., 1967. The influence of light and moisture on the growth of red pine seedlings in Minnesota. *Forest Sci.* **13**:183–191.

Stutzman, L. R., 1954. The Scotch pine color problem. Pa. *Christmas Tree Growers Assoc. Bull.* **41**:3.

Withrow, A. P., and R. B. Withrow, 1949. Photoperiodic chlorosis in tomato. *Plant Physiol.* **24**:657–663.

Woodwell, G. M., 1963. The ecological effect of radiation. *Sci. Amer.* **208**:40–49.

———, 1962. Effects of ionizing radiation on terrestrial ecosystems. *Science* **138**:572–577.

SELECTED REFERENCES

Cooper, A. J., 1964. Effect of light intensity, day temperature and water supply on fruit ripening disorders. *J. Hort. Sci.* **39**:42–53.

Decker, J. P., 1954. The effect of light intensity on photosynthetic rate in Scotch pine. *Plant Physiol.* **29**:305–306.

Francki, R. I. B., 1967. Effect of high light intensities on spontaneous and virus-induced local lesions in *Gomphrena globosa. Phytopathol.* **57**:329.

Huberman, M. A., 1943. Sunscald of Eastern white pine, *Pinus strobus* L. *Ecology* **24**:456–471.

Kozlowski, T. T., 1949. Light and water in relation to growth and competition on Piedmont forest tree species. *Ecol. Monogr.* **19**:207–231.

Kramer, P. J., and J. P. Decker, 1944. Relation between light intensity and rate of photosynthesis of loblolly pine and certain hardwoods. *Plant Physiol.* **19**:350–358.

Sparrow, A. H., and G. M., Woodwell, 1962. Prediction of sensitivity of plants to chronic gamma irradiation. *Radiat. Bot.* **2**:9–26.

Steemann-Nielson, E., 1952. High light intensity inhibiting photosynthesis. *Physiol. Plant.* **5**:334–344.

Tranquillini, W., 1964. The physiology of plants at high altitudes. *Ann. Rev. Plant Physiol.* **15**:345–362.

CHAPTER TEN

CLIMATIC EXTREMES: LIGHTNING, HAIL, ICE, AND SNOW

Lightning, hail, ice, sleet, and snow damage, associated with climatic extremes, is relatively common and sometimes causes devastating losses due to plant injury. These environmental components are discussed in the following pages.

10-1 LIGHTNING

Plant damage from lightning is remarkably common and widespread in many parts of the country. The extent of damage varies from area to area and from year to year but is often serious and worthy of recognition. Lightning has been blamed for about one-third of the total loss of ponderosa pine timber in northern Arizona (Wadsworth, 1943), and for an estimated 50 percent of the total volume of timber lost in the 1930s in the Southeast (Reynolds, 1940). Much of the loss was indirect, since injured trees were the first attacked by pine bark beetles and other insect pests. Once established in the weakened trees, the beetle population expanded enough to successfully infest the trees and cause extensive damage.

A survey of lightning damage in native Eastern deciduous forests by A. R. Thompson in 1943 over a seven-year period revealed that oak, elm, poplar, tulip trees, ash, maple, and pine were among the most prone to damage. Most of the trees hit, some 60 percent, were growing in groups consisting of a few trees. Only 24 percent of those hit were standing alone. Even fewer of the trees hit, some 9 percent were in a forest.

Certain types of trees appear more apt to be hit by lightning than others. Differences in susceptibility have been attributed to height, habitat, growth habit, chemical composition of the plant, and the unequal conductivity and water content of the wood.

Exceptionally tall trees, and those in exposed locations, are natural lightning rods and sooner or later will be hit by lightning. But often they escape for a considerable time, while shorter trees are damaged. In mixed oak-spruce forests oaks showed signs of having been hit by lightning with particular frequency, while the spruces, which were much taller, remained uninjured (Thompson, 1943). The reasons for this may be associated with the degree of hydration of the two species.

The fatty contents of plant cells have been reported to influence conductivity and subsequent tolerance to injury. Microscopic study showed that beech wood contains large amounts of oil, while oak wood is almost free from it and high in water content. This high degree of hydration predisposes the oak to lightning damage. The poor conduction and lightning resistance of such trees as birch, walnut, and linden are attributed to their high oil content. Oil content and conductance vary with the season of the year, so that damage is greatest from spring and summer storms when trees are high in sugars rather than oils.

When bark is smooth and constantly wet, as with beech trees, electric current is effectively conducted through the external water phase, and injury is mostly superficial. Superficial damage causes leaf scorching, bark fissures, and sometimes the killing of smaller branches. Trees usually recover from this. But when deepseated damage prevails, destroying larger limbs, splitting the bole, and shattering the trees, recovery is rare.

Effects of lightning are not always immediate and are sometimes expressed only after a year or two. Breakage may be immediately conspicuous, but the tree may not die for two or more years after a strike. This suggests that lightning may predispose the tree to biotic pathogens which are actually responsible for death. The continuous or discontinuous furrows ripped by the lightning charge provide an inviting entry for bacterial, fungus, and insect pests.

A single bolt of lightning often kills but a single tree, but many incidents have been reported where a single bolt was responsible for killing large numbers. One incident was described in Dumfries, Scotland (Murray, 1958), where about 100 Japanese larch trees were killed over some three-quarters of an acre. Yet only 3 of the 100 trees showed the scars of having been directly hit by lightning. It was believed that lightning caused this extensive damage by traveling through the trees from branch to branch and tree to tree, girdling the stems and roots. Another incident was recorded where lightning entered a planting of black locust seedlings (Jackson, 1940). The one-year-old trees were damaged over an area of approximately 50 by 100 ft. Most of the trees in the central part of the area were killed immediately. Those toward the periphery were severely damaged but not killed. Roots of these plants collapsed and turned brown. The following year the injured plants were dwarfed and chlorotic. Seedling blight over large areas of forests and nurseries may also be attributable to such lightning strikes.

The diffuse effects of lightning are further exemplified by an incident in which one larch tree had been struck and a considerable number of surrounding pines and spruces began to die (Murray, 1958). Affected trees showed no local injury but died in a semicircle 25 ft in diameter. It was postulated that such dying-back of large groups of trees was caused by "lightning spray," in which the lightning was scattered into a number of lethal rays. The current may also enter a plant from the ground, the electrical discharge moving up the tree and killing limbs and leaves. This may be followed by bark sloughing off from affected parts of the trees, with limbs freely cracking and wood splitting open. A. S. Rhoads (1943) reported two incidents where pines had been killed by lightning. In the first case, all the young pines in a circular area more than 100 ft in diameter in an open, glade-like region along the roadside had been killed. Damaged trees consisted chiefly of longleaf pine from 1 to 2½ in. in diameter and ranging up to 15 ft high. Death was so swift that the brown needles remained hanging on the tree.

The second incident was noted a few miles away in an area of shortleaf pine. In this case the trees were mostly 3 to 4 in. in diameter. The injury was older, and limbs on the affected side of the tree were completely defoliated.

Speculation concerning the path of lightning in the tree goes back to 1853 (Hartig), and the views have changed little. Hartig postulated that the main current of electricity, after breaking through the bark, passed down the tree through the cambium layer, since this provided the best conductor. Heat developed immediately, vaporizing the cell contents and bursting the bark, the wood being split along lines of the weakest union. The direction of the line of splitting depends on the structure of the wood. Since lightning follows the path of best conduction, it travels much further longitudinally than transversely in the wood. The electrical discharge may go up or down the tree, since discharge can be ground to cloud or cloud to ground.

While damage is more general on taller, woody plants, cane fruits and herbaceous species are not exempted. In 1949, N. A. Grubb reported lightning damage to raspberries where trellis wires were used to support the plants. The current was conducted along the wires and burned the leaves and canes which contacted it. Affected canes dried up and bore no fruit. Adam (1938) described damage to grapes in which pith cells became necrotic and the tissue collapsed.

Injury to herbaceous species lacks the mechanical splitting or tearing of the tissues which characterizes damage to woody plants. Rather, damage is more general and involves wilting and collapse similar to that caused by drought. The zone of injured plants tends to be circular, probably due to the radial conduction of the lethal current through the surface water. The size of the affected area depends on the soil and vegetation characteristics and their ability to conduct the electric current. Injury in dry, sparsely covered land will be minimal; damage in wet soils may extend 50 ft or more.

Lightning has been reported most frequently to damage the more succulent herbaceous crops, such as tomato, potato, and cabbage. Effects of lightning are not always immediately obvious. Sometimes plants decline or die gradually over a period of weeks; by the time symptoms are noticed, they can easily be confused with those produced by other pathogens. Symptoms of lightning damage on crops like tomato and potato consist of a general drooping of the tops and varying degrees of hollowing of the stem pith. Individual leaves become desiccated, and plants collapse. Leaf tips can be seen to wilt a few hours after a strike, but more obvious symptoms may not appear for several days. Small longitudinal or circular stem lesions together with irregular, necrotic burned areas on the stem, leaves, and fruit appear on plants near the periphery of the lightning strike. Affected plants also show varying degrees of growth retardation. The high temperatures from the lightning raise by many degrees the temperature of underlying tissues of the fruit, often cooking it and causing surface blisters.

Potato plants wither from the top down. The greatest amount of damage occurs at the point of the strike, where plants are killed. Beyond this, varying degrees of collapse are evident. In one incident (Weber, 1930) potato tops hit by lightning were broken and disheveled over a small area. Within a few days plants were dead and dried up over a circular area several feet in diameter. Injured, but still living, plants adjacent to this area were partially collapsed but slowly recovered. Because stems apparently serve to conduct the current from the ground, they are the most severely damaged. Leaves may remain normal at the periphery of the affected area, although they gradually collapse when the stem is damaged. Underground portions of the plant often escape injury but are not exempted. In one field in Lincolnshire, England, potato tubers were partly split open and showed both superficial and deep burns (Burr, 1933). Cross sections of the stem showed that the pith and parenchyma cells of the injured organs were collapsed, although vascular tissue characteristically remained intact. This is characteristic of lightning injury and helps distinguish symptoms from those caused by vascular wilt fungi.

Lightning damage to soybean plants also is characterized by blackened stems and blighted leaves over areas 40 to 50 ft in diameter (Johnson, 1943). Lightning was reported to have flattened growing kale plants in much the same way as might a strong wind. The mechanical force of the lightning is illustrated by a depression in the soil about 2½ ft across and 6 to 8 in. deep at the point of contact (Jones, 1917).

Banana plants have been damaged in areas up to 60 ft in diameter (Reinking, 1938). Plants were twisted, collapsed, and scalded near the strike; farther out leaves were partially cooked and drooping. Necrosis extended down the interior of the trunk into the rhizome. Damage was transient; soon new shoots arose from the underground parts and filled in the affected area.

Lightning damage to cotton plants affects both the reproductive and vegetative organs. One incident (Smith, 1943) was reported where all plants were killed over an area nearly 50 ft in diameter. Pith of the affected stem appeared to be the first tissue to collapse, turn brown, and die. Next, leaves of the plants wilted, died, and blackened. Bolls turned pale, yellowed, and dried up; the less severely damaged bolls failed to open. Surrounding this area of severe damage, injury gradually diminished. Peripheral plants clung to life for days or weeks, but slowly their color changed to yellow or brick-red, and finally the tissues blackened and died. In a few cases the inside of the stem was discolored, and vascular tissues appeared as though infected by parasitic wilt fungi. The same general kind of wilting expression has been reported on cabbage, sugarbeets, alfalfa, and onion (Munn, 1915; Walker, 1937).

10-2 HAIL

Hail literally can beat a crop to the ground and completely ruin it. Hail damage occurs suddenly, without warning, and little can be done to prevent it; staggering losses occur far more frequently than suspected. Hailstones can be as large as pigeon eggs or golf balls, capable of shattering windows and windshields, marring fruits, decimating foliage, pounding bark off trees, and breaking branches. In one Utah incident, hail completely destroyed one grower's peach crop two years out of three (Fig. 10.1); damage to his apple crop the same years was also extensive and necessitated intensive culling. Losses to nearby growers were only slightly less. Sugarbeets, corn, onions, and other field and vegetable crops take a similar severe beating annually (Fig. 10.2).

Dr. Larry Littlefield at Minnesota (1963) provided a detailed picture of hail damage on corn. Hail during the summer of that year caused moderate to severe damage to many crops in the southern half of the state. Hailstorms occurred in thirty counties. Surveys of 150,000 acres of corn in seventeen of these counties showed that yield reductions ranged from 0 to 100 percent with an average reduction of 10 to 15 percent. More serious, though, was the increased rate of stalk rot which developed in the moderate to heavily damaged fields. Every stalk hit by hail had bacterial and fungus rots developing in and around the hail wounds.

Cereal crops are also damaged by hail. Where the hail has not been severe enough to knock off the leaves or heads or to break down

FIGURE 10.1 *Hail injury and frost cracking on peach.*

FIGURE 10.2 *Hail injury of sugarbeet, showing lacerated foliage and petiole lesions.*

the whole stalk, whitish, bleached lesions can be found distributed over the exposed leaf surfaces. The lesions are caused by the mechanical rupture of clusters of mesophyll cells; even when the epidermis is not broken, air entering beneath it gives the spot a whitish coloration.

When cereals are in head, losses are serious and directly proportional to the amount of fruit which is shattered. Even when not knocked off, heads may be so badly bruised that only the most mature can free their tips and emerge from the outermost leaf sheathing. Heads hit directly are usually retarded in their subsequent development; kernels are lighter, lack uniformity, and often have black tips. Weight may be reduced 10 to 20 percent over those of uninjured heads.

Oats can endure severe injury if the panicles are still enclosed in the upper leaf sheath when hit. Sterile heads may be produced which give the appearance of having been damaged by thrips. The stippling and twisting of the heads can be caused by hail and thrips alike.

Hail injures such fleshy fruits as tomato both internally and externally. The more tender subepidermal tissue is killed, turns brown, and dries, leaving a hard cyst of corky tissue as the surrounding tissues mature.

Similar corky lesions have been observed on apples and pears. Even when no open wound is produced, the hail bruises the underlying parenchyma, producing depressed, brownish areas which are separated from the surrounding healthy tissue by a layer of cork cells. The earlier in fruit development hail occurs, the more serious are the consequences. Fruits hit within a few days of bloom will be permanently distorted or marred as the enlarging cells are killed. Simons and Lott (1964) studied the anatomical development of hail injury on Blackjon and Golden Delicious apples. Hail occurred ten days after full bloom. Within five days the injured tissue was dried and suberized. Cell division below the injury was first apparent as cork cells were initiated. Three weeks before harvest, injured areas were sunken even where the skin was not ruptured. Cells became suberized and many parenchyma cells disintegrated, while cork cells proliferated. Fruits were deformed and often unmarketable.

Even when the fruit itself is not marked, extensive leaf damage may cause production losses by reducing the effective photosynthetic area of crops such as onions. Dr. Leslie Hawthorn (1943) of the Texas Agricultural Experiment Station studied the importance of leaf area experimentally by cutting off various amounts of leaf tissue at different time intervals up to six weeks before harvest. The loss of foliage both retarded and reduced development of the onions. Removal of half the foliage two to six weeks before harvest caused an average loss in No. 1 yield of jumbo and medium-sized onion of 43.7 percent and loss in total yield of 34.7 percent. The same injury one week before harvest caused negligible loss.

Serious damage from hail has also been reported to such forest species as ponderosa pine (Marsden, 1951). Hail causes defoliation, debudding, bruising, stem lesions, and serious damage to newly forming cones. The characteristic symptoms of hail injury to forest stands initially are among the simplest to diagnose; yet a few years after its initial occurrence this type of injury is one of those most likely to puzzle the observer. The partially healed-over wounds are complicated by invasion of fungi, and the old symptoms or hail injury become obscured.

Diagnosis is simplified when one considers that the hail usually emanates from one particular direction, that is, downward. Therefore the bruising effect occurs on the upward and windward branches and on one side of the tree. These branches tend to shelter the ones on the lee side, which usually escape severe injury. One of the distinctions of hail damage, shared only by abiotic disorders like fire and frost, is that all species of trees are similarly affected.

The hail syndrome was described in detail by Dr. C. G. Riley (1953) in the Candle Lake Provincial Forest, Saskatchewan. The affected stands were 72- to 80-year-old white spruce, jack pine, and aspen. The width of the damage area was about 80 ft, but, as is characteristic of this type of injury, damage extended an undetermined distance in the direction followed by the storm.

The most conspicuous feature of the injury seven years after the hailstorm was the dead tops and one-sided crowns of larger trees, with the bare sides all facing the northwest direction, from which the hail had struck. On aspen, abrasions on the smooth white bark which gave rise to conspicuous black, rough callus bruises suggesting pockmarks always appeared on the same side of the tree. Looking southwest, only pock-dotted bark would be seen, while from the opposite direction the bark appeared smooth, white, and clean.

Injury on white spruce was most striking on the thin bark in the upper part of the trees. On the exposed side, down to where the stem was about 3 in. in diameter, patches of bark had been completely pounded off, leaving open wounds. The top was dead for several feet down from the tip. Below this the wounds were in all stages of healing, with rolls of callus at the margins. Small wounds were healed, leaving conspicuous scars. Still farther down the stem, where the hailstones had failed to break completely through the bark, numerous elongated scars 1 to 3 in. long appeared, indicating the severe bruising. Beneath each scar was a pitch pocket in the wood nearly overgrown with the normal number of annual rings.

In spruce, the effect of the one-sided reduction of the crown and wounding of the stem was strikingly evident in the growth-ring pattern. The annual rings formed prior to the date of injury were regular and quite broad; subsequent rings were narrow on the inside of the tree, even beneath the thick bark lower on the stem, where wounding was negligible. On the opposite side of the stem the regularity had been interrupted only by two narrow rings, which coincided with the years when injury occurred.

Symptoms in the jack pine stands were similar to those in the spruce, except that there were areas where the bark was completely stripped off the stems, and there was less evidence of healing. So many jack pine trees had died from the hail injury that a dense growth of brush and other understory vegetation had sprung up. Except for the lack of charring, the area probably looked very much as if it had been burned over. Much of the damage and mortality no doubt resulted from the drying of the sap that followed breaking of the bark and from the attack of insects and fungi which were favored by the wounds.

Hail wounds often bear a superficial resemblance to frost injury. On woody plants, hail wounds may be distinguished by the straight-lined normal wood with numerous vessels which soon appear again, while in the case of frost broad zones of parenchymatous tissue may be found due to the great extension of the adjacent edges. When hail injury is slight, the bark is not uniformly destroyed, and the cambium continues to grow with many gaps. Bark peels very unevenly and unsatisfactorily from the wood, rendering much of it unmerchantable and causing substantial economic loss.

Hail damage in forest stands is much more common than generally realized, although it is reported far less frequently than damage to agricultural crops. Relatively small geographical areas are affected by hail, and these may go unobserved in the large expanses of little-traveled forest, whereas similarly affected areas in agricultural land would not pass unnoticed.

Plants ranging in size from bluegrass to pine are damaged by ice sheets and heavy snows. Glazing, which is the coating of leaves or stems by ice, may cause damage by suffocation, the accumulation of toxic materials, oxygen deficiency, or breakage. Ice damage to forage species such as alfalfa has been basically attributed to internal accumulation of toxic byproducts of both aerobic and anaerobic respiration (Sprague and Graber, 1940). Ice contacting the crown and root inhibits CO_2 diffusion. Carbon dioxide and other respiratory products may accumulate rapidly in injurious or even lethal concentrations, causing much the same tissue injury and necrosis that flooding does.

Heavy accumulations of ice can strip twigs and branches from the trees and reduce growth for many years. Breakage is most common when ice storms are accompanied by strong winds. Broken tops cause permanent crooks or forks in the bole. These injuries also make trees more vulnerable to attack by insects and fungus pests. Because of their greater flexibility and manner of growth, conifers as a whole are more resistant to glaze injury than hardwoods. However, according to Downs (1943), even pines are not exempt. Glazing on ponderosa pine is most serious on young trees, causing bending, uprooting, and top and bole breakage. In stands of young, slender saplings of 2 to 6 in. dbh, bending is the principal type of damage; in stands 6 to 10 in. dbh, damage is attributed mostly to top breakage; over 10 in., damage is light and largely restricted to the weaker trees. Losses can be prevented by proper thinning of trees to obtain stronger limbs and boles.

J. D. Curtis (1936) described the effects of snow damage to Douglas fir, Norway pine, Colorado blue spruce, Scotch pine, pitch pine, northern white pine, jack pine, Austrian pine, Norway spruce, and red cedar. Trees were bent various degrees from the vertical, and stems and branches were often broken. Trees which were already one-sided were most prone to being thrown by snow than more symmetrical individuals. Snow damage was most prevalent in spring to Douglas fir trees under 3 ft high. After the trees were initially bent, snowfall contributed most to the final position of the tree. In the Midwest, cedars are particularly susceptible to breakage and are often damaged by heavy wet snows.

BIBLIOGRAPHY

Adams, D. B., 1938. The injury of grapevines by lightning strike. *J. Aust. Inst. Agr. Sci.* **4**:162–164.

Burr, S., 1933. Lightning damage to potatoes. *Gard. Chron.* **94**:48.

Curtis, J. D., 1936. Snow damage in plantations. *J. Forest.* **34**:613–19.

Downs, A. A., 1938. Glaze damage in the birch-beech-maple-hemlock type of Pennsylvania and New York. *J. Forest.* **36**:63–70.

Hartig, R., 1900. "Lehrbuch der Pflanzenkrankheiten," 3d ed. Springer-Verlag, Berlin,

Hawthorn, L. R., 1943. Simulated hail injury on yellow Bermuda onions. *Proc. Amer. Soc. Hort. Sci.* **43**:265–271.

Grubb, N. H., 1949. Lightning damage to raspberries. *A.R. East Malling Res. Sta.* 1948–49, A32, p. 81.

Jackson, L. W. R., 1940. Lightning injury of black locust seedlings. *Phytopathol.* **30**:1830184.

Johnson, H. W., 1943. Soybean diseases and their control. *U.S.D.A. Farmers Bull.* 1937, 24 pp.

Jones, L. R., 1917. Lightning injury to kale. *Phytopathol.* **7**:140–142.

Littlefield, L. J., 1964. Effects of hail damage on yield and stalk rot infection in corn. *Plant Dis. Rep.* **48**:169–170.

Marsden, D. H., 1951. Hail injury to trees. *Trees* **12**, 2 pp.

Munn, M. T., 1915. Lightning injury to onion. *Phytopathol.* **5**:197.

Murray, J. S., 1958. Lightning damage to trees. *Scot. Forest.* **12**:70–71.

Reinking, O. A., 1938. Lightning injury in banana plantations. *Phytopathol.* **28**:224.

Reynolds, R. R., 1940. Lightning as cause of timber mortality. *S. Forest. Exp. Sta. Notes* **31**:1.

Rhoads, A. S., 1943. Lightning injury to pine and oak trees in Florida. *Plant Dis. Rep.* **27**:556–557.

Riley, C. G., 1953. Hail damage in forest stands. *Forest Chron.* **29**:139–143.

Simons, K., and R. V. Lott, 1964. The morphological and anatomical development of apples injured by early season hail. *Proc. Amer. Soc. Hort. Sci.* **85**:60–73.

Smith, A. L., 1943. Lightning injury to cotton. *Phytopathol.* **33**:150–155.

Sprague, V. G., and L. F. Graber, 1940. Physiological factors operating on ice sheet injury of alfalfa. *Plant Physiol.* **15**:661–673.

Thompson, A. R., 1943. Lightning-struck tree survey. *Proc. 19th Nat. Shade Tree Conf.* 34–41.

Wadsworth, F. H., 1943. Lightning damage in ponderosa pine stands of northern Arizona. *J. Forest.* **41**:684–5.

Walker, J. C., 1937. Injury to cabbage by lightning. *Phytopathol.* **27**:858–861.

Weber, G. F., 1930. Lightning injury of potatoes. *Phytopathol.* **21**:213–218.

SELECTED REFERENCES

Beard, J. B., 1964. Effects of ice, snow, and water covers on Kentucky bluegrass, annual bluegrass and creeping bentgrass. *Crop Sci.* **4**:638–640.

Brown, H. D., and M. W. Gardner, 1923. Lightning injury to tomatoes. *Phytopathol.* **13**:147.

Carvell, K. L., E. H. Tryon, and R. P. True, 1957. Effects of glaze on the development of Appalachian hardwoods. *J. Forest.* **55**:130–2.

Duterme, C. J., 1965. Snow damage in pine stands. *Bull. Soc. Forest. Belg.* **72**:167–170.

Jones, L. R., and W. W. Gilbert, 1918. Lightning injury to herbaceous plants. *Phytopathol.* **8**:280–282.

Knowles, D. B., 1941. The effect of hail on wheat and other grain crops. *Univ. Sask. Agr. Res. Bull.* **102**.

Orton, C. R., 1921. Lightning injury to potato and cabbage. *Phytopathol.* **11**:96–98.

Samuel, G., 1940. Lightning injury to potato tubers. *Ann. Appl. Biol.* **27**:196–198.

Whipple, O. C., 1941. Injury to tomatoes by lightning. *Phytopathol.* **31**:1017–1022.

PART THREE

WATER AND SOIL RELATIONS

CHAPTER ELEVEN

WATER MECHANISMS

Water is the matrix of life, contributing as much to the properties of life as such complex molecules as carbohydrates, fats, and proteins. Water is directly or indirectly required for every life process and every chemical reaction; it is a vital component of most of these reactions, the medium in which they all take place, and the solvent for most of the reacting chemicals. It is the continuous phase between the substrate and the plant—within the plant, between the cells, and within the cell. Water is an essential part of every plant tissue, of the cells that comprise the tissues, and of the protoplasm that makes up the cell.

Life evolved in an aquatic environment, the sea, and can survive only in an aqueous medium. In the ancient seas that gave birth to life, the primitive simple, single cells were constantly bathed in fluids. The sea surrounding every cell not only provided the needed water but also bathed the cell in nutrients from which it synthesized the varied component carbohydrates, amino acids, and nucleotides.

When unicellular organisms aggregated into more complex multicellular forms which shared metabolic responsibilities, life became more complicated, but the basic needs did not change. Every living cell still required constant immersion in water. Cells within a tissue, unable to obtain water and nutrients directly, were dependent on the inward diffusion of water from the outer, absorbing cells.

Early life was adapted to this saturated environment. Drought, moisture stress, and wilting were unknown. These problems, and the adaptations required to avert them, arose only in response to a new environment as plants emerged onto the land.

Many problems had to be overcome for plants to survive a terrestrial environment and still more to endure the arid extremes. Success demanded remarkable physiological and morphological modification. How could the cell keep from drying out when the naked protoplast was exposed directly to the air? How could water be transported to every cell? How could the cell dispose of the toxic metabolic wastes? How could the cell obtain the essential mineral nutrients? How could a reasonable temperature be maintained, preventing the protoplast from literally "burning up" in the heat of the sun?

The key to these questions lies in the fact that plants have evolved mechanisms for providing every cell with a continuous supply of water.

11-1 WATER TRANSPORT

Mechanisms of water transport are discussed at length in plant physiology references and will not be enumerated here. Briefly, however, water passes into the plant through the root hairs, where it diffuses through the root cortex and endodermis and into the xylem vessels of the vascular tissue.

Water may also be absorbed from the atmosphere by leaves and stems. It has been postulated that atmospheric moisture, such as dew and fog, may provide some desert plants with their water needs (Stone et al., 1950; Duvderani, 1964). Thus water would pass from the atmosphere through the stomata into the intercellular spaces and mesophyll, through the vascular tissues, and presumably to the roots. However, evidence for this process is limited. If water is absorbed from the atmosphere, the quantity is slight and would rarely raise the leaf water content by more than a few percent. The primary effect of dew is probably exerted through its effect in reducing transpiration and creating a more humid environment for growth. Many species appear to require ample atmospheric moisture for normal development. The necessity of fog or high humidity for normal growth is exemplified by the distribution of the giant redwoods along the foggy shores of the California coast.

Water needs are more commonly met by *passive absorption* through the roots, largely initiated and controlled by transpiration from the shoot. Passive absorption is basically attributed to evaporation of water from the leaf mesophyll cells, which causes a reduced cell volume and results in a decrease of water potential in the mesophyll. This establishes a water potential gradient toward the leaf mesophyll cells from the xylem. As water molecules leave the leaf cells, additional water is needed to replace them. This water is provided through the xylem as the water column is essentially continuous, extending down through the stem and roots. Removal of water from the xylem reduces pressure in the xylem. This potential is transmitted to the roots, and a gradient develops across the root cortex to the epidermis and into the soil. Water is thus "pulled" into the cell by the pressure gradient initiated in the leaf. The cohesiveness of water molecules prevents the column of water in the xylem from breaking, maintaining a continuous supply to replenish the molecules lost through the leaf.

11-2 WATER NEEDS

Depending on the tissue and type of plant, approximately 80 to 90 percent of the plant is water. The figure varies considerably, though, from 10 percent in seeds in their dry state to over 90 percent in leaves of succulent and herbaceous species.

Water forms the greatest part of every cell, where it is the main constituent of protoplasm and of the proteins and nucleic acids which it comprises. Water is vital to these macromolecules since its properties of hydration facilitate reactivity between molecules in solution and between enzymes and substrates. The rate and physiological acitivity of many metabolic reactions depend on the degree of hydration of enzymes and other reactants. An ample water supply is also vital, since water is a major reagent for numerous biological reactions, including the many steps of respiration and photosynthesis.

Also, the large protein molecules depend for their configuration on the support of hydrogen bondings with the water matrix in which they are suspended. Bondings of water to the nitrogen and oxygen of proteins are vital and form a structural part of the protein. If the hydrogen bonding with water is lost, the proteins become denatured and collapse, destroying the protein structure. This dehydration is the fundamental basis for many disease symptoms.

Water is vital to every stage of plant development from seed germination to plant maturation. Germination proceeds only after seeds have imbibed sufficient water for rehydration. As water diffuses into the seed, the protoplast becomes rehydrated, food reserves are hydrolized, and the enzymes begin to function. Starch is digested to sugars, lipids to soluble materials, and storage proteins to amino acids. Energy is then liberated in respiration, foods are translocated, and embryo growth begins. Once growth is initiated, the water supply must be sustained to provide the hydrogen needed for carbohydrates to incorporate into the protoplasm of the new cells.

Water is further required to provide the constant pressure of turgor which supports the plant and facilitates cell enlargement after the new cells have been initiated. As the cells mature, water accumulates in the vacuole, building up a pressure which forces the protoplast against the cell wall, keeping it fully distended or turgid. The turgidity of the cells, tissues, and the entire plant depends entirely on an ample water content.

Water also provides the medium in which essential mineral nutrients are carried. Minerals dissolved in water enter the plant from a soil solution. From the root hair, the essential ions are transported throughout the plant. Without this nutrient solution the cells, tissues, and organs soon would starve.

11-3 WATER STRESS AND WILTING

Plant growth and survival depend on water availability as much as any single environmental factor. Yet the environment frequently fails to provide optimum water relations; a water deficit arises whenever water is used or lost, as in transpiration, faster than it is absorbed by the roots.

As water becomes deficient, the plant progressively undergoes incipient, transient, and permanent wilting. Even in the absence of any wilting the water balance may be poor and physiological stress may exist, limiting metabolic acitivity, growth, and production.

While wilting is the most obvious expression of moisture stress, it is by no means the most common or significant. Moisture stress influences such processes as water uptake, seed germination, stomatal closure, transpiration, photosynthesis, respiration, enzymatic activity, growth of shoots and roots, shrinkage of tissues, mycorrhizal development, and mineral nutrition. By altering chemical and physical composition of tissues, water stress also modifies the quality and taste of edible plants. All of these processes may be impaired in the absence of visible wilting symptoms.

Wilting refers to drooping of plant parts when they lose their turgor. Wilting most frequently effects herbaceous plants, especially the leaves and soft stems, but stress can be equally serious on woody species when metabolic processes are impaired. Woody tissues are more lignified, and deficits may exist long before any outward signs of stress appear. For instance, photosynthetic rate of apple trees was shown to be notably reduced as much as a week before leaves visibly wilted (cited by Kozlowski, 1964).

The earliest stage of a water imbalance, often called *incipient wilting*, arises before there is any persistent visible wilting or drooping of leaves. Incipient wilting occurs almost daily for varying periods of time during the summer, possibly several hours a day in warmer or drier regions. Turgor loss usually is slight and short-lived, with the plant recovering in the evening or as soon as the transpiration rate diminishes. But temporary water imbalance can also leave more lasting effects.

During the period of stress, protoplasmic dehydration reduces the physiological activity of the cells, particularly in young tissues, which require a constant water supply for development. According to Slavik (1963), the intensity of photosynthesis and respiration is related directly to differences in hydration of the leaf cells. Diminished metabolic activity means diminished growth and reduced productivity. Fruit size is also impaired, since growth is suppressed during hot days as water moves from the fruits to the leaves. This is partly offset at night, when fruits replenish the water supply. In the same way, tree trunks have been measured and found to shrink slightly in the afternoon as water is lost in transpiration.

Numerous examples can be cited illustrating this correlation of growth and water balance. One such study of shortleaf pine and white oak conducted by O. L. Copeland (1955) showed that 80 percent of the basal growth was completed by the end of June, after which the available moisture diminished. Three species of oak completed practically all seasonal height growth in just nineteen days during which water was ample. Later growth in the summer is often limited to periods when moisture is available.

The capacity of a plant to attain its expected, genetically inherited size potential is influenced by growth conditions in general and moisture relations specifically, even within its normal range of distribution. American linden trees attain heights of 80 to 90 ft in the most favorable parts of their range, where annual precipitation averages 33 in.; they average under 25 ft in height in regions of 22 to 23 in. of precipitation (Albertson and Weaver, 1945). The bur oak, which averages a 70-ft height in moist regions, reaches only 20 ft in areas with an annual rainfall under 23 in. Similar limitations are recognized in commercial timber and fruit crops as well as in ornamental park and street trees.

When turgor loss is more severe, visible wilting arises and persists until soil moisture is replenished. This temporary (or *transient*) wilting, from which the plant can recover, can be harmful even when only of short duration. Transient wilting occurs on hot, windy days, where turgor loss is pronounced and leaves are flaccid and "drooping" much of the day. Stomata close even before wilting is noticeable, and gas exchange is negligible throughout the day. Photosynthesis is correspondingly reduced, and carbohydrate reserves become depleted. If proper moisture relations fail to be restored, the chlorotic symptoms of starvation will soon accompany the wilted condition.

The most extreme form of wilting, known as *permanent* wilting, is irreversible; tissues are critically damaged and unable to absorb water even if made available. Tensions to the vascular tissues and parenchyma increase as drying progresses, eventually disrupting water flow; vascular and mesophyll cells collapse, subjecting protoplasm to increasing stress, and ultimately all structural organization is lost. Tissues wither, collapse, and die.

11-4 PLANT MODIFICATIONS

Xerophytes

Plants growing in their natural environments have adjusted to the prevailing moisture regime. Those species whose evolutionary changes kept pace with the environment have survived, while those failing to adapt have perished. Through continued evolutionary modifications over millions of years, plants became adapted to both excessively dry and excessively wet environments.

Responding to arid environments, plants have evolved numerous and remarkable mechanisms for obtaining water, retaining what they have, and surviving in unbelievably dry conditions. Desert species, collectively called *xerophytes* (dry plants), utilize a selection of one or more, often several, morphological and physiological characteristics enabling them to endure prolonged periods of unfavorable water balance. Modifications inherently possessed by xerophytic species for water conservation, such as a thick cuticle and dense cell structure, are much the same as those utilized to withstand temperature stress.

Even mesophytes, plants normally not well adapted to arid conditions of limited moisture, will develop xeric modifications when grown under conditions of moisture stress. For one thing, a lack of available water inhibits enlargement of newly forming cells. Thus, the stems and leaves are stunted. While cell enlargement is impaired, cell division remains unaffected. Cells are smaller than normal; the stomata are comparably reduced in size and closer together, and intercellular spaces are less extensive. Deposition of cell wall material is intensified, so that wall and cuticle are thicker and lignification is increased. The plant then is tough and less flexible; it possesses more mechanical tissue and is stronger and better able to withstand the mechanical stress of desiccation.

One of the most obvious morphological xerophytic modifications is the presence of a thick, well-developed waxy cuticle layer which reduces moisture loss from the epidermal cells. Such a cuticle characterizes many evergreens, as well as xerophytes in general.

The position, number, and size of stomata exert a still greater influence in regulating water loss. The depression of stomates below the general level of the plant surface characterizes many xerophytes, especially gymnosperms; stomata in others appear sunken when the cuticle covering the surrounding epidermal cells is particularly thick. The depression of the stomata provides a cavity or pit formed by the cuticle or epidermal cells. The water vapor must diffuse through this cavity in leaving the intercellular spaces of the mesophyll. The path of water is lengthened, and an area of saturated air resides in the cavity. Hence the diffusion pressure gradient is reduced, lowering the rate of transpiration and subsequent water loss.

Water vapor is also retained, and water loss reduced, when the leaf itself rolls up in such a way as to partly enclose the stomata.

Grasses often exhibit this characteristic. Tomatoes, sugarbeets, and many other broad-leaved plants can be observed to respond similarly to moisture stress.

Maintenance of a water vapor layer over the leaf surface, with a reduction in water loss, is facilitated in many plant species having epidermal hairs or pubescence. Dense hairiness often characterizes xerophytes. These hairs serve to trap a layer of moist air above the guard cells, reducing the gradient of diffusion. The hairs may also reflect appreciable amounts of solar radiation, preventing an excessive leaf temperature increase; this further reduces transpiration and conserves moisture.

Location, position, and shape of the leaves themselves may further aid the plant in conserving moisture as well as minimizing temperatures. Heat absorption, and with it moisture loss, may be minimized by orienting leaves vertically with the sun's rays. Plants may grow initially with this orientation or become oriented only when they are exposed to bright light. The fine leaflets of the locust tree turn their edge to the sun's rays after a few minutes of bright light. The wilting of leaves is a similar response. Plants in many desert and dry prairie or mountain regions possess leaves growing in close, rosette-like formations near the ground. Moisture is conserved, as it tends to accumulate within these rosettes.

Minimizing the leaf surface area exposed to the air is also important to survival in dry areas. Gates et al. (1968) found that temperatures of leaves measuring 1 by 1 cm or less will remain close to air temperature because of the greater amount of convection. The leaf surface area (or surface/volume ratio) also tends to be minimal in xerophytic plants. Leaves may even be reduced to scales or dispensed with altogether. The photosynthetic chores must then be assumed by the stems (spines) or flattened petioles (phyllodes) of the plant. Where leaves are retained, they are characteristically circular in outline so the exposed surface is minimal. Even where leaves are normally present, the plant may drop them to escape desiccation. California buckeye, creosote bush, almond, and locust are but a few plants which exhibit this defoliating characteristic.

Internal, cellular adaptations also aid the plant in conserving water. Succulent plants store water in large thin-walled cells and tissues during periods when water is available. Plants able to retain water in this way possess an extremely thick and effective cuticle which minimizes evaporation and a minimal surface/volume ratio in both leaves and stems.

Many succulents also possess the unique feature of keeping the stomates closed during the heat of the day and opening them only at night, further minimizing water loss. This mechanism works because the CO_2 needed for photosynthesis is absorbed and combined with pyruvic acid by a dark-fixation process. The acids formed are later decomposed in daylight to release CO_2 internally, which is then used for photosynthesis.

Extensive root systems, quite logically, aid many xerophytes in maintaining adequate water relations. Xerophytes typically possess such systems, enabling them to absorb maximum amounts of water. Often the root systems are shallow but have a tremendous lateral spread, enabling plants to take advantage of the brief downpours or shallow rains typifying desert climes. Other desert plants tap the deeper soil strata for water, with roots penetrating depths of 20 ft or more.

When all else fails, the ability of protoplasm to resist drought becomes vital. The cellular contents of xerophytes are particularly tolerant of desiccation. Their ability to become desiccated far below the normal survival level without apparent harm appears due to the presence of large amounts of solutes and mucilages within the protoplast and vacuole. These substances may in some way help restrict evaporation from the cell walls. More important, though, they are able to hold large quantities of water, which increases cellular resistance to desiccation. Ultimately, however, when moisture stress becomes excessive, the cell will dry out and die.

Henckel (1964) discusses the physiology of drought resistance in detail. During dehydration cells die as a result of the mechanical disruption of the protoplasm. More basically, desiccation, by loosening specific molecular connections in the protoplasm network, breaks down the structure of protein molecules. This is accompanied by changes in viscosity, permeability, hydration, and electric charge, and by deactivation of enzymes.

Death may also be due to mechanical injury associated with drying and rehydration. As the vacuole shrinks during drying, an inward pull is exerted on the protoplasm; simultaneously, an outward pull is exerted by the cell wall to which the protoplasm remains attached. Hence, protoplasm is exposed to injurious tensions during desiccation; if it should survive, it may be subject to further stresses in rehydration.

Plants may survive in arid environments by being physiologically active only during moist periods. These drought-evading—or drought-escaping—plants simply complete their life cycle during the relatively brief periods when moisture is available. Such plants are not true xerophytes, since they require greater amounts of water and look more like mesophytes. Dry periods are escaped only by rapid production of seeds which remain dormant during the periods of drought. The world's semiarid areas abound with species, often annuals, which flourish and bloom during the brief periods of favorable moisture relations. Perennial species that must endure unfavorable moisture relations most of the year require xerophytic morphological or physiological modification if they are to survive.

Hydrophytes

Plants have successfully colonized the earth's surface from the driest to wettest extremes. In contrast to the xerophytes, many species of flowering plants, the *hydrophytes*, grow and reproduce with their roots continually submerged in water, or at least inundated for prolonged periods. The modifications and adaptations possessed by hydrophytes are essentially the opposite of those found in xerophytes.

Hydrophytes grow partially or wholly immersed in open bodies of water, near the banks of open water, or in low-lying situations where the soils are continually moist and often waterlogged. Hence, the limitation imposed by the environment is not lack of water but inadequate oxygen, and hydrophytic modifications are directed toward providing every cell with adequate amounts of oxygen.

Oxygen is only slightly soluble in water. The small amount of oxygen available to submersed plants diffuses into the cells slowly and is in demand by every organism in the water. How then can plants obtain an adequate supply? In most cases the oxygen passes down the plant tissue from the leaves through a highly modified aeration system. Large, gas-filled intercellular spaces are present throughout the leaves, stem, and roots, forming extensive passageways within the submerged tissues. Also, during periods when the plant is photosynthesizing, the oxygen released by the green cells is stored in the intercellular spaces for later use in respiration.

The extensive, highly organized intercellular spaces are surrounded by a dense epidermal layer through which water cannot penetrate. The cuticle, on the other hand, is poorly developed or absent, so that dissolved gases and minerals can diffuse freely. Where the leaves are aerial, their morphology is characteristic of mesophytes.

Leaves are typically thin and flexible but may be of any shape or size. Roots serve more for support than for absorption and are far less extensive than on mesophytes. Root hairs are nearly always absent.

Plants lacking hydrophytic modifications are unable to obtain adequate amounts of oxygen for respiration in a saturated environment; they are unable to produce adequate usable energy for metabolic reactions and slowly starve as the energy supply becomes exhausted.

Mesophytes

Between the extreme modifications of xerophytic and hydrophytic plants lie the far more numerous mesophytic species possessing intermediate characteristics. These lack the modifications which would enable them to endure either limited or excessive moisture and are readily damaged by either extreme. These mesophytic species include nearly all our agronomic crops and most woody ornamental and native species. Mesophytes are poorly adjusted to moisture deficits or excesses, yet they are subjected to both. Excessive rains or irrigation, or poor soil drainage, provide too much water (or more accurately, inadequate oxygen); growth diminishes, foliage yellows, and plants gradually die.

In the arid environments found over most of the Western United States, a moisture deficit is essentially a normal feature of plant growth. Moisture stress is constantly limiting to plant development and production even in the absence of gross morphologic symptoms. Thus, what we have come to accept as normal growth and productivity may be far from optimal.

BIBLIOGRAPHY

Albertson, F. W., and J. E. Weaver, 1945. Injury and death or recovery of trees in prairie climate. *Ecol. Monogr.* **15**:393–433.

Copeland, O. L., 1955. The effects of an artificially induced drought on short leaf pine. *J. Forest.* **53**:262–264.

Duvderani, S., 1964. Dew in Israel and its effect on plants. *Soil Sci.* **98**:14–21.

Gates, D. M., R. Alderfer, and E. Taylor, 1968. Leaf temperatures of desert plants. *Science* **159** (3818):994–995.

Henckel, P. A., 1964. Physiology of plants under drought. *Ann. Rev. Plant Physiol.* **15**:363–386.

Kozlowski, T., 1964. "Water metabolism in plants." Harper, New York, 227 pp.

Slavik, B., 1963. In A. J. Rutte and R. H. Whitehead (ed.), "Water relations of plants." Wiley, New York.

Stone, E. F., F. W. Went, and C. L. Young, 1950. Water absorption from the atmosphere by plants growing in dry soil. *Science* **111**:546.

SELECTED REFERENCES

Billings, W. D., 1964. "Plants and the ecosystem." Wadsworth, Belmont, Calif., 154 pp.

Briggs, G. E., 1967. "Movement of water in plants." Davis, Philadelphia, 160 pp.

Knight, R. O., 1965. "The plant in relation to water." Dover, New York, 147 pp.

Kramer, P. J., 1944. Soil moisture in relation to plant growth. *Bot. Rev.* **10**:525–559.

———, 1949. "Plant and soil water relationships." McGraw-Hill, New York, 347 pp.

Letey, J., and G. B. Blank, 1961. Influence of environment on vegetative growth of plants watered at various soil moisture suctions. *Agron. J.* **53**:151.

Philip, J. R., 1966. Plant water relations: Some physical aspects. *Ann. Rev. Plant Physiol.* **17**:245–268.

Rutter, A. J., and F. H. Whitehead (eds.), 1963. "The water relation of plants." Blackwell, Oxford, 394 pp.

Slatyer, R. O., 1967. "Plant-water relationships." Academic, New York, 366 pp.

Steward, F. C., 1959. "Plant Physiology. II. Plants in relation to water and solutes." Academic, New York, 758 pp.

Subramanian, D., and L. Saraswathi-Devi, 1959. Water is deficient, in J. G. Horsefall and A. E. Dimond (eds.), "Plant pathology," vol. 1, chap. 9, The diseased plant. Academic, New York.

Taylor, A., 1951. A continuous supply of soil moisture to the growing crop gives highest yield. *Utah Farm and Home Sci.* **12**:50, 51, 61.

U.S. Dept. Agr. Yearbook, 1955. Water. Government Printing Office, Washington, D.C., 751 pp.

Vaadia, Yoash, F. C. Raney, and R. M. Hagan, 1961. Plant water deficits and physiological processes. *Ann. Rev. Plant Physiol.* **13**:265–292.

Zelitch, Israel, 1963. Stomata and water relations in plants. *Conn. Agr. Exp. Sta. Bull.* **664**:116.

CHAPTER TWELVE

DISORDERS ASSOCIATED WITH ADVERSE WATER RELATIONS

12-1 WATER DEFICIENCIES

The health and survival of a plant depends on a proper water balance. Any degree of water imbalance will produce a proportionately deleterious deviation in physiological activity, growth, and reproduction. Soil moisture levels are rarely optimal during the growing season. Moisture relations might be technically adverse when they are not optimal, but the point of imbalance at which stress becomes measurably meaningful, or economically significant, is hard to delimit. Actual growth losses may be tremendous, but they are impossible to calculate since we cannot estimate how much more growth might have occurred had conditions been optimal. The decreased vigor and productivity of plants subjected to a slightly disturbed water relation generally pass unnoticed.

Metabolism and Growth

Photosynthesis and related metabolic processes dependent on gas exchange are quickly impaired by water deficiency, as the deficit induces closure of the stomatal aperture, stopping carbon dioxide uptake. Carbohydrate synthesis is then suppressed and the growth potential limited.

A more fundamental mechanism by which water stress inhibits growth is based on the impaired synthesis of the proteins needed for growth (Vaadia et al., 1961). Water stress was found to inhibit amino acid utilization and protein synthesis. While amino acid synthesis was not impaired, the cellular protein levels decreased. Water stress, in blocking amino acid utilization, induced a ten- to one hundred fold accumulation of free asparagine. Valine levels increased, and glutamic acid and alanine levels decreased. Water shortage also impairs auxin production so that cell enlargement and growth is suppressed. Limited cell enlargement produces a reduced leaf area, shortened internodes, and short, stunted, rosetted plants.

Productivity is at least as sensitive to moisture stress as growth. The necessity of ideal moisture conditions for optimum production is illustrated in studies by Dr. Sterling Taylor of Utah State University (1951). Sugarbeet, alfalfa, and potato plots were irrigated with varying frequency during the growing season. All plots were kept moist and thoroughly soaked at each irrigation, but test plots were irrigated more frequently. Maximum yields could be obtained only by irrigating adequately before the plants showed any apparent wilting. Any decrease in water availability reduced production. The more pronounced the wilting before irrigation, the greater the yield reduction. Sugarbeet yields were reduced 4 to 13 percent for each increase of 1 atm of mean integrated soil moisture tension. Second-crop alfalfa plants allowed to reach 4 atm of soil moisture tension before being irrigated turned dark green in color and wilted on hot days; yields were reduced 5 to 11 percent for each increase in 1 atm tension.

Potatoes were even more affected by soil moisture, and yields responded to moisture tension changes of less than 1 atm. Plants were characterized by dark green color and slight wilting prior to irrigation, and yields decreased about 20 percent for every increase of 1 atm tension.

The initial fruit set also can be affected by water availability. Experiments with Sultana grape vines suggested that fruit set was significantly less when water stress was imposed within four weeks of flowering (Alexander, 1965). Not only is water imbalance responsible for reduced yields; it can also reduce fruit quality, especially size, to an extent where the crop is unmarketable.

Fruit Symptoms

Blossom end rot of tomato is the classic example of a drought-related fruit disorder. The disease develops when plants have grown rapidly and luxuriantly during an early favorable season and then subjected to moisture stress. The first expression of blossom end rot is a water-soaked spot at the stylar end of the fruit. The spot turns whitish to deep brown and can affect up to half the fruit. The basic cause of the disease is actually a calcium deficiency, or an excess of ammonia, potassium, magnesium, or sodium salts which affects the calcium ratio, but water imbalance provides the stress which evokes the deficiency response. The basic disease process is treated in Chap. 13.

Analogous symptoms have been observed on fruits of sweet and sour cherries, peach, and pear trees. In all cases the tip of the fruit becomes depressed, and a brown to black lesion develops (Fig. 12.1). Less severe symptoms on sweet cherry involve only a shriveling of the stylar end of the fruit. A related disorder, *black end of pears* (Tufts and Davis, 1930), is characterized by hard, brownish-black areas ranging in size from isolated flecks to 2 in. or more across at the base of the fruit. Black speckling often occurs around the central lesion. Up to 50 percent of the fruit in some orchards may be affected, although losses rarely exceed 5 to 10 percent. A similar disorder, *drought spot* of apples and prunes (Fig. 12.2, Burrell and Heinicke, 1938), is caused by a sudden, pronounced water deficit occurring near the end of a normal season. Symptoms consist of dark-colored lesions on the fruit surface and an underlying fine network of brownish vascular tissue. The outer cortical cells are frequently necrotic in close proximity to the vein endings. Tiny drops of clear cell sap exudate appear at the surface of the skin of both apples and prunes. When fruits fail to drop, they become wrinkled, or the lesions may develop into rough or cracked areas.

When the moisture stress has been more severe, lesions are correspondingly deeper and characterized by brown, corky or pithy, somewhat sunken areas. These lesions are pronounced on apples but

FIGURE 12.1 *Stylar end blackening of Montmorency sour cherry associated with drought.*

tend to be obscured on prune, where the normal dark purple skin color masks all but the most severe expression.

Internal decline of lemon, also known as *endoxerosis, blossom end decay, dry tip,* or *yellow tip* (Bartholemew, 1926), is characterized by the loss of luster of the stylar end or the premature yellowing of this region while the rest of the fruit remains green. Also, the stylar end of the thin-skinned fruits may be slightly sunken. Internally, small cylindrical openings develop within the ring of vascular bundles at the stylar end. These openings gradually fill with brownish gum deposits. Surrounding vascular tissue breaks down, with further deposition of gums and resins and development of local brownish lesions in the stylar end.

The extent of internal decline depends on water relations during fruit development. When water is deficient, the leaves draw it from the fruit, creating an acute water stress and causing collapse and death of fruit tissues, particularly in the stylar end.

Foliar Symptoms

BROAD-LEAVED PLANTS Foliar symptoms of moisture stress are far more common than fruit symptoms. Broad-leaved species such as maple and horse-chestnut are native to moist environments and ill adapted to drought. But the plant undergoes substantial stress before the characteristic leaf-burning symptom appears. As water becomes deficient, the guard cells lose their turgor and the stomates close, reducing further water loss. Water loss through the cuticle of the leaf blade and petiole now becomes significant, and as the underlying tissues continue to lose water, tissues become flaccid and the leaves wilt. Wilting is the earliest readily visible expression of moisture stress; it is a normal, defensive mechanism serving to orient the leaf blade parallel to the sun's rays, thus minimizing the solar energy directly hitting the leaf surface.

Leaves in a wilted state are relatively nonfunctional, and if wilting

FIGURE 12.2 *Drought spot of Italian prune fruit, showing discoloration of vascular tissue.*

is prolonged and tissues remain desiccated, they will ultimately die. When the leaves lose too much moisture, margins of affected leaves are characteristically the first tissues to become desiccated and necrotic. Necrosis ranges from straw-colored to deep brown or reddish, depending on the species and the condition under which necrosis developed. Three separate types of drought-induced necrosis may develop: (1) Necrosis most often is confined to the leaf tip, and the boundary is sharply delimited. Such is often the case with aspen, willow, elm, maple, and conifers such as pine, spruce, or fir (Fig. 12.3). (2) The margin of the necrotic tissue tends to be diffuse, and the tissue between the necrotic and healthy parts of the leaf becomes chlorotic (Fig. 12.4). Necrosis tends to extend from the margins inward between the secondary veins toward the midrib. This expression usually characterizes a chronic moisture stress. (3) Necrosis may appear as dull brown or bronzed interveinal lesions, irregular in outline. In most instances high temperatures are primarily instrumental in causing this symptom (Fig. 12.5).

The first type of necrosis is exemplified by *terminal bleach* of cereals, a drought-related disease in which temperature plays a key

FIGURE 12.3 *Moisture stress of quaking aspen associated with thin soil and an exposed site.*

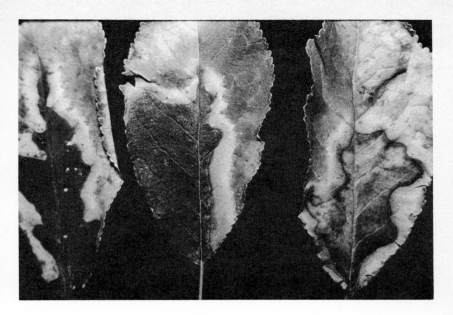

FIGURE 12.4 *Drought-induced marginal necrosis of Italian prune leaves.*

FIGURE 12.5 *Intercostal necrosis associated with moisture stress.*

role (Treshow, 1957). Terminal bleach denotes the sudden appearance of whitened leaf tips of cereals when such plants as barley, wheat, or oats are exposed to drying winds. The tallest, deepest-green, most rank-growing plants are most seriously affected. Developing in local areas of a field where soil moisture and nutrition have been ample, such plants have a "soft" growth and are especially sensitive to sudden water demands. Sudden termination of a long cool spring can be disastrous. When warm winds arise, the plants are unable to maintain a sufficient water balance and become desiccated. Subjected to these conditions, the same response can be expected by other plants, e.g. onion, iris, and gladiolus.

On woody plants, leaf desiccation is followed by drought injury to twigs and limbs. Summer drought injury and mortality have been described on deciduous trees throughout their normal distributional range. The effects of one widespread drought in the 1930s were reported in the midwest (Alberston and Weaver, 1945). From June 8 to September 8 only 2.8 in. of rain fell in southeastern Nebraska, Kansas, Missouri, and Iowa. Native species including hackberry, American elm, and Carolina poplar came into fruit prematurely, leaves matured early, and the most severely affected trees had dropped all their leaves by the end of July.

Moisture stress has also been considered responsible for such specific diseases as *maple decline,* in which leaves become progressively yellow, burn, color prematurely, and drop early (Banfield, 1967). Each year foliage becomes more sparse and remaining leaves more chlorotic as more and more dead twigs and branches appear. Banfield finds that "decline is most common along roadways in groves on southern or western slopes when removal of trees or grazing has transformed the normal forest to a savannah or orchard formation." Trees on shallow soils, or on sites where much of the root area has been covered by pavement or otherwise compacted, are most severely affected.

CONIFERS   Needle-leaved species, despite their xerophytic characteristics, are also intolerant of desiccation. Stresses in evergreen, needle-leaved forests are most frequent during winter months, when the needles continue to transpire. As the water in the soil and stem freezes, less is available for transpiration, and the trees are subjected to a slowly increasing moisture stress. Some evidence has been provided which indicates that water may move through frozen tissues, but even though this may be the case, the amounts are still inadequate to fill the demands under all conditions.

The etiology of winter drought, winter injury, and drought injury involves moisture stress in each case. One of the most striking winter needle-burn diseases, known as *red belt*, was discussed briefly under low-temperature effects, but desiccation plays an equal role in its development (Henson, 1923). The reddish-brown foliage discoloration characterizing red belt is limited to trees growing in distinct altitudinal bands or belts often along the sunlit slopes. Damage is greatest on more exposed trees and the most exposed foliage of each tree, particularly the previous year's needles, but all the foliage may be discolored in the most severe cases. Where injury is most severe, the new buds may be killed, so that the next year's crop of needles is sparse or wanting.

Death of the needles is attributed to their desiccation when they are unable to obtain water through the root and stem tissues fast enough to compensate for its loss through the needles. Conifers depend on late fall precipitation, coming from August through November, to provide the reserve water required to sustain them over the winter months. When soil is dry from inadequate precipitation during the fall months, needle blight of conifers, early defoliation, twig dieback, and sparse foliage may be expected the following spring. Such drought damage has been reported to native populations across the United States and is common over extensive areas of the West as well as in the Plains and Lake States. Widespread droughts in the 1930s were responsible for the death of many newly established forest plantations in the glaciated thin or eroded soils of the Great Lakes region (Stoeckler and Rudolf, 1949). Many hemlocks in Connecticut were killed, and the growth of the others was suppressed, following a five-year drought (Stickel, 1933). Differences in the severity of damage were attributed largely to the depth of the root systems. Slightly damaged and surviving trees were thought to have deeper, more extensive root systems than trees which were killed.

The successful growth and survival of native and introduced plants alike throughout the world depends on their ability to withstand prolonged droughts; trees on gravelly or sandy soils are most severely affected. The effect of drought on conifers in the Pacific Northwest was described by Charles Leaphart (1959). Following an abnormally hot, dry summer in 1958, needles on western larch trees of all ages began to turn yellow and shed. Drought symptoms appeared on western white pine by September. Foliage on particularly the leader and upper crown branches began to wilt and turn light straw color; the pale discoloration darkened, turning reddish-brown by mid-October. Uppermost and outermost limbs were most severely damaged, and some entire trees were killed to the ground. In the same region, lodgepole pine needles turned reddish-brown and wilted slightly. Seedlings and saplings were most severely damaged; needles were burned, and trees were often killed. Notable drought symptoms also appeared on grand fir, Douglas fir, and ponderosa pine saplings and trees.

A similar, but more chronic, disorder of western white pine known as *pole blight* also predominates on shallow, sandy soils or soils having an available moisture storage capacity of less than 5 in. (Leaphart and Gill, 1955; Wellington, 1954). Pole blight is characterized by a general "blighting" of affected trees. Needles turn yellow and many drop, growth declines each year until it is zero, and the tree dies slowly from the top down. Elongated lesions or necrotic streaks infiltrated with resin and running longitudinally in the bark and sapwood develop. Radial and terminal growth are markedly reduced. Pole blight extends from British Columbia through eastern Washington, Idaho, and western Montana, where it has been recognized since about 1933. Pole-sized trees, generally forty to a hundred years old, are most often affected; at this stage water demands probably reach their peak.

White pine, the most important timber species where it grows, is more seriously affected than the associated species, possibly because of its inability to compete successfully for water. White pine trees lack the profuse system of fine roots characterizing some of the associated species (such as Douglas fir and lodgepole pine) with which they must compete for water. Furthermore, white pine trees have absorbing structures only at the tips of the roots, while competing species absorb water along their entire root length. On shallower soils capable of holding less than 6 to 8 in. of available moisture, the incidence of rootlet and root-tip mortality associated with pole blight is high in drier years. The inability to compete successfully for water may lead to root mortality and death of the tops.

Similar needle burning has been reported from Wisconsin on Scotch pine, red pine, and jack pine (Voigt, 1948). While this type of injury had been attributed to frost, sunscald, or drought, winter drought was most frequently blamed. Necrosis was most serious on trees on thinner, dry, upland soils most exposed to the warm southerly winds which stimulated transpiration. Fall drought in the 1947–1948 season had produced a soil moisture deficiency to a depth of nearly 5 ft. When southerly winds in January arose, the combination led to extensive necrosis. Foresters wondered why Scotch pine, usually an extremely hardy species, was the most severely damaged. Anatomical studies revealed the answer: Jack pine and red pine have a slightly thicker cuticle layer protecting the leaf, which may have reduced transpirational water loss. Scotch pine also had more stomata, a greater leaf surface/volume ratio, and, perhaps most significantly, a shallower root system.

The *little-leaf disease* of shortleaf pine is a similar disorder involving a different species. In an effort to determine its cause, Otis L. Copeland, Jr. of the U.S. Forest Service conducted a quantitative study in which he artificially induced drought by diverting all the rainfall away from the study trees (1955). Lower branches died one by one, until the remaining crowns were only about half their original size. Needle retention diminished so that only sparse tufts remained at the tips. Branch dying began at the base of the crown and slowly proceeded upward. While trees were not always killed, their growth rate diminished to nothing, and their appearance was shabby.

12-2 WATER EXCESS

The normal, mesophytic land plant must have oxygen for the liberation of energy required to maintain life processes. When the pore spaces of the soil are filled with water, the plant root suffocates. The immediate effect of suffocation is suppression of growth. Studies with hybrid tea roses showed that total linear growth in an unaerated soil-peat mix was only half that of plants in aerated plots (Biocourt and Allan, 1941). Plants may survive brief periods of oxygen deficiency where as little as 0.5 percent oxygen is available in the soil, but roots need 2 to 8 percent for optimum growth. Below this, leaves become chlorotic, growth ceases, no new roots develop, shoots die back, and death ensues. This decline and death may take place in a few days in some plants, or months or even years in others.

In addition to reducing the available oxygen, flooding may decrease transpiration over 90 percent, limiting water absorption and photosynthesis.

Decline symptoms resulting from the suffocation and death of roots were described and first called *root rot* by Robert Hartig in 1878. He applied the term root rot to denote that death of roots was from lack of oxygen rather than fungus infection.

There are several situations which give rise to excessive water. The poor drainage afforded by heavier soils is probably the most common; water fills the pore space of the soil for prolonged periods, preventing the roots from obtaining adequate oxygen. Changes in the water table from alteration in the soil line, heavy rains, or diking an area so that water becomes impounded all have the same effect. Land also becomes inundated from flooding due to excess rains, runoff, or irrigation; and unless the soil is extremely well drained, water may persist long enough to cause marked damage. But whatever the reason for excessive water, the effect is much the same and depends on the sensitivity and tolerance of the species.

Fruit Crops

The overall syndrome of excess water is exemplified by an insidious decline, called simply *the death,* which has plagued fruit growers in England (Furneaux and Kent, 1937). Leaves of fruit trees remain small, gradually fading in color, becoming yellow, and wilting. Growth of terminal shoots is short or wanting. The feeble terminals and laterals often wilt, shoots gradually collapse and die, and the entire tree may ultimately succumb. Susceptible plants included apple, pear, plum, cherry, raspberry, gooseberry, blackberry, and loganberry.

Much has been written on the damage to submerged apple roots showing a close relation between temperatures, the time of flooding, and tolerance. McIntosh roots could be submerged from late fall to late spring without injury (Heinicke and Boynton, 1941). But if trees remained in water after being exposed to high temperatures, severe damage was inevitable. If flooding came after the leaves had emerged, but while temperatures were still cool, foliage might remain normal for more than a month; but as soon as there was a single day with high temperatures, serious injury would be expected to appear within a few hours.

Low soil oxygen then is most damaging when soil or air temperatures are high. Sunflower plants grown under low oxygen levels made weak, stunted growth at 13 and 24°C but died if they were exposed to 34°C. Plants require less oxygen at low temperatures, so are better able to endure low oxygen levels at low temperatures. At higher temperatures, respiration is accelerated and the demand for oxygen great. The identical response has been reported for blueberry (Bailey and Franklin, 1950).

When aeration in the summer is marginal, whether from inadvertent flooding or excessive irrigation, cankers which are entirely free from parasitic organisms develop at the soil line, and twigs and limbs of most fruit trees can be expected to die out. The resulting disease, known as *collar rot*, can be arrested if detected in its early stages by providing adequate subsoil drainage or aeration close to the tree trunk.

Stone and pome fruit trees, grapes, cane fruits, and many ornamental, woody species are sensitive to poor aeration. A single week of flooding has been reported to kill apricot trees (Cunningham, 1920). Peach is nearly as sensitive. Water may be from excess precipitation or irrigation or may arise from a high subsoil water table. When the water table in fruit orchards along the Nile rose to within 3 ft of the ground surface, peach and other stone fruit trees developed gummosis, shoot and twig dieback, leaf shedding, and root rot and finally died (Fikry, 1947). A prolonged high spring water table even in such arid areas as Utah causes decline of apple, peach, cherry, and raspberry. Cankers, gumming, and bark splitting have all been observed on young pie cherry, sweet cherry, and peach trees growing in poorly drained sites. If affected trees survive to an age of ten to twelve years, cankers may heal over and the trees produce normally if parasitic fungi have not meanwhile invaded the weakened tissues.

Citrus too becomes chlorotic and develops dieback with excessive irrigation (Burgess and Pohlman, 1928). Elongation of tap and lateral roots of sour orange seedlings was entirely suppressed when oxygen in the soil was below 1 to 2 percent. Some retardation was evident even at 5 to 8 percent oxygen.

Avocado is another species sensitive to poor aeration (Parker and Rounds, 1944; Haas, 1940). Root fungi are always important under such conditions, but even in their absence heavy subsoils alone can cause decline. Even on better soils, excessive fruit abscission follows heavy rains, which reduce the oxygen content of soil solution.

Even bog-type plants such as cranberry can be suffocated (Bergman, 1943). Cranberry bogs are flooded intentionally in many areas to protect them from winter killing, but losses from suffocation may be experienced especially when bogs are frozen over with a cover of snow or ice, further reducing oxygen diffusion. In such cases, leaves may all drop and terminal fruit buds may be killed, with a corresponding crop loss.

While the most significant long-range damage from poor aeration is to the plant, losses to the fruit crop can also be expected.

Bitter pit of apple (also known as Baldwin spot, fruit spot, and fruit pox in the United States, Stippen in Germany, and liège in France) illustrates the consequences of a disturbed water relation to the fruit. Water is not the primary pathogen but is closely associated with development of the disease.

The first indication of bitter pit is the appearance of small, circular, water-soaked spots on the fruit. These usually develop near harvest but may appear any time after the fruit is about half grown. Spots are dark, dull green on green fruits and dark red on red fruits. Gradually the spots darken, becoming brown after the fruits are in storage. By this time the tissues underlying the lesions die, become pithy, and present a prime target for invasion by decay fungi. The areas of dead cells are closely associated with the vein endings which ramify just beneath the fruit surface. This cellular collapse and death occurs on sensitive varieties when a period of excessive moisture or a late, heavy irrigation is preceded by a period of hot dry weather. More basically, bitter pit is caused by calcium deficiency. It is discussed in more detail under mineral nutrition.

The general nature of oxygen deficiency affecting so many tissues complicates recognition of the disorder. Grapefruit, for instance, may show no symptoms of water excess other than chlorosis of lower leaves and failure to make satisfactory new growth. The same symptoms of chlorosis from lack of aeration also appear on grapes and other small-fruit crops growing in heavy soils. The general dieback and decline is equally deceptive since so many factors can be responsible. Knowledge of soil moisture conditions and soil texture and structure then are vital to an accurate diagnosis.

Vegetable and Field Crops

The resemblance of water imbalance symptoms to those caused by other pathogens makes diagnosis difficult. Take the case of *white spot.* Symptoms of this disease of alfalfa, sweet clover, and black medic closely resemble sulfur dioxide injury. White spot is characterized by sharply delimited, bleached, necrotic spots with an irregular outline and ranging in diameter from a few to many millimeters. Foliar lesions can be marginal, intercostal, or both (Fig. 12.6). Affected tissues pass from a water-soaked, dull, dark green color to yellowish gray-green and finally pale tan or bleached within forty-eight to sixty hours. In its mildest form, chlorosis and bleaching involve but a few cells of the leaflet tip or margin; in its severest form, the entire leaflet is bleached, leaving only the green skeleton of the largest veins.

The disease is distributed from New York to California, wherever alfalfa is grown. While rarely causing significant losses, as much as a 35 percent yield reduction has been reported from retarded growth, leaf destruction, and defoliation. The most vigorous, rank-growing plants and those on the heavier soils are most commonly and seriously affected.

White spot develops when plants which have been subjected to near-drought conditions suddenly receive ample moisture over a period of one or two days. B. L. Richards (1929), one of the few to conduct extensive research on white spot, suggests that the cell mortality is caused by the sudden hydration of the tissues. Hydration results from the rapid absorption of water accompanied by increased humidity and reduced transpiration, which decrease the oxygen level in the plant and interfere with respiration. Further, the byproducts of incomplete respiration could then accumulate in toxic amounts, killing the cells.

Potatoes are also sensitive to excess water, and adequate aeration

FIGURE 12.6 *White spot of alfalfa showing distribution of interveinal and marginal necrosis virtually indistinguishable from SO₂ injury.*

is essential to maximum production (Bushnell, 1935). Results of field drainage studies showed that roots were consistently most abundant in aerated zones, and insufficient aeration limited potato yields in soils heavier than a silt loam.

Aeration continues to be important after harvest; *blackheart,* which was most common under older types of storage, is representative of the diseases associated with oxygen deficiency. The disease results from high respiratory activity and the failure of gas exchange to keep pace with the respiration rate (Davis, 1926). When the affected potato is sliced open, irregular, diffuse areas of necrotic black cells, involving as much as half the cross-sectional area, are found toward the center of the tuber. The etiology of blackheart is tied to the interrelation between temperature and oxygen demand. High temperatures stimulate respiration and the need for oxygen by cells in the tuber; the inner cells cannot obtain adequate oxygen and are killed.

Poor aeration can directly affect fruit formation and production. Shedding of young flower buds of cotton can be caused by poor aeration from heavy rains or excessive irrigation (Albert and Armstrong, 1931). Fruit is directly affected in the case of tomatoes. Cuticle cracks follow heavy rains at relatively high temperatures. *Tomato fruit pox,* a disease characterized by small dark-green specks up to 3 mm in diameter scattered over the fruit surface, has been attributed to a wet early growing season followed by dry conditions near harvest (Ivanoff and Young, 1940). Pox is most common in the southeast, where such conditions prevail.

The similar diseases known as *tomato puffs* or *pockets* have also been attributed to water imbalance (Taubenhaus and Alstat, 1939). Affected fruits are light in weight, misshapen and often flat-sided, and more or less hollow. Puffing begins in the embryonic stages and progresses as the fruit matures. The mature fruit is unmarketable. Symptoms result from large air spaces which arise under sunken parts of the fruit. When cut open, large openings can be seen where seeds failed to develop. The pockets result from ovule abortion, which prevents normal development of the tissue surrounding the affected ovule. Almost any factor interfering with pollination or preventing fertilization and normal ovule development will cause pockets. Water imbalance, either a deficit or an excess, at the time of fertilization is one of the most common causes, but sustained temperatures of 13 to 16°C, or lack of fertilization, have the same effect.

Under natural field conditions the retarded root development in poorly drained soils or during periods of prolonged saturation in the spring leads to extensive infection by soil fungi. Presence of secondary pathogens further complicates the etiology of decline as well as providing a further threat to the development and survival of the species.

Forest Species

The response of plants to water excesses is much the same whether they are cultivated or growing naturally. Growth increment studies have demonstrated that water impounding impairs growth of plants including river birch, silver maple, and pin oaks even when no chlorosis or necrosis appears on leaves (Yeager, 1949). No species can survive excess water for long; even the rare species or individual trees able to endure a few years of water impoundment cannot survive indefinitely. River birch and silver maple lose vigor within a year, and dieback begins during the second.

Raising the soil levels has the same deleterious effect on altering water balance and aeration (Thompson, 1944). Species including birch, beech, oak, maple, and linden and most conifers rarely tolerate more than 2 to 3 ft of fill over established root systems. Sensitive trees become progressively weaker over a period of years and are soon characterized by drooping leaves, short twig growth, an abnormal number of dead twigs and branches, excessive suckering, and loosening of the bark. Thriftiness can usually be restored by removing the fill, installing drainage tiles, or putting wells around the trunk.

The decline symptoms from suffocation can be misleadingly similar to those caused by biotic pathogens. *Hemlock canker* is a good example of this (Elliott and Baldwin, 1961). Symptoms would lead the observer to look for a parasite: Resinous, gummy substances exude from the lower trunk, cankers underlie the bark, bark frequently splits over the cankered area, wood is discolored, and twigs die back; gradually, over a period of years, the tree is girdled and killed. The disease appears whenever hemlocks are grown on heavy, poorly drained soils, in areas where there is a high water table, or in areas of frequent flooding. Such soil contrasts sharply from the well-drained loamy sites in which the hemlock is native and where their far-spreading shallow root systems remain aerated.

Observation of suffocation symptoms and the relative sensitivity of other plant species were reported by C. E. Ahlgren and H. L. Hansen at the University of Minnesota School of Forestry (1957).

Reduced terminal growth was the most noticeable symptom of flooding. Growth suppression was most serious in the smaller, 1- to 2-ft size class but was also significant in larger trees. Flooding for only twenty days reduced growth nearly as much as twice that long. The flooding came during the early summer when most of the new growth would normally have taken place. In the long run, this effect on growth might be significant to the success and survival of a species. Only 10 percent of the trees of all size classes were actually killed by twenty-eight days or less of flooding, while 90 percent of the trees over 5 ft high survived forty-eight days. Close to 100 percent of the small balsam fir and black spruce survived forty-eight days of flooding. Where flooding exceeded sixty days, mortality was much higher. Up to 25 percent of the 3 -to 4-ft white pine died following thirty-eight and forty-eight days of flooding. White spruce was nearly as sensitive. Fifty-five percent of the jack pine 5 to 12 ft high flooded for 123 days died. One hundred percent of the smaller trees and 70 percent of the white pine trees 15 to 30 ft high flooded over 60 days died.

Flooding is one of the many, ever-occurring environmental factors altering the environment and competitive advantage of plant species so that conditions are made less favorable for some while becoming more desirable for others. Sensitivity to submergence is an important ecological factor influencing the ability of a species to survive in areas subjected to intermittent or prolonged flooding.

BIBLIOGRAPHY

(Water Stress)

Albertson, F. W., and J. E. Weaver, 1945. Injury and death or recovery of trees in prairie climate. *Ecol. Monogr.* **15**:393–433.

Alexander, D. McE., 1965. The effect of high temperature regimes or short periods of water stress on development of small fruiting sultana vines. *Aust. J. Agr. Res.* **16**:817–823.

Banfield, W. M., 1967. Significance of water deficiency in the etiology of maple decline. *Phytopathol.* **57**:338.

Bartholomew, E. T., 1926. Internal decline of lemons. III. Water deficit in lemon fruits caused by excessive leaf evaporation. *Amer. J. Bot.* **13**:102–117.

Burrell, A. B., and R. M. Heinicke, 1938. First progress report on prune drought spot. *Proc. Amer. Soc. Hort. Sci.* **36**:275–278.

Copeland, O. L., Jr., 1955. The effects of artificially induced drought on short leaf pine. *J. Forest.* **53**:262–264.

Henson, W. R., 1923. Chinook winds and red belt injury to lodgepole pine in the Rocky Mountains Park area of Canada. *Div. Forest. Biol. Sci. Ser. Dept. Agr.* **28**:62–64.

Leaphart, C. D., 1959. Drought damage to western white pine and associated tree species. *Plant Dis. Rep.* **43**:7.

―――― and L. S. Gill, 1955. Lesions associated with pole blight of western white pine. *Forest Sci.* **1**:3.

Stoeckeler, J. H., and P. O. Rudolf, 1949. Winter injury and recovery of conifers in the upper midwest. *Lake States Forest. Exp. Sta.* no. 18, St. Paul, Minn.

Taylor, S. A., 1951. A continuous supply of soil moisture to the growing crop gives highest yield. *Farm and Home Sci.* **12**:50–51, 61.

Treshow, M., 1957. Terminal bleach of cereals. *Plant. Dis. Rep.* **41**:118–119.

Tufts, W. P., and L. D. Davis, 1930. Hard end or black end of pears in California. *Proc. Wash. State Hort. Assoc.* **25**:108–115.

Vaadia, Y., F. C. Raney, and R. M. Hagan, 1961. Plant water deficits and physiological processes. *Ann. Rev. Plant Physiol.* **12**:265–292.

Voigt, G. K., 1948. "Causes of injury to conifers during the winter of 1947–1948 in Wisconsin." Wisconsin Academy of Science.

Wellington, W. G., 1954. Pole blight and climate. *Bimon. Prog. Rep. Div. Forest. Biol. Dept. Agr. Can.* 10, 6, pp. 2–4 (R.A.M. 1956, p. 132).

SELECTED REFERENCES

(Water Stress)

Allmendinger, D. G., A. L. Kenworthy, and E. L. Overholser, 1943. The carbon dioxide intake of apple leaves as affected by reducing the available soil water to different levels. *Proc. Amer. Soc. Hort. Sci.* **42**:133–140.

Bartholomew, E. T., 1923. Internal decline of lemons. II. Growth rate, water content and acidity of lemons at different stages of maturity. *Amer. J. Bot.* 117–126.

Faucett, H. S., 1926. Endoxerosis or internal decline of the lemon, in "Citrus diseases and their control," pp. 421–426. McGraw-Hill, New York.

Fisher, D. V., J. E. Brittan, and S. W. Porritt, 1950. Some horticultural aspects of black end of pear. *Proc. Amer. Soc. Hort. Sci.* **55**:207–223.

Graham, D. P., 1958. Results of pole blight damage surveys in the western white pine type. *J. Forest.* **56**:9.

Hepther, M. J., 1928. Pear black and its relation to different root stocks. *Proc. Amer. Soc. Hort. Sci.* **38**:27–31.

Hubert, E. E., 1918. A report on the red belt injury of forest trees occurring in the vicinity of Helena, Montana. *Mont. State Forest. Bien. Rep.* **5**:33–38.

Kozlowski, T. T., 1958. Water relations and growth of trees. *J. Forest.* **56**:498–502.

Leaphart, C. D., and E. C. Wicker, 1966. Explanation of pole blight from response of seedling grown in modified environments. *Can. J. Bot.* **44**:121.

Reisch, Kenneth W., 1958. Effects of drought on plant growth. *Proc. Int. Shade Tree Conf.* **25**:11–23.

Shirley, H. L., 1936. Lethal high temperatures for conifers and the cooling effect of transpiration. *J. Agr. Res.* **53**:239–258.

—— and L. J. Meuli, 1939. Influence of moisture supply on drought resistance of conifers. *J. Agr. Res.* **59**:1–21.

Stickel, P. W., 1933. Drought injury in hemlock-hardwood stands in Connecticut. *J. Forest.* **31**:573–577.

Subramanian, D., and L. Saraswathi Devi, 1959. Water is deficient, in J. G. Horsefall and A. E. Dimond (eds.), "Plant pathology," vol. 1, chap. 9, The diseased plant. Academic, New York.

Toole, E. R., 1949. White pine blight in the southeast. *J. Forest.* **47**:378–382.

BIBLIOGRAPHY

(Water Excess)

Ahlgren, C. E., and H. L. Hansen, 1957. Some effects of temporary flooding on coniferous trees. *J. Forest.* **55**:9.

Albert, W. B., and G. M. Armstrong, 1931. Effects of high soil moisture and lack of soil aeration upon fruiting behavior of young cotton plants. *Plant Physiol.* **6**:585–591.

Bailey, J. S., and H. J. Franklin, 1950. Blueberry culture in Massachusetts. *Mass. Agr. Exp. Sta. Bull.* 358.

Bergman, H. F., 1943. The relation of ice and snow cover on winterflood cranberry bogs to vine injury from oxygen deficiency. *Mass. Agr. Exp. Sta. Bull.* 402, 3–24.

Biocourt, A. W., and R. C. Allen, 1941. Effect of aeration on the growth of hybrid tea roses. *Proc. Amer. Soc. Hort. Sci.* **35**:315–319.

Burgess, P. S., and G. G. Pohlman, 1928. Citrus chlorosis as affected by irrigation and fertilizer treatments. *Ariz. Agr. Exp. Sta. Bull.* 124.

Bushnell, J., 1935. Sensitivity of the potato to soil aeration. *J. Amer. Soc. Agron.* **27**:251–252.

Cunningham, G. H., 1920. Mortality among stone-fruit trees of central Otago. *J. Agr. N. Z.* **20**:359–364.

Davis, W. B., 1926. Physiological investigation of blackheart of potato. *Bot. Gaz.* **81**:323–338.

Elliott, E. S., and R. E. Baldwin, 1961. Control hemlock canker. *W. Va. Univ. Agr. Exp. Sta. Bull.* 460.

Fikry, A., 1947. Water table effects. V. Peach functional disorder. *Minn. Agr. Egypt. Tech. Sci. Serv., Plant Pathol. Ser. Bull.* 245.

Furneaux, B. S., and W. G. Kant, 1937. "The death": A trouble of fruit trees due to root suffocation. *Sci. Hort.* 567–77. (R.A.M. 1937, p. 540).

Haas, A. R. G., 1940. The importance of root aeration in avocado and citrus trees. Calif. Avocado Assoc. Yearbook, pp. 77–84.

Hartig, R., 1879. Zersetzungserscheinungen des Holzes der Nadelbaume und der Eiche, Die Wurzelfaule: 75–81.

Heinicke, A. J., and D. Boynton, 1941. The response of McIntosh apple trees to improved subsoil aeration. *Proc. Amer. Soc. Hort. Sci.* **38**:27–31.

Ivanoff, S. S., and P. A. Young, 1940. Tomato fruit pox. *Phytopathol.* **30**:343–344.

Parker, E. F., and M. B. Rounds, 1944. Avocado tree decline in relation to soil moisture and drainage in certain California soils. *Proc. Amer. Soc. Hort. Sci.* **44**:71–79.

Richards, B. L., 1929. Whitespot of alfalfa and its relation in irrigation. *Phytopathol.* **19**:125–141.

Taubenhaus, J. J., and G. E. Alstat, 1939. Some factors contributing to tomato puffing. *Plant Physiol.* **14**:575–581.

Thompson, R. A., 1944. Protection of trees in grade changes. *Amer. Nurseryman* **79**:9–11, 35–37, 42.

Yeager, L. E., 1949. Effect of permanent flooding in a river bottom, timber area. *Bull. Ill. Nat. Hist. Surv.* 25, Art. 2.

SELECTED REFERENCES

(Water Excess)

Bergman, H. F., 1959. Oxygen deficiency as a cause of disease in plants. *Bot. Rev.* **25**:418–485.

Bolton, J. L., and R. E. McKenzie, 1946. The effect of early spring flooding on certain forage crops. *Sci. Agr.* **26**:99–105.

Bratley, C. O., 1930. Notes on flooding injury to strawberries. *Phytopathol.* **20**:685–686.

Chester, K. Starr, 1944. A cause of "physiological leaf spot" of cereals. *Plant Dis. Rep.* **28**:497–499.

Childers, N. F., and D. G. White, 1942. Influence of submersion of the roots on transpiration, apparent photosynthesis and respiration of young apple trees. *Plant Physiol.* **17**:603–618.

Conway, Verona M., 1940. Aeration and plant growth in wet soils. *Bot. Rev.* **6**:149–163.

Cooley, J. S., 1948. Collar injury of apple trees in water-logged soil. *Phytopathol.* **38**:736–739.

Degman, E. S., 1946. Anjou pear growing on heavy soils in the Medford, Oregon area. *Proc. Wash. State Hort. Assoc.* **41**(1945):139–142.

Denyer, W. B. G., and C. G. Riley, 1964. Dieback and mortality of Tamarack caused by high water. *Forest. Chron.* **40**:3.

Dorsey, M. H., and W. A. Ruth, 1930. A record of an unusual flood in an apple orchard. *Proc. Amer. Soc. Hort. Sci.* **27**:565–569.

Foster, A. C., and E. C. Tatman, 1937. Environmental conditions influencing the development of tomato pockets or puffs. *Plant Physiol.* **12**:875–880.

Green, W. E., 1947. Effect of water impoundment on tree mortality and growth. *J. Forest.* **45**:118–20.

Heinicke, A. J., 1932. Effect of submerging roots of apple trees at different seasons of the year. *Proc. Amer. Soc. Hort. Sci.* **29**:205–207.

Kienholz, J. R., 1946. Performance of pear orchards with flooded soil. *Proc. Amer. Soc. Hort. Sci.* **47**:7–10.

Kramer, P. J., 1951. Causes of injury to plants resulting from flooding of the soil. *Plant Physiol.* **26**:722–736.

Letey, J. L., H. Stolzy, N. Valoras, and T. E. Szuskiewics, 1963. Low soil oxygen most damaging to plants during hot weather. *Calif. Agr.* **17**:15.

Parker, J., 1950. The effects of flooding on the transpiration and survival of some southeastern forest tree species. *Plant Physiol.* **25**:453–460.

CHAPTER THIRTEEN

DEFICIENCIES OF MINERAL ELEMENTS

When plant growth is weak and production not up to expectations, one or more of several pathogens may be responsible, but malnutrition is one of the first factors to consider. Plants require ample, but not excessive, amounts of at least sixteen different nutrient chemicals for normal, optimum growth. Nutrients form the structural components of the plant, the nucleic acids directing plant processes, the enzymes and catalysts regulating metabolic activity, and the carbohydrates providing energy reserves; they are further necessary for maintaining osmotic balance and absorbing ions from the soil solutions.

Most of these elements are provided by the soil, but hydrogen, a component of every organic plant constituent, is provided by water; and oxygen and carbon, which comprise over half of the dry matter in plants, are obtained from the air. Oxygen is combined in various oxides as well as in plant proteins, fatty acids, carbohydrates, and many other plant constituents. The oxygen content of the atmosphere is high and varies little; consequently, it is seldom a limiting element for above-ground tissues. Carbon, providing the basic building block of all organic molecules, is the backbone of the cell and is needed in tremendous quantities, but ample amounts are provided by the carbon dioxide in the atmosphere. Carbon, oxygen, and water comprise roughly 95 percent of the total plant weight. The remaining 5 percent —consisting of such major elements as nitrogen, phosphorus, potassium, sulfur, magnesium, and calcium, and trace elements (needed only in minute amounts) including iron, manganese, boron, zinc, copper, molybdenum, and chlorine—are obtained from the soil. The plant will also pick up trace amounts of every other element found in the soil whether beneficial or not. Elements may be beneficial, sometimes even essential, to certain plant species but not to others. For instance, sodium is required by marine algae and improves the growth of many plants native to saline soils, but it is not essential to other species. Other elements may stimulate growth, but since the plant can survive in their absence, they are not considered essential.

The essential elements are often divided into two groups: the macro elements, which are required in large amounts, and the micro or minor elements, required in trace amounts. All are equally vital to normal plant development. For the most part, the macro elements form part of the plant structure. The bulk of carbon, hydrogen, and oxygen, together with calcium, makes up the cell walls and membranes. Nitrogren, phosphorus, and sulfur are part of the amino acids and proteins forming the structural framework of the protoplast. Magnesium is part of the chlorophyll molecule.

The trace elements are part of the enzymes or coenzymes regulating plant development and have largely a functional role. The role of a third group of chemicals, which stimulate growth but are not essential to it, is not known. While each essential element has one or more specific roles in the structure and function of the plant, their activity is closely integrated, and a shortage of one element will soon affect the activity of others. Interaction of nutrients is illustrated by deficiencies of potassium, phosphorus, or calcium which can induce deficiency of iron. High phosphorus greatly accentuates the symptoms of iron and potassium deficiencies. Conversely, potassium deficiency symptoms are more severe in iron-deficient plants than in those supplied with adequate iron. At normal phosphorus levels the severity of iron deficiency is determined mainly by the potassium supply. Interaction of calcium and phosphorus is illustrated in a study by Greenwood and Hallsworth (1960). A phosphorus level of 40 ppm, which is normally favorable, was found to be toxic at a low calcium level of 8 ppm but beneficial at a high calcium level of 64 ppm.

Cases have also been reported of one element substituting for another, as where strontium may partially substitute for calcium and rubidium for potassium. Strontium was beneficial only when the calcium supply was low; when calcium was absent, growth of barley was markedly improved (Walsh, 1945). Another example shows that selenium can replace sulfur in certain amino acids such as selenomethionine or selenocystine.

Nutrient ion interaction may affect the absorption of elements from the soil, so that a chemically similar ion may be absorbed rather than the essential element. In this way arsenate may interefere with phosphate absorption, selenate with sulfate, bromide with chloride, and rubidium with potassium.

The interaction of nutrients, where a deficiency of one element may cause visible deficiency symptoms of another, makes visual diagnosis not only difficult but uncertain. Diagnosis is further complicated by the similarity of deficiency diseases to those caused by chemical excesses, viruses, air pollutants, and other pathogens.

When the nature of chemical deficiencies is thoroughly understood, an accurate diagnosis is not only possible but probable. Symptoms of mineral deficiencies are often distinctive for each element although the detailed expression may vary among different plant species. Many symptoms are so definitive that the disease has a well-established, descriptive name.

Early diagnosis is vital to maximum production; if a nutrient deficiency is suspected and no visible symptoms are apparent, chemical methods of diagnosis can be utilized for more immediate results. Several chemical tests are available which can be used to analyze the nutrient status of plants. Tissues can be chemically analyzed for their mineral composition, or spot tests may be used where a small portion of petiole or leaf tissue is macerated in a drop of suitable reagent and a color change observed. The intensity of the color is proportional to the amount of the essential chemical sought. Another test consists of spraying the plant with the element in question and noting the response.

Deficiencies arise when the supply of some element becomes

limiting to the continued activity of a chemical reaction or becomes inadequate to provide the substrate needed for the continued development of the cell wall or protoplast. When other conditions such as temperature or moisture are not limiting, the chemical nutrient status becomes critical. When deficiencies become acute, not only are growth and production limited, but such clearly recognizable disease symptoms as distorted growth, chlorosis, necrosis, fruit and stem malformation, dieback, and death appear. The reasons why plants need certain elements, where they get them, how much they require, and what happens when they are deficient are discussed in the following pages. More comprehensive descriptions of symptoms, together with excellent color illustrations, are given in *Hunger Signs of Crops,* edited by H. B. Sprague (1964).

13-1 NITROGEN

Nitrogen is required in large amounts and is the most universally deficient element. It is found in every amino acid, and as such it is a major component of the proteins forming the matrix of the protoplast and comprising from 40 to 50 percent of the plant dry matter. It is required in enzymes and for the synthesis of enzymes, and a deficiency limits every enzymatic reaction. Nitrogen also forms part of the chlorophyll molecule and is required for both its synthesis and its structure. With such dependence of the chlorophyll on nitrogen, it is no wonder that plants low in nitrogen can contain very little chlorophyll, and a loss of green color is one of the early visible expressions of nitrogen deficiency. The ubiquity of nitrogen in the plant also explains why so much is needed for normal development. In the absence of an adequate nitrogen supply little growth can take place, and plants remain stunted and undeveloped.

An average of 100 to 200 lb nitrogen per acre is required by a typical agricultural crop such as sugarbeets, corn, wheat, or potatoes, and nitrogen must be continually provided throughout the growing season.

Vegetable crops, which make a tremendous amount of growth during a short period, utilize several pounds of nitrogen per acre each day, and several applications are necessary during a growing season. The same is true for forage crops and grasses, including home lawns, which must be fertilized every few weeks to maintain good growth and color.

Although in a natural plant community the total nitrogen in the soil may appear adequate, only a fraction of the total is available to the plant. The bulk is tied up in proteins in various stages of degradation. As nitrate is gradually released, it is absorbed and utilized by the plant. As long as no plant materials are removed, the amount available is usually balanced by the nitrogen utilized, and no acute deficiencies arise. However, the net nitrogen supply might be low and limit growth even in an apparently healthy community. Growth in a natural community is often limited by nitrogen, and addition of nitrogen would greatly accelerate growth. However, the resulting lush plants would be highly sensitive to moisture and heat stress and greatly predisposed to destruction by insect and fungus pathogens.

In the mildest visible form, nitrogen deficiency is characterized by pale green leaves, often smaller than normal, and stunted, spindly shoot growth. More specifically, chlorosis appears first on the older leaves as nitrogen is translocated from them to the younger leaves. Gradually the nitrogen supply is depleted even in the younger leaves; chlorophyll fades, unmasking the xanthophylls and carotenes and exposing areas of orange and red tints. These hues arise first on the older leaves and gradually on progressively younger leaves as the deficiency becomes more acute. Anthocyanin production is often accelerated, and leaves take on a purplish coloration. Stems and twigs also become reddish-yellow, with numerous red and brown spots arising at the base of the twigs. Bark of nitrogen-deficient apple trees becomes reddish-brown. Because of their small leaves, early defoliation, and suppressed shoot growth, nitrogen-deficient plants are conspicuously thin, sparse-foliaged, and weak.

While leaf chlorosis is a general symptom, the details of response vary among species. In corn, yellowing is most prominent along the leaf margins. In stone fruits, apple, and cauliflower, orange and red tints are particularly prominent, with veins on the lower leaf surface becoming prominently purplish. In cereals, tillering is poor and stems tinted reddish or purplish. The most severely deficient leaves may develop necrotic areas along the margins, and in pecans and walnuts the leaflets may drop in mid-season. If the deficiency in tung trees is severe enough to induce leaf symptoms, fewer pistillate flowers differentiate, and the succeeding year's crop is reduced. The current year's crop suffers from poorly filled nuts.

Adequate nitrogen is essential for normal flowering and fruit setting of all species; a deficiency will reduce fruit quality and size, and maturation will be earlier than normal. Where nitrogen deficiency is acute, blossoming is seriously reduced and yields proportionately limited. Flower buds often turn yellow and shed rather than mature to form marketable fruit. Apple fruits are small, hard, and highly flushed; background color is pale green or white.

13-2 PHOSPHORUS

Phosphorus is a vital structural component of the nucleic acids, nucleoproteins, phytin, phospholipids, adenosine triphosphate, and numerous phosphorylated compounds. As a component of nucleic acids, phosphorus is built into the DNA of the chromosomes and the RNA of the nucleus and ribosomes, where it is vital to nuclear and cell division and regulation of every cellular process. Phosphorus is vital also as a structural component in the phospholipids of the cell membranes, regulating movement of materials in and out of the cells and organelles. Phospholipids also may act as storage material in seeds, although phosphorus in seeds is stored principally as phytin, which is hydrolized as the seed germinates, releasing the phosphate to serve in the energy-bearing ATP.

Phosphate also occurs in certain enzymes catalyzing metabolic reactions. Phosphoglucomutase, vital to sugar metabolism, is but one example. Phosphorus also occurs in the primary cell walls as the enzyme acid phosphotase. Phosphorus is further involved in the initial reactions of photosynthesis, where it is found in the 5-carbon sugar with which CO_2 initially reacts.

Phosphorus is likely to be deficient in any soils other than those formed from parent materials high in phosphorus or those in which phosphorus has accumulated through years of fertilization. The available phosphorus is absorbed by the plant roots and utilized by the plant in rather large quantities. The amount present in a crop may range from 15 lb/acre for apples and grapes to 130 lb in celery. Such quantities may be present in virgin soils, but the supply soon becomes depleted if phosphate is not added. Fertilization can be overdone, however, and in some soils continued applications of phosphates have led to toxic accumulations.

Phosphorus content is greatest in the most active growing tissues of the plant, such as meristematic regions and young, developing fruits, and in seeds; but, on the average, the phosporus content of a healthy plant is considered to be about 0.25 percent of the dry weight. Should the content drop below about 400 ppm, deficiences are likely. The plant responds to phosphorus deficiency in much the same way as to nitrogen deficiency. Even before visible symptoms appear, growth of shoots and roots will be greatly restricted. Shoots are short and thin, and growth is upright and spindly; leaves are small, and defoliation follows, starting with the oldest leaves. Lateral buds often die or remain dormant, so the shoots are sparse. Flower bud development is also inhibited, reducing the bloom with corresponding yield reductions. Opening of buds is sometimes delayed, so that growth is uneven and fruits mature over a longer period.

While the general syndrome is similar to that of nitrogen deficiency, specific differences exist, particularly in the leaf expression. Phosphorus-deficient leaves tend to lack luster more and are more bluish than yellow-green in color due to the abnormally excessive formation of anthocyanin. Most distinctively, they tend to remain dark green, older leaves becoming darkest green first. Where it occurs, the purple pigmentation is most prominent on the underside of leaves or along the veins. Tips of some leaves, as on cabbage, take on a purplish cast. Unfortunately for diagnosis, purple coloration may also characterize nitrogen deficiency and other stress conditions. Tints from phosphorus deficiencies are more purple than yellow or red and may have a dull bronze cast with purple or brown spotting. A purplish tinge is particularly prominent on leaves of grasses. Leaf margins of some plants (e.g. potato) are more apt to become necrotic, and margins are apt to roll or curl under.

Fruit symptoms are also distinct. Fruits are not only sparse, but they are small and their color variable, with green background color. Flavor may be acidy and keeping quality poor.

The suppression of root development in general, combined with the paucity of a fibrous root system in phosphorus-deficient plants, is especially limiting to normal plant growth. The weak root systems further render phosphorus-deficient plants highly sensitive to damage by root-rotting fungi. Young plants are particularly vulnerable to destruction when their root growth is too slow to keep ahead of advancing fungi.

13-3 POTASSIUM

Potassium is not a structural component of the plant, nor is it a component of nucleic acids or enzymes. Its function is primarily regulatory. Much is known about its distribution in the plant and what happens in its absence, but the specific role of potassium remains a mystery.

Experiments over the years have implicated potassium in virtually every metabolic process studied. Deficiencies produce water imbalance, decreased photosynthetic activity, disturbed carbohydrate metabolism, increased respiration, decreased chlorophyll synthesis and protein content, and visible leaf injury. The most specific role which has been established is in protein metabolism. Potassium is essential to activating the enzymes that synthesize certain peptide bonds and incorporate amino acids into protein (Webster, 1953). Potassium also is associated with maintenance of a positive ion balance to satisfy negative charges on the protein and thereby to stabilize it (Krantz and Melsted, 1964).

On a general basis, potassium is first necessary for formation of sugars and starch, then it is required for their transport throughout the plant. It is necessary also for cell division, growth, and, along with calcium, neutralization of organic acids. In some way, potassium is essential to maintaining cellular organization, permeability, and hydration.

Potassium improves the rigidity of straw and stalks, increases resistance of plants to disease, and helps plants withstand such environmental stresses as adverse water and temperature relations and poor soil. Also, an ample supply enhances fruit size, flavor, and color.

Potassium is present in all parts of the plant in fairly large quantities and is especially concentrated in meristematic regions and in leaves. The normal concentrations of potassium of about 1 to 2 percent of the dry matter of new growth is higher than for any element except nitrogen and calcium.

In order for the plant to accumulate such relatively large amounts of potassium, the soil content must be high. An average crop may use somewhere around 150 lb of potash per acre. A fair crop of celery, for instance, removes 470 lb of potash from the soil. Wheat, cotton, and apples require close to 50 lb/acre.

When the available potassium supply diminishes, plant growth becomes progressively weaker. Shoot growth is restricted, stems are thinner, leaf symptoms appear, and shoots may die back. Weak growth and poor production associated with a slight deficiency may be difficult to distinguish, but the leaf discoloration and necrosis reflecting a more acute shortage are distinct. General symptoms of potassium deficiency are similar for many plants. The first feature to appear is the dull, bluish-green discoloration, particularly in the interveinal areas of the leaves. A dull, general chlorosis, particularly at the leaf tips and margins, is characteristic on some species. This is usually followed, beginning with the older leaves, by tip burning, marginal scorch, and sometimes the development of bleached or brown spots (usually most numerous along the margins). The blades of many broad-leaved species tend to curl either upward or under. By this time, growth is severely stunted, production is markedly reduced, and root systems poor.

Histologically, potassium deficiency decreases cambial activity except at the stem apex. Cells become thin-walled and unable to support the plant (Nightingale et al., 1930). Cell walls of tracheids are less lignified and parenchyma cells abnormally large. The size of phloem cells is reduced, and pith cells often disintegrate. Potassium persists in cambial cells but disappears from most other tissues.

Symptoms on different types of plants differ in detail. Vegetable crops which characteristically grow from a crown without a definite stem develop an acute rosette habit. In other species—for example, flax and cereals—apical dominance appears to be suppressed so that the growth habit is especially bushy. Tillers in barley may be nearly prostrate. Under severe deficiency, the terminal bud is killed, producing dieback with dominant basal growth.

A lack of potassium in cotton results in a striking symptom and disease commonly known as *cotton rust* (Oaks, 1938). At first a yellowish-white mottling appears on the foliage, changing the leaf color to pale yellow-green. Yellow spots then develop along the margin and between the veins; the center of these dies and becomes necrotic and reddish-brown in color, and leaves drop early. Potassium deficiency limits carbohydrate production, so that the main stem and branches often wither and die prematurely.

On legumes, brown spotting is most distinctive. In peas, the seeds develop a thickened, tough coat which gives the frozen or canned product a poor quality. Spots on alfalfa begin as small, whitish flecks along the margins of the leaves. As the spots develop progressively toward the midrib, the tissue between veins becomes yellow and turns brown, and the edges of the leaves become ragged. The disease is sometimes called *alfalfa yellows*.

Potato plants lacking potassium are stunted and spreading. Leaves are dark green but become dull and often bronzed, with numerous dark brown spots on the lower leaf surface. Eventually they become brown and wither. Tuber size is reduced, and the crop is negligible. Tomatoes, which also have a high potassium requirement, are similarly affected, but marginal leaf scorching is more prevalent. Fruits ripen unevenly, and yellowish-green patches may persist when fruits ripen, producing a "blotchy" expression.

Scorching of older leaves is the outstanding symptom on fruit trees. In stone fruits, an upward, lateral curling appears before the leaf burn. The necrosis may be any shade of brown, or even black in the case of pears and grapes. Stunted growth and excessive twig dieback are common. Even when blossom set is ample, yields may be low due to excessive drop. When the crop sets, the fruit is small and fails to mature evenly. On grape leaves, marginal chlorosis and necrosis, gradually extending toward the midrib, is prominent. Leaves are noticeably smaller than normal; fruit clusters are dense, fruit small, and maturity uneven and late.

13-4 SULFUR

As a component of the amino acids cysteine, cystine, and methionine, sulfur is vital to protein structure and required in rather large amounts. Sulfur is also found in the plant hormones thiamin and biotin, also in such volatile compounds as mustard oil. Sulfur aids in the synthesis of oils and appears to be associated with chlorophyll synthesis. It is also essential to the development of root nodules and subsequent nitrogen fixation by legumes.

The required sulfur is obtained from the soil, which may contain anywhere from 15 to several thousand lb/acre-foot. Sulfur is absorbed from the soil solution as sulfate, with upwards of 40 lb/acre needed by crops having a high requirement. Fifteen to 20 lb is adequate for others, including legumes, tobacco, cotton, and alfalfa; 10 lb/acre adequately fills the needs of small grains, grasses, and corn (Jordon and Reisenauer, 1957).

Deficiencies arise in areas of the Pacific Northwest, California, and most of the Southeastern states, where soils have been formed from parent materials low in sulfur. The amount of sulfur in the leaves should be 500 to 14,000 ppm by dry weight. Concentrations below 250 ppm are considered critical and give rise to visible deficiency symptoms.

Symptoms of sulfur deficiency are much like those caused by nitrogen deficiency. The reduced leaf size, stunted growth, chlorosis, development of orange to purple pigmentation, and delayed maturity are characteristic of both. But in contrast to nitrogen deficiency, symptoms of sulfur deficiency are usually most striking on new growth. Chlorophyll production is retarded, so that new leaves of plants such as citrus, tobacco, cotton, soybean, and tomato remain pale green. Leaves become pale yellow to bleached and turn faint pink in later stages (Kramer and Shrader, 1942).

Chlorosis from sulfur deficiency in corn may be limited to interveinal regions. With the veins remaining green, symptoms may be mistaken for iron or zinc deficiency. However, the latter are more common on alkaline soils, while sulfur deficiency is more likely on acid soils.

Sulfur deficiency in tea was described by Storey and Leach (1933) as the cause of *tea yellows*. Affected plants were characterized by extremely chlorotic, yellow leaves of much reduced size. This was followed by necrosis and rolling of tips and margins of young leaves, and defoliation and death of the shoot apex. Lyon and Garcia (1944) have described the anatomical effects of sulfur deficiency on tomato as consisting of an increase in the thickness of the fiber, xylem, and collenchyma cells. Such cellular changes make the plants hard and woody.

Chemically, protein synthesis is inhibited, causing a decrease in protein nitrogen and an accumulation of total soluble nitrogen, amides, free amino and ammonia nitrogen, and nitrate. This is accompanied by increased sucrose and reduced sugar levels.

13-5 CALCIUM

Calcium has many roles in the structure and metabolism of plants. The bulk of calcium occurs as calcium pectate, a structural component of the middle lamella of the primary cell walls and a regulator assisting in controlling the amount and development of new growth. Calcium is also found in the calcium phosphates of the protoplasmic membranes and may strongly influence their properties of structure and ionic permeability. Calcium enters the constitution of proteins as one of the important mineral bases, helping to regulate the hydration of cytoplasm. It is also thought to aid in translocating carbohydrates, but the mechanism is not known. In the absence of calcium large amounts of starch accumulate, and new cell walls are imperfectly formed. Calcium may reduce the toxicity of such inorganic elements as sodium and magnesium should they accumulate in excessive concentrations, and it is important in neutralizing or precipitating excesses of organic acids formed as by-products of metabolism which might otherwise become injurious to plant cells. Calcium oxalate crystals, which are formed as excesses of oxalic acids, are neutralized. Calcium is important in assimilating nitrogen and phosphorus in proteinaceous constituents, and a lack of calcium produces more carbonaceous vegetation.

The minute amounts of calcium required for normal meiosis, and the close relation found between calcium deficiency and chromosome abnormality, suggest that calcium has a specific function in the organization of chromatin or the mitotic spindle.

Calcium exists in a delicate balance with magnesium, potassium, and boron. Any imbalance in the ratio of these elements to each other will produce abnormal plant responses. An apparent deficiency of calcium may in reality be due to an excess of magnesium, potassium, or boron, and an excess of boron or potassium may produce symptoms similar to those produced by calcium deficiency.

Most soils are sufficiently high in calcium to fulfill the plant requirements, but much of it is unexchangeable and not available to the plant. The calcium supply and uptake is further limited by competition with other ions. For instance, hydrogen and sodium ions are absorbed in preference to calcium, so that in highly acid soils calcium uptake will be impaired. In highly acid soils, calcium deficiency symptoms are complex and often exist in combination with magnesium deficiency and manganese or aluminum toxicities. Similarly, calcium uptake will be restricted in high-sodium soils. As the exchangeable sodium percentage increases, there is a comparable increase in alkalinity (pH) which induces deficiencies of calcium along with other elements such as magnesium.

Calcium deficiency symptoms appear first in the young leaves and the shoot apex. The newly emerging leaves become severely distorted, with tips hooked back and the edges curled. Margins are often irregular and ragged and may show brown scorching or spotting. In other instances leaves are pale and chlorotic, and much of the mesophyll tissue is collapsed. Leaf growth is uneven, and lateral development is restricted to where the margin is concave in the chlorotic regions.

Chlorosis, tip curling, and marginal curvature become more pronounced with decreasing leaf age. As the leaves enlarge, the areas of restricted growth and chlorosis become necrotic and blackened. The youngest, most severely affected leaves may fail to develop a blade, so that only the blackened tip of the petiole appears. In more severe instances, the youngest leaves and shoot apex may wither and die without expanding.

Growing points are often killed, as in *blackheart of celery,* a common disease appearing wherever celery is grown. The disease is not confined to celery but may develop on carrot, fennel, parsley, and other umbelliferous crops. Identical death of the growing point has also been observed on sugarbeets growing in waterlogged soils. Symptoms usually appear when plants are about half grown, developing first on the youngest leaves in the center of the stalk. The margins of the smallest leaves fade in color and turn pale brown and ultimately black. The dead tissue of the young leaves is quickly invaded by soft-rot bacteria which cause a slimy, watery decay and render the stalks unmarketable.

Blackheart is basically caused by a calcium deficiency, but predisposing factors, primarily adverse moisture relations, are important in its development. Plants on heavy, waterlogged soil are particularly sensitive, regardless of the soil calcium content. The disease is quick to develop when plants have been allowed to become dry or subjected to water stress and then irrigated heavily. Blackheart is readily controlled by applications of .05 to 0.25M sprays of calcium sulfate or calcium nitrate to the celery heart (Geraldson, 1954, 1956).

Blossom end rot of tomatoes, another disease associated with calcium deficiency and water stress, can also be prevented by calcium sprays. The disease is widespread and annually causes serious crop losses in major tomato-growing regions. Fruits making rapid growth have a constant and high calcium demand; if the requirements are not met, the newly developing cells will be unable to divide and enlarge normally. The primary cell walls will be thin, succulent, and readily subject to collapse. Blossom end rot first appears on a small water-soaked spot at or near the blossom scar of one-third to half-grown green fruits. Gradually the lesion enlarges, and the affected tissues dry out and become light to dark brown, finally black in color. The remainder of the fruit enlarges normally, leaving a collapsed, leathery, sharply delimited, sunken area at the base. Spots vary greatly in size, affecting as much as half the fruit.

The disease develops most seriously when plants have made rapid growth and are most sensitive to high temperature and moisture stress. Extreme moisture fluctuations further increase the likelihood of disease. Plants which are high in nitrogen are especially susceptible, as are those grown under high-salt conditions. Raleigh and Chucka (1944) found that the disease was greatest in plants deficient in calcium and appeared when the calcium content of the fruit fell below 0.2 percent. The ratio of calcium to such elements as boron, nitrogen, sulfur, magnesium, and potassium is equally important to disease development, and plants containing excessive amounts of these will be predisposed.

Carrots grown on soils low in calcium or high in soluble salts may produce rough-surfaced roots that develop pitting or *cavity spot*. This disorder occurs when calcium content of the root is less than 0.25 percent of the dry weight.

Bitter pit of apple, which was described briefly in Chap. 12 under improper water relations, is fundamentally caused by a calcium deficiency. Symptoms consist of small, sunken, corky or spongy, circular dark brown lesions appearing on the nearly mature fruit. Pitting is induced in the orchard, although it may not show up until after fruit is in storage.

Anatomical studies have shown that symptoms are produced by cell proliferation induced in cortical cells near the vascular bundle endings. Development of the newly divided cells, unable to compete with older cells for nutrients, is impaired and abnormal. Pectic protuberances extend from these cells into the intercellular spaces, causing the bitter-pit symptom. Interaction of other elements is also important, and bitter pit may develop even though calcium is "adequate" if the ratio of magnesium or potassium to calcium is too high or nitrogen levels excessive. The incidence of bitter pit has been recognized to be related to low calcium levels in the fruit since 1937 (Delong, 1937). Subsequently, workers have found that bitter pit is also associated with high potassium levels, which depress calcium absorption. The heavier fertilization practices in recent years have caused an increased incidence of bitter pit. Up to an 80 percent fruit loss is reported in some German orchards where susceptible varieties are grown. The problem is less severe in the Northwest, since strains of the Delicious varieties most frequently grown there are not especially sensitive.

The disease is most important on soils of low calcium availability and in areas such as South Africa and Australia where the major varieties grown are particularly sensitive. In such areas, bitter pit is routinely controlled by applications of calcium sprays such as $CaCl_2$ or $CaNO_3$ at 2 to 3 lb per 100 gal water. However, some varieties are sensitive to the calcium sprays, especially $CaNO_3$, and are readily injured; leaf symptoms often occur when over 3 lb per 100 gal is used.

Withertip of flax, another calcium-related disease, is characterized by the withering or drooping of the shoot tips. Necrosis appears at the point of bending when plants are about 1 ft high, and the tip dies. Shoots emerge from buds below the injury, producing considerable lateral branching and a bushy habit. Fiber cell formation is inhibited, weakening the stems and reducing their market value. Moderate to severe calcium deficiency causes a similar wilting in other species including crucifers, tulips, and legumes. Major veins, midribs, and petioles become involved successively; the stem tips wilt next, and the leaves topple over.

Calcium-deficient potato plants may appear essentially normal above ground while tubers are profuse, smaller than normal, and malformed. Internally, tubers may show vascular breakdown and necrosis.

Roots are also sensitive to calcium deficiency. Absence of calcium has been found to exert a profound influence on root cell behavior (Sorkin and Sommer, 1940). Mitosis was abnormal, and, while nuclei separated into two parts, the small, newly formed cells were unable to differentiate. Often cell walls were not formed, so that cells contained two nuclei. Many incomplete stages of nuclear division were observed which showed partial spindle formation, incomplete separation of chromosomes, and aggregation of chromatin into granules and lumps. Roots of cereal crops are especially sensitive; they become stunted, translucent, and gelatinized and die back from the tips. Apple, peach, and tomato roots become bulbous and proliferate just behind the primary root, which ceases growth and dies, much like the shoot apex. Calcium deficiency results in a lack of hardening of the cell wall of root hairs as well as roots, also in gross expansion of the weak walls with consequent swelling and branching (Cormack, 1949). Loss of roots increases the top-to-root ratio, causing further complications from moisture and other stresses imposed by the inadequate root system.

13-6 MAGNESIUM

As the only metal .ion in the chlorophyll molecule, magnesium is of obvious significance to the structure of this pigment and the initial absorption of light energy for which it is responsible. Magnesium ions are also thought to bind together the ribosomal particles containing ribonucleic acid (RNA) and protein. Magnesium is also vital in the enzyme systems concerned with all aspects of phosphorus metabolism, in the uptake of phosphorus into organic combinations, and in one or more reactions of photosynthetic phosphorylation (Arnon et al., 1958).

Deficiencies of magnesium arise regularly in the Southeastern coastal states, where soils have been derived from parent materials containing calcium to the nearly complete exclusion of magnesium, in the cotton soils of southern Appalachia, and in the Florida citrus soils.

When plants become deficient in magnesium, synthesis of chlorophyll is impaired, and chlorosis, sometimes accompanied by brilliant tints of orange and red, develops on the older leaves. As symptoms progress to younger leaves, the more severely affected older leaves may wither and drop. The veinal area may remain green, so that the deficiency is characterized by an interveinal orange, chlorotic mottle.

Chlorosis in corn is light yellow to almost white, with veins remaining fairly green, giving the leaf a striped appearance. Chlorosis in cereals is accompanied by dwarfing, and yields may be greatly reduced. Starting at the older leaf tips, chlorosis becomes increasingly severe, and tissues break down and become necrotic, leaving dead tips and margins. Symptoms on potatoes appear first on the terminal leaflets of the oldest leaves, with minute brown flecks often developing in the most chlorotic areas. Symptoms on broad-leaved plants such as cotton are difficult to tell from aging or maturity.

In vegetable crops, magnesium influences earliness and uniformity of maturity, size of roots and fruits, and quality of the edible organs. Brown lesions, irregular in outline, are characteristically associated with the interveinal chlorosis. In tomato, the chlorotic, mottled leaves become brittle and tend to curl upward.

Interveinal islands of necrosis and chlorosis, together with marginal chlorosis, also characterize magnesium-deficient stone and pome fruit leaves. Necrotic areas in pear are nearly black. Lesions subsequently drop out, leaving holes; leaf margins become necrotic and torn in much the same way as when leaves are deficient in potassium.

Magnesium deficiency is extremely important in blueberry culture, due largely to the acid soils on which this crop is often grown. The disease is recognized by the bright red leaves of affected plants. Tissue along the midrib and larger veins remains green, giving the leaf a "Christmas tree" appearance (Smith et al., 1964).

Magnesium deficiency of tobacco has been given the common name of *sand drown* (Garner et al., 1923). The older, lowest leaves become pale green to nearly white toward the tip, while veinal tissues tend to remain normal. Plants are dwarfed, with corresponding yield reductions. More serious is the effect of the deficiency on reducing leaf quality by producing a dark ash when used as cigar tobacco.

Magnesium deficiency of citrus in Florida, where it has been significant for many decades, is known as *bronzing*. Symptoms consisting of irregular chlorotic blotches along the leaf midrib appear most frequently in late summer or fall when the crop is maturing (Camp and Peech, 1939). Intensity of chlorosis increases as the magnesium is translocated from the leaves to the fruit, and if the deficiency is severe, defoliation follows. The crop is reduced, fruits are low in soluble acid and vitamin C, color of the fruit is pale, and trees become predisposed to cold injury (Sites, 1947).

Magnesium deficiencies can be readily corrected by foliage sprays of magnesium nitrate at 10 lb per 100 gal water applied to the spring flush of growth (Embleton and Jones, 1959), but heavy applications of magnesium salts ($MgSO_4$) to the soil are more lasting.

13-7 IRON

Iron in living cells occurs chiefly in porphyrins, or hemes, which are required to catalyze a number of reactions. The peroxidases and catalases of plants are iron porphyrin–containing enzymes that catalyze reactions in which hydrogen peroxide is an electron acceptor. It is also postulated that the respiratory energy required for salt uptake and accumulation in plants involves the heme-containing enzymes (Nason and McElroy, 1963).

Iron is considered to be involved primarily in forming the chloroplast protein in leaves (Gauch, 1957). As a catalyst for the production of chlorophyll, iron is essential for the synthesis of chlorophyll pigments. Iron also affects the iron-porphyrin protein complex which acts as an oxygen carrier, transporter of electrons, and activator of oxygen. A deficiency of iron causes a decrease in the size of chloroplasts, reduced chlorophyll, and corresponding reductions in photosynthesis.

The kind of iron in soils and its availability to plants depend largely on the parent materials from which it was derived. Iron is usually available in acid soils except where phosphates are high. When the pH is below 5, complex phosphates of iron are formed which have a very low solubility. Both phosphates and iron become unavailable to the plant. The form of iron is important not only to uptake but to its utilization after it gets into the plants. Iron is most deficient in the alkaline, lime-containing soils of arid and semiarid regions where soluble, available iron is low. The exchangeable iron in calcareous soils and in others with a pH of 8 or more may be so low that plants cannot absorb enough for normal growth.

Iron-deficiency symptoms more often result from a high pH or mineral imbalance rendering the iron unavailable than from an actual lack of iron in the soil.

Deficiencies of available iron exist in most major fruit-growing areas of the country. When the available iron content in the leaf is inadequate, a disease condition develops commonly called *iron chlorosis* or *lime-induced chlorosis*. Symptoms have been reported on over 250 different species or varieties of plants in the Western and Plains states. Trees, shrubs, field crops, flowers, grasses, and many vegetable crops are affected (Locke and Eck, 1965). General symptoms consist of yellow foliage, poor vigor, and unproductiveness. But the veins, which remain bright green and stand out in vivid contrast to the pale yellow blade, are the most striking and distinctive feature. Even when the young leaves are nearly bleached, the large veins retain their green color. As chlorosis intensifies, the leaf margins may become chlorotic and shoot tips die. Often chlorosis is limited to the younger leaves of a few shoots, but as the supply of available iron in the soil is used, the disease progresses throughout the entire tree. The most severely affected leaves drop, and affected shoots die from the tip.

Fruit crops are particularly sensitive. Trees in the lower, most heavily watered parts of an orchard or in sites where drainage is poor are most seriously affected. In some heavy, alkaline soils, sensitive plants such as stone and cane fruits cannot be grown. There may be sufficient iron for a young tree to grow for a year or two, but as the supply is depleted, leaves become increasingly yellow, growth diminishes to nothing, shoots die back, and the young plants succumb.

Vegetable and field crops are only slightly more tolerant of iron shortage. Affected plants become chlorotic to ivory or white, especially at the leaf tips; veins of cereal crops remain green, giving a striped appearance, and plants become progressively weaker and die. P. A. Young (1967) described a peanut chlorosis in Texas which was caused by iron deficiency and reduced yields. Moderate light green to yellow chlorosis was associated with 20 percent yield reduction. Severe chlorosis, when the upper half of the leaf was white to light green, with some necrosis and defoliation, reduced yields about 50 percent.

Pine, spruce, fir, and juniper trees also show iron chlorosis. Symptoms consist of a general needle or leaf yellowing which could resemble decline from improper moisture relations or nitrogen deficiency, except that when iron is involved, the upper and youngest (rather than the older) leaf clusters are most severely affected.

Iron chlorosis of lawns is infrequent, but treatment may greatly improve their vigor in more alkaline soils.

Symptoms occasionally appear on fruits. Fruits on chlorotic tomato plants become silvery-green and tend to be orange rather than red when ripe (Hewitt, 1945). Apple and pear fruits have an abnormal reddish coloration due to the production of anthocyanins rather than carotenoids (Wallace, 1961). Such fruits are highly flushed but have poor, pale background color.

Several approaches have been used to overcome iron deficiency. Such cultural methods as controlling soil moisture and improving drainage are helpful when the condition is not serious. Planting tolerant species or varieties may also be practical; but in highly calcareous soils, such approaches are inadequate. Likewise, the soil application of iron in the form of iron sulfate is usually inadequate since the iron is soon rendered insoluble and of little value to the plant. The only successful approach to controlling lime-induced chlorosis is to use iron chelates, organic reagents which bind the iron ion into their structure. The iron is held in such a way that it does not combine with common precipitating agents in the soil such as phosphate or hydroxide. Yet the iron is readily soluble and available to the plant.

The chelate concentration required to correct an iron deficiency depends on the soil characteristics and sensitivity of the plants to iron deficiency. Ten pounds per acre may be adequate for vegetable crops which are relatively shallow-rooted, while 100 lb/acre may be required to benefit fruit trees. Lesser amounts of only 1 lb/100 gal of water, or 1 to 5 lb/acre, will be sufficient if the chelate is sprayed directly on the leaves. When iron is supplied to the rapidly expanding leaves in the spring, or to flushes of growth later in the season, the leaves may respond within five days. Treatment is rarely successful once the chloroplasts have developed abnormally and the leaves have expanded.

13-8 MANGANESE

Manganese, along with iron, is intimately associated with controlling oxidation-reduction reactions in plants and the synthesis of chlorophyll. Catalytic, regulatory, and enzymatic roles have also been ascribed to manganese. It may further have a specific role in bringing ribonucleoprotein molecules together in ribosomes and helping suppress their dissociation (Littleton, 1960).

Manganese is a constituent of respiratory enzymes and has been found to stimulate respiration (Ruck and Bolas, 1954). Carbon dioxide production is stimulated by manganese, but the precise mechanism is not known. It is also apparently essential for photosynthesis, although its role is not explained. The enzymatic action of manganese and its role in nitrogen and carbon assimilation are also vague, although as a constituent of enzymes responsible for protein or amino acid synthesis, it would be expected to have some effect. Manganese was also found to promote nitrate reduction in roots (Burnstrom, 1950).

When the supply of available manganese is deficient, varied symptoms will appear. In a general way, symptoms on fruit trees, cane fruits, vegetable crops, and many ornamentals consist of interveinal, mottled-appearing leaf chlorosis. Leaves become increasingly lighter green and finally yellow, and small necrotic lesions typically develop within the most chlorotic areas. Symptoms resemble those of iron chlorosis except that more of the tissue along the veins remains green, and the border between chlorotic and healthy tissue is more diffuse. When the smaller veins remain green, interveinal islands of chlorotic tissue appear, giving the leaf a mottled appearance. Growth is weak, little or no blossoming takes place, and no fruit forms.

More specific manganese deficiency symptoms on various crops are illustrated by diseases which have been known for many years and are sufficiently distinctive to have popular names.

Gray Speck of Oats

Gray speck, also called gray stripe, gray spot, dry spot, and halo blight, characterizes manganese deficiency not only on oats but on other cereals, corn, and grasses. It is serious in Australia and Europe and widespread in the United States. Symptoms consist first of one or more grayish lesions developing on the base or lower half of the leaf. The lesion increases in size, gradually turning bright yellow or orange at the edge of the leaf. Tissue within the lesion dies and dries. Soon the remainder of the leaf becomes chlorotic, ultimately brown, and dies. The disease first appears on young plants in the third or fourth leaf, and when the disease is severe, plants are killed early. Less severely affected plants may produce flowers, but little grain develops (Gallagher and Walsh, 1943; Sherman, 1957).

Pahala Blight of Sugarcane

Pahala blight is characterized by the fading of color toward the leaf tip and between the veins. The long white streaks developing become reddish and finally necrotic as the disease intensifies. As lesions increase in number and coalesce, the leaf may split along the necrotic streaks. The disease develops on plants grown in calcareous or alkaline soils or where the ratio of available iron to manganese is excessively high (Lee and McHargue, 1928).

Speckled Yellows of Sugarbeet

Yellows appears most often on plants growing in sandy, light-textured soils that have been limed. Symptoms consist first of a mottling of new leaf growth. As chlorosis intensifies, brownish lesions develop in the mottled areas. Affected tissue dies and drops out, leaving small holes. Veins and adjacent tissues remain green, and leaf margins curl upward toward the upper surface. The absence of marginal chlorosis helps distinguish this disease from potassium deficiency.

Marsh Spot of Peas

Marsh spot consists of brownish lesions or cavities on the center of the cotyledons of pea and certain varieties of bean, and of dark-colored spots on the seeds of sensitive legumes (Piper, 1941). One or more of the seeds may show symptoms while still in the pod. Leaf symptoms on pea may be lacking entirely, and the plants themselves may look quite healthy, while on the beans chlorosis may develop strongly. The first symptom on snap beans is chlorosis of the trifoliate leaves; the leaves turn a golden yellow within a few days, with brown spots developing along the larger veins. Affected leaves never attain normal size.

Frenching of Tung Trees

Frenching exemplifies another type of manganese deficiency response in which crimping of the leaf margin is the most distinctive feature. Leaves are narrower than normal and the edge prominently scalloped; chlorotic areas develop interveinally, and brown spots usually develop within these. Leaves are smaller than normal, so that photosynthetic activity and production are reduced (Reuther and Burrows, 1942).

Manganese deficiencies of all types are most commonly corrected by applying manganese salts to the soil. Fifty to a hundred pounds of manganese sulfate or chloride per acre is usually adequate, but the amount depends on the degree of soil acidity and the amount of other ions, such as iron, which may be present.

13-9 ZINC

Zinc is vital to plant growth through its association in auxin formation, also as a component of specific enzymes. Its interrelationship with auxin appears to be its most prominent role. Tsui (1948) reported that zinc is apparently required for synthesis of molecule tryptophan from which auxin is produced. Indirectly, then, zinc regulates stem elongation and cell enlargement. This relation is consistent with a correlation between accumulation of auxin tryptophan and zinc in leaf tips reported by Hoagland (1944). Plants deficient in zinc were low in auxin, and applications of zinc increased the auxin level. The dependence of auxin on zinc means that zinc is important to normal development of flowers and inflorescences.

Zinc is also a necessary component of several enzyme systems regulating various metabolic activities (Nason and McElroy, 1963) and forms part of the dehydropeptidase and glycoglycine dipeptidase enzymes functioning in protein metabolism. The activity of triosephosphate dehydrogenase is also dependent on zinc. This enzyme is concerned with the oxidation and further phosphorylation of phosphoglyceric aldehyde and with the production of diphosphoglyceric acid, an essential step in the glycolysis of carbohydrate in respiration.

Zinc deficiencies are most frequent in the calcareous soils in regions of limited rainfall where the soil reaction and other soil factors serve to render the zinc unavailable. Low zinc availability has also been associated with excess soil phosphate in major fruit regions in the West, where the phosphate radical is thought to form an insoluble, relatively unavailable zinc phosphate complex. More agricultural soils in the West are high in native phosphate, and deficiency symptoms are frequent on crops grown in such soils.

Zinc deficiency can also be associated with highly organic soils or soils high in nitrogen in which it becomes tied up in organic compounds (Camp, 1945). Deficiency symptoms are common around barnyards or corrals where manure has been stored.

As with so many deficiencies, the first sign of zinc deficiency is interveinal chlorosis. The first leaves to emerge in the spring remain small, often no more than one-twentieth of their normal size. The shoots fail to elongate, and the internodes are so short, sometimes less than ¼ in. in length, that the leaves appear to be in whorls or rosettes. These symptoms were described long before the cause was known, and early workers gave the disease such names as *rosette, mottle leaf, little leaf, yellows, frenching, sickle leaf, bronzing,* or *white bud,* depending on the response of a particular plant species.

Fruit trees are highly sensitive and frequently affected by zinc deficiency. The characteristic dark green veins and bright yellow chlorosis is particularly striking in citrus and in many stone fruit species. Chandler (1937) ranks fruit trees in order of decreasing sensitivity as sweet cherry, apple, plum, peach, walnut, apricot, avocado, citrus, and grape. Field and vegetable crops are more tolerant and not as widely affected.

While symptoms on trees appear first on the young terminal flushes of growth in the spring, symptoms on such crops as tomato, tobacco, peas, and beans appear mainly on older leaves. In addition to the prominent chlorosis, which often reminds one of iron deficiency, shoot elongation is inhibited, and leaf margins may be severely distorted, twisted, wavy, corregated, or even curled.

Chlorosis may be followed by development of irregular, necrotic areas in interveinal or vascular tissues which provide an excellent court of entry for biotic pathogens.

Seed production is impaired in beans and peas (Reed, 1942, 1944). Riceman and Jones (1958) found that seed and flower production was increased about 100 times when the zinc supply was raised to normal levels. At low levels only a few inflorescences were formed, and these were mostly aborted; all but 2 percent remained seedless.

Anatomical and histological studies of zinc deficiency have revealed further abnormalities (Reed, 1933; Carlton, 1954). Palisade and spongy mesophyl tissues were abnormally compact, and intercellular spaces were practically absent. The number of palisade cells was decreased, and they were three to four times longer and nearly twice as wide as normal. Cells often failed to differentiate or became atrophied. Mature root cells of tomato were isodiametric instead of columnar. Intracellular changes were even more striking in that plastids were smaller and fewer in number than normal. Plastids in green areas of mottled avocado leaves remained normal, but in the chlorotic areas they became agglutinated and segregated into groups at the inner ends of the cells. Chloroplasts disintegrated or dissolved in a manner resembling lysis. Nuclei were more resistant to breakdown but often migrated to one end of the cell, and nucleoli tended to be smaller than normal. Phenolic compounds, oil droplets, and tannins accumulated in many woody species.

The different symptoms which appear on various plant species can be exemplified by the named disorders given by early workers:

Little Leaf or Rosette of Deciduous Fruit Trees

Little leaf is widespread throughout the Western United States and Canada on apple, pear, cherry, peach, almond, apricot, and prune trees. The disease is most clearly characterized by the terminal rosettes of small, chlorotic leaves emerging in the spring. Shoots bearing normal leaves may develop below the rosettes later in the season, but the new leaves are progressively smaller and mottled and often abnormal in shape. If the affected trees are not treated, the new leaves become progressively smaller and more chlorotic each year; within a few years buds fail to emerge and branches die back (Wann and Thorn, 1950; Hoagland et al., 1935).

In pear, buds are especially late to open, and instead of a little-leaf condition leaves are elongated and more uniformly yellow. Affected branches are distributed irregularly over the tree, and leaves on one limb may show severe symptoms while leaves on adjacent branches remain normal.

Fruit symptoms may appear even when leaf symptoms are lacking. Zinc-deficient apple fruits are smaller than normal, and (more significantly) they are misshapen enough to be unmarketable.

Pecan and Walnut Rosette

Abnormal, yellow leaf mottling characterizing *rosette* becomes apparent as the leaves begin to unfold at the tips of a branch. Leaves at the top of the trees are the first affected, remaining small, usually crinkled, misshapen, and brittle. Chlorotic areas are thin and gradually become reddish-brown and necrotic (Finch and Kinnison, 1933).

The lack of internode development is most striking. Terminal shoot growth is severely stunted, so that the leaves emerge close together and appear rosetted. Below this buds often fail to develop, so that the shoots are bare. Dieback follows as in little leaf, but the trees are rarely killed. Fruit production is reduced long before symptoms become this severe. Alben (1932) reported that before the cause was recognized, hundreds of acres of pecans were abandoned in the southeastern states because of this disease.

Rosette of Pinus radiata

Rosette has also been described in Australia, where plantations of Monterey pine have been established on poor soils (Kessell and Stoate; 1936, 1938). The first symptom of the disease was a decrease in the rate of growth. Trees gradually developed a flat-topped appearance because of the lower activity of the apical meristem. Apical buds were bunched together, secondary needles stopped growing, so that they appeared short, thick, and stiff, and bundles failed to spread open. Gradually small yellow dots appeared near the tip of needles, followed by browning, which gave the tips of the branches a bronzed appearance.

Mottle Leaf or Frenching of Citrus

The disorder has been called *mottle leaf* in California and *frenching* in Florida, but both are caused by zinc deficiency (Johnson, 1933). Symptoms do not differ greatly from those described for stone fruits. Leaves are narrow and deformed, with interveinal chlorosis giving affected leaves a mottled appearance. As the disease intensifies, chlorosis overlaps more and more of the smaller veins, so that ultimately most of the veins are also yellow. Leaves become nonfunctional, buds fail to emerge, and branches die. Shortened growth gives the tree a bushy appearance. In severe cases, the entire tree may be killed either by the disease or secondary pathogens. Fruits are small and thick-skinned.

Sickle Leaf of Theobroma cacao

Leaves affected with *sickle leaf* or *narrow dented leaf* of cacao are dull and leathery and soon express a bluish-green vein banding

(Greenwood and Haybron, 1951). Leaves developing subsequently are slightly corrugated. After several weeks, small chlorotic to reddish spots appear on the crests of these convex interveinal areas. Leaves of the next growth flush are deformed and smaller. They may exhibit a lateral curvature, giving the leaf a sickle shape, or leaf tips may be severely constricted, forming a long point which is often curled and crinkled. When the leaf has developed, veins are dark green and interveinal areas chlorotic, often with translucent streaks near the midrib. Chlorosis becomes more pronounced and leaf size smaller on progressive growth flushes. Necrosis intensifies, leaf tips die, and defoliation becomes more prevalent. Shoots ultimately wither and die.

In Ghana, where the disease was first described, affected plantations were found chiefly on soils containing rotting husks of cacao pods. The high potassium and phosphate content of these alkaline soils seriously limit the zinc availability.

Bronzing of Tung Trees

Bronzing was first described on trees in Florida growing on soils high in phosphate but is not limited to such soils (Newell et al., 1930). Terminal leaves are deformed and become bronzed as they enlarge. As the bronze color darkens, necrotic lesions often develop which later drop out, giving the leaf a ragged appearance. When necrosis becomes severe, the whole leaf drops. New leaves are successively smaller and more deformed. Internodes fail to develop, producing a rosette symptom. Symptoms appear first on scattered branches but soon affect the whole tree. Within three years after bronzing begins, affected trees may be conspicuously smaller than normal, branches almost bare of foliage, and the tree appears half dead.

White Bud of Maize

Like many zinc deficiency diseases, *white bud* of corn is most prevalent on overcropped soils or poor soils newly brought under cultivation (Berger, 1962). The first signs of the disease are light yellow streaks appearing between the veins of older, lower leaves, followed by the rapid development of necrotic, white spots. Tissues by the midrib and along the margins remain green. In severely deficient areas, the symptoms may appear within two to three weeks of seedling emergence. The new leaves emerge pale yellow to white, giving the disease its name (Barnette and Warner, 1935). Actually, however, the name is not appropriate, since the buds and leaf tips usually remain green, only the lower portions being bleached. Shoot elongation is suppressed, causing severe stunting over the more chlorotic areas. The disease is most severe during cool, wet weather and may be evident in fall or winter crops while absent during the summer.

Symptoms on related crops such as sorghum, sugarcane, and grains are somewhat similar except that bleaching is not as definitive and can more readily be confused with other deficiencies.

Other crop plants frequently stunted and reduced in yield by zinc deficiency include bean, cotton, onion, tomato, hops, potato, and flax. Symptoms are essentially the same as previously described, with poor leaf size and interveinal chlorosis the most distinctive features. Yellowing tends to be more intense on the lowest leaves of these crops and is followed by dull bronzing, necrosis, and premature death. A "fern leaf" symptom occurs on potato in which leaflets are chlorotic, less than one-fifth of normal size, and crinkled to a degree where the symptom resembles 2,4-D injury.

Zinc deficiency can be corrected by applying zinc as zinc sulfate or zinc chelate to the plants or soil. Deficiencies in tree crops can be remedied by spraying 3 to 5 lb of zinc sulfate per 100 gal of water per acre. Since the zinc is used up during the season and lost when the leaves drop, sprays should be applied every year. However, continued use may cause an accumulation of zinc in excessive, toxic quantities.

Zinc sulfate sprays have also been effective in controlling deficiencies in field and vegetable crops. Applications of zinc sulfate to the soil at the time of seeding are still more effective, since zinc is then available as soon as the plants begin to grow.

13-10 BORON

Plants require so little boron that it is remarkable that deficiencies of it are so common and widespread. One-quarter pound of common borax per acre is all that is required to produce a normal crop. Less than 0.06 lb is removed by an acre of alfalfa. A ton of alfalfa hay contains but 1 oz of boron, and a ton of sugarbeets but 2.5 oz. Yet deficiency symptoms have been reported in forty-one states on over ninety crops.

Boron serves many roles. Gauch and Duggar (1954) reviewed some fifteen possible ways in which boron may affect plant growth. In summary, they point out that boron affects flowering and fruiting processes, pollen germination, cell division, metabolism, photosynthesis, active salt absorption, hormone movement and action, the metabolism of pectic substances, and the water relations in plants. Boron is also reported to be a constituent of membranes, to serve in precipitating excess cations, to act as a buffer, to be necessary in the maintenance of conducting tissues, and to exert a regulatory effect on other elements. Still, the role of boron in the living cell has not been clarified.

One hypothesis concerning its role holds that boron may be essential for the translocation of sugar in the plant (Gauch and Duggar, 1954). Movement of highly polar compounds such as sugar through cell walls and membranes requires energy. Boron may reduce the energy required for translocation by reducing the polarity of sugar. Boron has the outstanding property of complexing with various compounds, including several common sugars, and is thought to be a loosely bound constituent of the membrane which forms temporary union with the sugar while it passes through the membrane. This hypothesis is supported by the comparatively high concentrations of sugar and starch found in leaves of boron-deficient plants. The role of boron in facilitating sugar translocation would reconcile and explain the many varied roles proposed for it. The effect of boron on meristem activity, salt absorption, hormone translocation, and protein synthesis may all reflect direct or indirect effects of impaired carbohydrate translocation.

Boron has been implicated in regulating cell wall formation, and there seems to be a direct relationship between lignification of the cell wall and boron nutrition. A deficiency of boron often results in a collapse of the meristematic cells, but this may be a manifestation of sugar deficiency.

Symptoms of boron deficiency are difficult to generalize since individual plant species respond in many different ways. Some of the important diseases caused by boron deficiency include: *Heart rot of sugarbeet, canker of table beet, brown heart of swedes and turnips, hollow stem of cauliflower, snakehead of walnut, crinkle leaf in cherry, cracked stem of celery, bark measles of apple, yellows and rosetting of alfalfa, corky core and dieback of apples, hard fruit of citrus, top sickness of tobacco,* and *fruit pitting and dieback of olive.*

The most consistently characteristic and prominent symptoms of boron deficiency are injury to the meristem and actively dividing tissues and death of the apical meristem. Death of the growing point results in the development of numerous axillary buds which produces a bushy growth habit. Often, though, the buds are killed before developing, so that no new growth is regenerated.

Less striking symptoms usually appear before the shoot apex is killed. Interveinal tissues and smaller veins of the younger leaves may be crinkled, and the lower part of leaf margins of the youngest leaves often becomes chlorotic and then necrotic. The greater chlorosis on the lower half of the leaf is important to diagnosis. The leaf tip tends to remain green since boron is primarily translocated to it from the basal area.

Semimature leaves often develop reddish or purple tints and tend to be thicker than normal, turgid, brittle, and rolled. Leaf shape is distorted, and leaves may blacken or shrivel before they expand more than a few millimeters. Petioles are often brittle and crack or split, producing corky ridges. Stems are similarly affected, with corky lesions appearing on kale and cauliflower stems and potato tubers. Corky splits arise in the epidermis and cortex; internally, the pith parenchyma is killed, producing irregular brown areas and large stem cavities. Apple and olive stems may develop excrescences from the lenticels.

Roots become severely necrotic, with the growing point becoming enlarged, blackened, and killed. Numerous lateral, secondary roots may then develop, forming a rosette. Storage roots break down internally.

Flowering is often totally suppressed, but even when flowers are formed, they may drop without producing seed. The abnormal fruits formed may be obvious in the absence of other symptoms. Apples are especially sensitive and become malformed, with numerous irregular sunken areas on the surface and clusters of cork cells throughout the flesh. Citrus fruits also become hard and misshapen. Olives are malformed and pitted.

At the cellular level, the lack of boron stimulates excessive cell division in the cambium areas, but the new cells are much enlarged, and the cell walls remain thin. Phloem and xylem differentiation is inhibited; cells initiated in the vascular cambia are largely parenchymatous, forming clusters of "callus"-type cells which disrupt the normal flow of food and water. The new cells are weak, the walls soon rupture due to pressure from surrounding rapidly dividing cells, and the protoplast dies, leaving streaks of necrotic, cracked tissue which gradually coalesce. Cells are abnormal in size and shape. Phloem cells fill with dense materials, and brownish deposits appear between the xylem cells. This is the beginning of lesions which soon become visible. Breakdown of areas along the vascular cambium occurs in root storage organs, stems, and petioles, producing the visible necrotic streak or heart rot symptoms of boron deficiency (Walker, 1944; Jolivette and Walker, 1943).

Cell division at the apical meristem of the shoot of broad bean is reported to be suppressed rather than stimulated (Whitington, 1957). Cells are irregular in shape and abnormally enlarged; nuclei sometimes fused into giant nuclei, and cells finally disintegrated and died.

Sensitivity of plants to boron varies greatly. A concentration of 10 ppm in a culture solution which is optimal for sugarbeet is toxic to plants such as persimmon and kidney bean, which have a lower tolerance and requirement (Eaton, 1944). Deficiency symptoms are most common in such plants as sugarbeet, table beet, and radish, which have a high boron requirement. Where boron is deficient, yields can generally be increased by boron supplements, even when there is no visible indication of a disease situation. The more common diseases caused by boron deficiency are discussed below.

Heart Rot of Sugarbeet and Mangold

Heart rot, also known as *crown rot* or *dry rot,* is widely distributed in the United States and Europe, especially on lime soils, where losses of up to 30 percent of the crop have been reported (Walker, 1943). Symptoms first appear on the youngest leaves of the crown. These heart leaves spread outward and show dark brown, scurfy patches on the inner, concave surfaces of the petioles. Veins turn from white to yellow from the base upward, and the heart leaves become curled, wilt, and turn yellow. New shoots develop in the axils of the dead leaves, but these soon die. When the deficiency is more acute, the shoot apex and flower buds are killed. After the apical leaves are damaged, necrotic lesions appear on the crown. The necrotic areas gradually extend into the root, increasing in size until ultimately a large part of the root tissues are destroyed, producing the "heart rot" symptoms.

Internal Black Spot of Table Beet and Chard

Internal black spot, or *canker,* as it is known in England, is very similar to heart rot but may be even more important since affected roots are often unmarketable (Walker, 1939, 1944). The first symptoms appear after midseason; the youngest leaves in the center of the crown become misshapen, often unilaterally developed, and elongated. Leaves are dwarfed and more intensely red than normal, except for the base, which tends to become chlorotic. The young leaves tend to die early, leaving a rosette of dead, stunted leaves. Older leaves usually appear normal. The most prominent symptoms are the internal, hard, black, necrotic masses of tissue which render the beet unfit for canning. The necrotic areas are most obvious between the vascular rings formed by the secondary cambium zones in the pericycle. This dead tissue appears anywhere in the root; when it is completely internal, no outside symptoms are evident. When lesions arise in the peripheral region, an excellent court of entry is provided for soil microorganisms.

Brown Heart, or Raan, of Swede, Turnip, or Rutabaga

During the growing season, the major symptom of *brown heart* is a mottled brownish discoloration of cambium tissue in proximity to the xylem in the inner, central part of the root. In severe cases, the central tissue may break down, producing a "hollow" heart. Affected turnips are unfit for human consumption, and even when symptoms are mild, the sugar content is low and the roots taste bitter.

Cracked Stem of Celery

Cracked stem is widely distributed in the United States and Canada. It caused serious losses to celery grown in alkaline, calcareous, and organic soils (Kendrick et al., 1954) until the cause was recognized and boron was routinely applied to deficient soils. Early symptoms begin in midseason as light tan, sunken, greasy-looking lesions appear on the inner surface of the petioles. As the tissue dies and dries, the spots turn dark brown, and horizontal cracks develop in the area over the associated vascular bundles. Lesions occasionally also develop on the outer ridges of the stalk. Brown to blackish spots and longitudinal streaks develop, rendering the stalks unmarketable.

Browning of Cauliflower

The first sign of *browning* is the appearance of pale brownish, water-soaked areas on the developing head. Affected patches become darker brown and hard. The lesions progress into the curd or head, causing a general discoloration. Pith tissues may break down, producing a hollow stem. Leaves around the curd are thicker than normal, brittle, often curled downward, and chlorotic, particularly at the tips (Dearborn, 1942).

Alfalfa Yellows and Rosetting

Boron deficiency in alfalfa and some clovers appears as a uniform yellowing of the terminal leaves, or bronzing or reddening of interveinal areas; poor development of internodes; and death of the growing points. Even in the absence of visible symptoms, growth and flowering may be impaired. As the shoot apex is killed, new, weak shoots arise from the lateral buds, producing a rosetted or witches' broom appearance (McLarty et al., 1937; Chandler, 1941).

Internal Cork, Rosette, and Dieback of Apple and Pear

This disease, also called *corky pit, corky core, spot necrosis, drought spot,* and *crinkle,* has been reported primarily from New York and British Columbia but is also important in the calcareous soils of the Northwest.

Leaves are affected only when the boron deficiency is acute; more often symptoms are restricted to the fruit (Burrell, 1940). Light brown areas of dead cells may appear at any point in the flesh of the fruit. Affected areas dry out and become corky and either hard or spongy, depending on the maturity of the fruit when symptoms first arise. A drought-spot expression, consisting of fairly large, superficial necrotic areas which ultimately become russeted and crack, may or may not accompany corking. Most of the affected fruits drop, but those remaining are bumpy and distorted, especially at the calyx end. When the deficiency is more acute, terminal twig dieback and subsequent rosetting may develop. This is accompanied by development of leaf yellowing, reddening, scorching, curling, and dropping. A few brittle, thick leaves may remain at the tips of the twigs, forming a rosette.

Brown Spotting and Dieback Disorders

Brown spotting of apricot fruits was described by Askew and Williams (1939) in one of the only reports of the disease. Masses of brown spots developed in the flesh, especially near the stem end, and the texture around the pit was dry and spongy. Terminal buds may abort or fail to open, and twig tips may die. Terminal leaves are dwarfed, often curled upward, and sometimes chlorotic. Affected trees have a general dieback appearance, with a dense, stunted, bushy habit. Trees may be unilaterally affected, so that symptoms appear only on part of the tree.

A similar leaf curling and *crinkle-leaf* expression has been observed on sweet cherry. Leaves have a crinkled, wavy margin and are strap-shaped. Symptoms resemble the viruslike crinkle disease of bing cherry but can be controlled by applications of borax.

Symptoms of the same nature has been reported on several other crops. One, *dieback* of raspberries, described by Askew et al. (1951), is characterized by retardation of growth or failure of the buds to open and failure of fruit formation. Leaves on older canes are long, thin, and deeply incised and "fernlike."

Ivy leaf of hops was described by Cripps (1956). Internodes were short, growing points killed, and lateral shoots poorly developed, giving the plant a bushy appearance. Most definitively, leaves were small, distorted, and deeply toothed and had chlorotic margins.

Snake head of walnut is characterized by weak, deformed twig and leaf growth and short internodes. Shoots possessing leaves only at the tips develop in the tops of trees. These sharply curved shoots may die during the winter.

All of these diseases can be readily controlled by applying adequate amounts of available boron to the plants or soils. Borax (sodium tetraborate) containing 10 to 13 percent boron is most commonly used. The amount to apply depends on a number of factors such as the needs of the crops, the soil, the season, and the method of application. The optimal amount of borax may vary from as little as 1 lb to correct cracked stem of celery in certain soils to over 50 lb for alfalfa. Since borax is an effective herbicide, and excesses are harmful, care must be taken to avoid using more than the crop needs.

Because boron availability is intimately interrelated with other elements, boron content alone may not be the determining factor in the appearance of deficiency symptoms. Calcium and potassium levels are particularly significant, and normal growth can be expected only when a balance of calcium and potassium to boron exists. Excessive amounts of either will cause boron deficiency symptoms even though ample amounts of boron are present. A calcium/boron ratio in the plant in the range of 80:1 to 600:1 results in normal growth. Above this, boron deficiency symptoms appear. High potassium levels, particularly in the external substrate, are also conducive to producing boron deficiency symptoms.

13-11 COPPER

Copper is essential as a component of a number of different plant enzymes, including polyphenol oxidase, monophenyl oxidase, lactase, ascorbic acid oxidase, and cytochrome oxidase (Nason and McElroy, 1963). One of the vital roles of copper salts is in catalyzing the oxidation of various organic substrates to form water as the end product. In this process copper is essential to the enzymatic transfer of electrons from the substrate to oxygen.

Indirect evidence has been obtained by Neish (1939) that copper also functions in photosynthesis. He found that 75 percent of the total copper in clover leaves was localized in the chloroplasts. Arnon (1950) speculated that if a quinone-like substance were the primary hydrogen acceptor in photosynthesis, a copper enzyme may be responsible for its regeneration. Evidence for this was that copper enzyme–inhibiting chelating agents inhibited part of the photosynthesis reactions. When more copper was added, the inhibition was reversed.

Copper may also be involved in chlorophyll formation. Elvehjen (1931) postulated that copper is required for the synthesis of the iron porphyrin precursor of chlorophyll.

Copper deficiency occurs in crops growing in the sandy soils of the Southeast, on newly reclaimed peats in Europe, and in some sandy and gravelly soils in Australia and South Africa.

The symptoms of copper deficiency vary considerably among different species but generally consist of lack of vigor, smaller-than-normal leaves, and a bluish-green cast to the leaves. Microscopically, palisade cells contain many large, intensely deep green plastids. These ultimately degenerate, becoming aggregated at one end of the cell. The upper ends of the palisade cells separate, forming cavities between them. Soon the protoplast lyses, and cells collapse with the appearance of necrotic areas on the leaf (Reed, 1939).

Specific symptoms are illustrated by the named diseases caused by copper deficiency:

Reclamation Disease

Copper deficiency affects farm crops in Holland and Denmark, where the disease is known as *reclamation disease,* and in Australia, where it is called *yellow tip* and *wither tip.* Cereals, beets, and legumes are affected. Leaf tips become chlorotic, especially along the margins

(Stiles, 1961). Gradually the pale green leaf tips roll, wither, and die. Heads are dwarfed and distorted, and the chlorotic tips remain undeveloped. Grain formation is severely inhibited. The bases of plants remain green and bushy in appearance.

Exanthema or dieback of citrus and other fruit trees is also caused by copper deficiency (Pittman, 1936). The disease has variously been called red rust, multiple bud, and ammoniation. The latter name was used because symptoms were most severe when ammonia fertilizers had been used. In orange trees, the more vigorous water shoots tend to bear abnormally large leaves, while the shoots form an S-shaped curve rather than growing straight. Small, blister-like, gummy swellings appear on the young shoots and develop into longitudinal ruptures bordered by reddish-brown ridges from which yellowish-red gum exudes in wet weather. The gum may glaze the outside of the twig with a reddish-brown exudate. Affected shoots lose their leaves and die back. The lateral shoots which develop from the base produce a witches'-broom symptom. Leaves become yellow-veined and drop, leaving a bare yellowish or brown-stained twig which soon dies back. Fruit is small and frequently marked with irregular brownish or reddish blotches which may dry out and split open.

In apple trees the same condition is known as *wither tip* or *summer dieback* (Dunne, 1938). This symptom, plus bark cracking and gummy exudation in more severe cases, is characteristic on plum, apricot, apple, peach, pear, and olive.

Copper deficiency has also been reported on tung and walnut. The most characteristic symptom in tung is the cupping of the terminal leaves produced by the upward curling of their margins. This is followed by interveinal chlorosis and necrotic spotting in the chlorotic areas. This cupping symptom is especially noteworthy in its resemblance to fluoride toxicity in cherry trees.

13-12 MOLYBDENUM

Molybdenum is intimately associated with nitrogen metabolism in higher plants, being required for nitrate reduction and assimilation. Molybdenum has many varied roles in plant metabolism. The chief role is possibly in bacterial nitrogen fixation by *Azobacter* species in leguminous plants.

The amount of molybdenum required is infinitesimal, yet definite diseases are caused by its absence. Deficiency symptoms first involve the older leaves and progress toward the apex. Symptoms in many plants begin as bright yellow-green or pale orange interveinal mottling distributed rather uniformly over the leaf. Mottling differs from that caused by nitrogen or magnesium deficiency in that it lacks the bright red or purple tints. Marginal wilting, inrolling, and cupping often follow the chlorosis. Flower formation is suppressed, together with seed development. Often the few seeds produced shrivel before they mature (Hewitt and Jones, 1947).

More detailed symptoms are illustrated by the following specific diseases caused by molybdenum deficiency:

Whiptail of Cauliflower and other Brassicas

The *whiptail* symptom is among the more striking expressions of molybdenum deficiency (Stiles, 1961). It begins with the inrolling, cupping, and interveinal chlorosis of the younger leaves. As the leaves expand, the midrib twists abnormally and leaf tissue matures irregularly, producing an undulated to incised margin. The epidermis collapses, and then the mesophyll. Chloroplasts shrivel and disintegrate. Pale, chlorotic patches develop between the veins of brussels sprouts, giving leaves a mottled appearance.

Strap Leaf of Hibiscus

Strap leaf also involves the collapse of the massive central portion of flowers and subsequently the fruit, so that it is unfit to use for food or making jelly. Seed production may also be abnormal in cauliflower, with seeds failing to develop after fertilization. Seed coats remain green, and seeds shrivel before they mature.

Yellow Spot or Orange Spot of Citrus

Yellow spot was described on the acid soils of Florida in 1908 (Floyd), but the cause was not established until 1952 (Stewart and Leonard). Symptoms develop first on the new growth after leaves of the early summer flush mature; oval water-soaked–appearing lesions develop which gradually turn yellow and finally necrotic. The chlorotic areas are irregular, often in rows between the main veins, and develop mostly along the margin. During fall, the affected areas become impregnated with a resinous gum which exudes through the lower surface of the leaf, forming reddish-brown deposits up to ¼ in. across in mandarins and to ½ in. on grapefruit. When the deficiency is more severe, spots increase in number and leaves drop early. Spotting and defoliation progress during the growing season; by fall, the trees are thin and unthrifty in appearance.

Bean Scald and Yellows of Legumes

Molybdenum deficiency in legumes is closely tied to the suppression of nitrification, which causes symptoms of a nitrogen-deficiency type. Symptoms on bean appear first as a chlorotic, interveinal mottling followed by necrosis of interveinal and marginal tissues (Wilson, 1949). Alfalfa, clover, and peas become stunted and pale green. Chlorosis is most striking interveinally but soon spreads over the leaf, which then dies and falls off prematurely. Molybdenum deficiencies can usually be controlled by applying 1 oz of sodium or ammonium molybdate to 100 gal of water per acre. Liming also provides a cure and has been especially effective in poorly drained, acid soils, where the disease is more severe.

BIBLIOGRAPHY

(General)

Greenwood, E. A. N., and E. G. Hallsworth, 1960. Studies on the nutrition of forage legumes. II. Some interactions of calcium, phosphorus, copper and molybdenum on the growth and chemical composition of *Trifolium subterraneum* L. *Plant and Soil* **12**:97–127.

Sprague, Howard B. (ed.), 1964. Hunger signs in crops: A symposium, 3d ed. McKay, New York, 461 pp.

Walsh, T., 1945. The effect on plant growth of substituting strontium for calcium in acid soils. *Proc. Roy. Irish Acad.* **B50**:287–294.

(Potassium)

Krantz, B. A., and S. W. Melsted, 1964. Nutrient deficiencies in corn, sorghums and small grains, in Howard B. Sprague (ed.), "Hunger signs in crops," 3d ed., pp. 25–58. McKay, New York.

Nightingale, G. T., L. G. Shermerhorn, and W. R. Robbins, 1930. Some effects of potassium deficiency on the histological structure and nitrogenous and carbohydrate constituents of plants. *N. J. Agr. Exp. Sta. Bull.* 499.

Oaks, J. Y., 1938. The effect of potash fertilizer on cotton in Louisiana. *La. Agr. Exp. Sta. Bull.* 291, 11 pp.

Webster, G. C., 1953. Enzymatic synthesis of gamma-glutamyl-cystine in higher plants. *Plant Physiol.* 28:728–730.

(Sulfur)

Jordan, V., and H. M. Reisenauer, 1957. Sulfur and soil fertility. U.S.D.A. Yearbook of Agr. Soils, pp. 107–111.

Kramer, A., and A. L. Shrader, 1942. Effect of nutrients, media, and growth substances on the growth of the Cabot variety of *Vaccinium corymbosium. J. Agr. Res.* 65:313–328.

Lyon, C., and C. R. Garcia, 1944. Anatomical responses of tomato stems to variations in the macronutrient anion supply. *Bot. Gaz.* 105:394–405.

Storey, H. H., and R. Leach, 1933. A sulphur deficiency disease of the tea bush. *Ann. Appl. Biol.* 20:23–56.

(Calcium)

Cormack, R. G. H., 1949. The development of root hairs in Angiosperms. *Bot. Rev.* 15:583–612.

Delong, W. A., 1937. Calcium and boron related to blotch cork. *Plant Physiol.* 2:552–556.

Geraldson, C. M., 1954. The control of blackheart of celery. *Proc. Amer. Soc. Hort. Sci.* 63:353–358.

———, 1956. Calcium spray control blackheart of celery. *Market Grow. J.* 85:8–9.

Raleigh, S. M., and J. A. Chucka, 1944. Effect of nutrient ratio and concentrations on growth and composition of tomato plants and on the occurrence of blossom-end rot of the fruit. *Plant Physiol.* 19:671–678.

Sorkin, H., and A. L. Sommer, 1940. Effect of calcium deficiency upon the roots of *Pisum sativum. Amer. J. Bot.* 27:308–318.

(Magnesium)

Arnon, D. I., F. R. Whatley, and M. B. Allen, 1958. Assimilatory power in photosynthesis. Photosynthetic phosphorylation by isolated chloroplasts is coupled with TPN reduction. *Science* 127:1026–1035.

Camp, A. F., and M. Peech, 1939. Magnesium deficiency in citrus in Florida. *Proc. Amer. Soc. Hort. Sci.* 36:81–85.

Embleton, T. W., and W. W. Jones, 1959. Correction of magnesium deficiency of orange trees in California. *Proc. Amer. Soc. Hort. Sci.* 74:280–288.

Garner, W. W., J. E. McMurtrey, Jr., C. W. Bacon, and E. G. Moss, 1923. Sand drown, a chlorosis of tobacco due to magnesium deficiency, and the relation of sulphates and chlorides of potassium to the disease. *J. Agr. Res.* **23**:27–40.

Sites, J. W., 1947. Internal fruit quality as related to production practices. *Fla. State Hort. Sci. Proc.* **60**:55–62.

Smith, C. T., N. Shaulis, and J. A. Cook, 1964. Nutrient deficiencies in small fruits and grapes, in Howard B. Sprague (ed.), "Hunger signs in crops," 3d ed. McKay, New York, 461 pp.

(Iron)

Gauch, H. G., 1957. Mineral nutrition of plants. *Ann. Rev. Plant Physiol.* **8**:31–64.

Hewitt, E. J., 1945. Experiments in mineral nutrition. *Prog. Rep. 2, Long Ashton Res. Sta. Ann. Rep.* (1944), 50–60.

Locke, L. F., and H. V. Eck, 1965. Iron deficiency in plants: How to control it in yards and gardens. *U.S.D.A. Home and Garden Bull.*, pp. 102–107.

Nason, A., and W. D. McElroy, 1963. Modes of action of the essential mineral elements, in "Plant physiology," vol. 3, Inorganic nutrition in plants, pp. 451–456. Academic, New York.

Wallace, T., 1961. The diagnosis of mineral deficiencies in plants (A colour atlas and guide), 3d ed. H.M. Stationery Office, London.

Young, P. A., 1967. Peanut chlorosis due to iron chlorosis. *Plant Dis. Rep.* **51**:464–467.

(Manganese)

Burstrom, H., 1950. The action of manganese on roots, in "Trace elements in plant physiology." Chronica Botanica, Waltham, Mass., 144 pp.

Gallagher, P. H., and T. Walsh, 1943. The susceptibility of cereal varieties to manganese deficiency. *J. Agr. Sci.* **33**:197–195.

Lee, H. A., and J. S. McHargue, 1928. The effect of manganese deficiency on the sugar cane plant and its relationship to pahala blight of sugar cane. *Phytopathol.* **18**:775–786.

Littleton, J. W., 1960: Stabilization by manganous ions of ribosomes from embryonic plant tissue. *Nature* **187**:1026–1027.

Piper, C. S., 1941. Marsh spot of peas: a manganese deficiency disease. *J. Agri. Sci.* (Eng.) **31**:448–453.

Reuther, W., and F. W. Burrows, 1942. The effect of manganese sulfate on the photosynthetic activity of frenched tung foliage. *Proc. Amer. Soc. Hort. Sci.* **40**:73–76.

Ruck, H. C., and B. D. Bolas, 1954. The effect of manganese on the assimilation and respiration rate of isolated rooted leaves. *Ann. Bot.* (*London*) (*N.S.*) **18**:267–297.

Sherman, D. G., 1947. Manganese deficiency of oats on alkaline organic soils. *J. Amer. Soc. Agron.* **33**:1080–1092.

Stiles, W., 1961. "Trace elements in plants," 3d ed. Cambridge University Press, London, 249 pp.

(Zinc)

Alben, A. O., J. R. Cole, and R. D. Lewis, 1932. Chemical treatment of pecan rosette. *Phytopathol.* **22**:595–601.

Barnette, R. M., and J. D. Warner, 1935. A response of chlorotic corn plants to the application of zinc sulfate to the soil. *Soil Sci.* **39**:145–156.

Berger, K. C., 1962. Micronutrient deficiencies in the United States. *Agr. Food Chem.* **10**:178–181.

Camp, A. F., 1945. Zinc as a nutrient for plant growth. *Soil Sci.* **60**:157–164.

Carlton, W. M., 1954. Some effects of zinc deficiency on the anatomy of the tomato. *Bot. Gaz.* **116**:52–64.

Chandler, W. N., 1937. Zinc as a nutrient for plants. *Bot. Gaz.* **98**:625–646.

Finch, A. H., and A. F. Kinnison, 1933. Pecan rosette: soil, chemical, and physiological studies. *Tech. Bull. Ariz. Agr. Exp. Sta.* no. 47, pp. 407–442.

Greenwood, M., and R. J. Haybron, 1951. Iron and zinc deficiencies in cacao in the Gold Coast. *Emp. J. Exp. Agr.* **19**:73–86.

Hoagland, D. R., W. H. Chandler, and P. L. Hibbard, 1935. Little leaf or rosette of fruit trees. V. Effect of zinc on the growth of plants of various types in controlled soil and water culture experiments. *Proc. Amer. Soc. Hort. Sci.* **33**:131–141.

———, 1944. Inorganic plant nutrition. Prather lectures. Chronica Botanica, Waltham, Mass.

Johnston, J. C., 1933. Zinc sulfate promising new treatment for mottle leaf. *Calif. Citrogr.* **18**:107, 116–118.

Kessell, S. L., and T. N. Stoate, 1936. Plant nutrients and pine growth. *Aust. Forest.* **1**:4–13.

——— and ———, 1938. Pine nutrition. *Bull. W. Aust. Forest. Dept.* no. 50.

Nason, A., and W. D. McElroy, 1963. Modes of action of essential mineral elements, in "Plant physiology," vol. 3, Inorganic nutrition of plants. Academic, New York.

Newell, W., H. Mowry, and R. M. Barnette, 1930. The tung-oil tree. *Bull. Fla. Agr. Exp. Sta.* no. 221.

Reed, H. S., 1938. Cytology of leaves affected with "little-leaf." *Amer. J. Bot.* **25**:174–186.

———, 1942. The relation of zinc to seed production. *J. Agr. Res.* **64**:635–644.

———, 1944. The growth of ovules of *Pisum* in relation to zinc. *Ann. Bot.* **31**:193–199.

Riceman, D. S., and G. B. Jones, 1958. Distribution of zinc and copper in subterranean clover (*Trifolium subterraneum*) grown in culture solution supplied with graduated amounts of zinc. *Aust. J. Agr. Res.* **9**:73–122.

Tsui, Chen, 1948. The role of zinc in auxin synthesis in the tomato plant. *Amer. J. Bot.* **35**:172–178.

Wann, F. B., and D. W. Thorne, 1950. Zinc deficiency in the western States. *Sci. Mon.* **70**:180–184.

(Boron)

Askew, H. O., and W. R. L. Williams, 1939. Brown spotting of apricots, a boron deficiency disease. *N. Z. J. Sci. Tech.* **ZA21,** 103–106.

——, E. T. Chittenden, and R. J. Mark, 1951. Dieback in raspberries. *J. Hort. Sci.* **26**:268–284.

Burrell, A. B., 1940. The boron deficiency disease of apple. *N.Y. Agr. Exp. Sta. Bull.* 428.

Chandler, F. B., 1941. Boron deficiency of alfalfa in western Washington. *Wash. Agr. Exp. Sta. Bull.* 396.

Cripps, E. G., 1956. Boron nutrition of the hop. *J. Hort. Sci.* **31**:25–34.

Dearborn, C. H., 1942. Boron nutrition of cauliflower in relation to browning *Bull. Cornell Univ. Agr. Exp. Sta.* no. 778.

Eaton, F. M., 1944. Deficiency, toxicity and accumulation of boron in plants. *J. Agr. Res.* **69**:237–279.

Gauch, H. G., and W. M. Duggar, Jr., 1954. The physiological action of boron in higher plants; a review and interpretation. *Univ. Md. Agr. Exp. Sta. Coll. Park Bull.* A80.

Jolivette, J. P., and J. C. Walker, 1943. Effect of boron deficiency on the histology of garden beet and cabbage. *J. Agr. Res.* **66**:167–182.

Kendrick, J. B., Jr., R. T. Wedding, J. T. Middleton, and B. J. Hall, 1954. Some factors affecting development and control of adaxial crack stem of celery. *Phytopathol.* **44**:145–147.

McLarty, H. R., J. C. Wilcox, and C. G. Woodbridge, 1937. A yellowing of alfalfa due to boron deficiency. *Sci. Agr.* **17**:515–517.

Walker, J. C., 1939. Internal black spot of garden beet. *Phytopathol.* **29**:120–128.

——, J. P. Jolivette, and J. G. McLean, 1943. Boron deficiency in garden and sugar beet. *J. Agr. Res.* **66**:97–123.

——, 1944. Histologic-pathologic effects of boron deficiency. *Soil Sci.* **57**:51–54.

Whittington, W. H., 1957. The role of boron in plant growth behavior. I. The effect on general growth, seed production and cytological behavior. *J. Exp. Bot.* **8**:353–367.

(Copper)

Arnon, D. I., 1950. Functional aspects of copper in plants, in W. O. McElroy and H. B. Glass (ed.), "Copper metabolism," pp. 89–110. Johns Hopkins Press, Baltimore.

Dunne, T. C., 1933. Wither-tip or summer dieback. *J. Agr. W. Aust.* (2nd ser.) **15**:120–126.

Elvehjem, C. A., 1931. The role of iron and copper in the growth and metabolism of yeast. *J. Biol. Chem.* **90**:111–132.

Nason, A., and W. D. McElroy, 1963. Modes of action of essential mineral elements, in "Plant physiology," vol. 3, Inorganic nutrition of plants. Academic, New York.

Neish, A. C., 1939. Studies on chloroplasts. II. Their chemical composition and distribution of certain metabolites between the chloroplasts and the remainder of the leaf. *Biochem. J.* **33**:300–308.

Pittman, H. A., 1936. Exanthema of citrus, Japanese plums and apple trees in Western Australia. *J. Agr. W. Aust.* (2d ser.) **13**:187–193.

Reed, H. S., 1939. The relation of copper and zinc salts to leaf structure. *Amer. J. Bot.* **26**:29–33.

Stiles, W., 1961. "Trace elements in plants." Cambridge University Press, London, 249 pp.

(Molybdenum)

Floyd, B. G., 1908. Leaf spotting of citrus. *Fla. Univ. Agr. Exp. Sta. Ann. Rep.* 1908, 91.

Hewitt, E. J., and W. E. Jones, 1947. The production of molybdenum deficiency in plants grown in sand cultures, with special reference to tomato and brassica crops. *J. Pomol. Hort. Sci.* **23**:254–262.

Steward, I., and C. D. Leonard, 1952. Molybdenum deficiency in Florida citrus. *Nature* **170**:714–715.

Wilson, R. D., 1949. Molybdenum in relation to the scald disease of beans. *Aust. J. Sci.*, 209–211.

SELECTED REFERENCES

(General)

Black, C. A., 1957. "Soil-plant relationships." Wiley, New York, 332 pp.

Ballard, E. G., and G. W. Butler, 1966. Mineral nutrition of plants. *Ann. Rev. Plant Physiol.* **17**:77–112.

Childers, N. F. (ed.), 1954. "Fruit nutrition." Somerset, Somerville, N.J., 907 pp.

Evans, H. J., and G. J. Sorger, 1966. Role of mineral elements with emphasis on the univalent cations. *Ann. Rev. Plant Physiol.* **17**:47–76.

Gauch, H. G., 1957. Mineral nutrition of plants. *Ann. Rev. Plant Physiol.* **8**:31–64.

Gilbert, F. A., 1949. "Mineral nutrition of plants and animals." Univ. of Oklahoma Press, 131 pp.

Reuther, W. (ed.), 1960. Foliar analysis and fertilizer problems. *Amer. Inst. Biol. Sci. Publ.* 8, Washington, 454 pp.

Russell, J. E., 1950. "Soil conditions and plant growth." Longmans, London, 635 pp.

Steward, F. C. (ed.), 1963. "Plant physiology," vol. 3, Inorganic nutrition of plants. Academic, New York, 811 pp.

Stiles, W., 1961. "Trace elements in plants," 3d ed. Cambridge University Press, London, 249 pp.

Wallace, T., 1961. The diagnosis of mineral deficiencies in plants (A color atlas and guide), 3d ed. H.M. Stationery Office, London.

(Nitrogen)

Allison, F. E., 1957. Nitrogen and soil fertility, in U.S.D.A. Yearbook of Agr. Soils, pp. 85–94.

Blaser, R. E., and N. C. Brady, 1950. Nutrient competition in plant association. *Agron. J.* **42**:128–135.

Bosemark, N. O., 1954. The influence of nitrogen on root development. *Physiol. Plant.* **7**:497–502.

McKee, H. S., 1962. "Nitrogen metabolism in plants." Clarendon, Oxford, 728 pp.

(Phosphorus)

Neller, J. R., 1947. Mobility of phosphates in sandy soils. *Soil Sci. Soc. Amer. Proc.* **11**:227–230.

Olsen, S. R., 1953. Inorganic phosphorus in alkaline and calcareous soils. *Agronomy* **4**:89–122.

——— and M. Fried, 1957. Soil phosphorus and fertility, in U.S.D.A. Yearbook of Agr. Soils, pp. 377–396.

Pierre, W. H., and A. G. Norman (eds.), 1953. Soil and fertilizer phosphorus in crop nutrition. *Agronomy* **4**.

(Potassium)

Bolle-Jones, E. W., 1955. The interrelationships of iron and potassium in the potato plant. *Plant and Soil* **6**:129–173.

International Potash Institute, 1954. Potash symposium, International Potash Inst., Zurich.

Reitmeier, R. F., 1957. Soil potassium and fertility, in U.S.D.A. Yearbook of Agr. Soils, pp. 100–106.

Richards, F. J., 1956. Some aspects of potassium deficiency in plants. Potassium symposium, International Potash Inst., Zurich.

Shaulis, Nelson, 1960. Associations between symptoms of potassium deficiency, plant analysis, growth, and yield of Concord grapes, in W. Reuther, "Foliar analysis and fertilizer problems," pp. 44–57. *Amer. Inst. Biol. Sci. Pub.* **8**

(Sulfur)

Benson, N. R., E. S. Degman, I. C. Chmelin, and W. Chenhaull, 1963. Sulfur deficiency in deciduous tree fruits. *Proc. Amer. Soc. Hort. Sci.* **83**:55–62.

Chapman, H. D., and S. M. Brown, 1941. The effects of sulfur deficiency in citrus. *Hilgardia* **14**:185–201.

Gilbert, S. G., 1951. The place of sulphur in plant nutrition. *Bot. Rev.* **17**:671–691.

Thompson, J. F., 1967. Sulfur metabolism in plants. *Ann. Rev. Plant. Physiol.* **18**:59–84.

(Calcium)

Baxter, P., 1960. Bitter pit of apples. Effect of calcium sprays. *J. Agr. (Victoria)* **58**:801–811.

Davis, D. E., 1949. Some effects of calcium deficiency on the anatomy of *Pinus taeda. Amer. J. Bot.* **36**:276–282.

Foster, A. C., 1934. Blackheart disease of celery. *Plant Dis. Rep.* **18**:177–185.

————, 1937. Environmental factors influencing the development of blossom end rot of tomatoes. *Phytopathol.* **27**:128–129.

Maynard, D. N., B. Gersten, E. F. Black, and H. F. Vernell, 1961. Effects of nutrient concentrations and calcium levels on the occurrence of carrot cavity spot. *Proc. Amer. Soc. Hort. Sci.* **78**:339–352.

Millikan, C. R., 1944. "Withertop" (calcium deficiency) disease in flax. *J. Victoria Dept. Agr.* **42**:79–91.

Muttus, G. E., 1953. Cork spot and bitter pit of apples. *Va. Fruit* **51**:35–40.

Oberly, G. H., and A. L. Kenworthy, 1961. Effect of mineral nutrition on the occurrence of bitter pit in Northern Spy apples. *Proc. Amer. Soc. Hort. Sci.* **77**:29–34.

Simon, R. K., 1962. Anatomical studies of the bitter pit areas of apples. *Proc. Amer. Soc. Hort. Sci.* **51**:41–50.

Spurr, A. R., 1959. Anatomical aspects of blossom-end rot on the tomato with special reference to calcium nutrition. *Hilgardia* **28**:269–295.

Stiles, W. C., 1964. Influence of calcium and boron tree sprays in York spot and bitter pit of York Imperial apples. *Proc. Amer. Soc. Hort. Sci.* **84**:39–43.

(Magnesium)

Lott, W. L., 1952. Magnesium deficiency in muscadine grape vines. *Proc. Amer. Soc. Hort. Sci.* **60**:123–131.

Moon, H. H., et al., 1952. Early-season symptoms of magnesium deficiency in apple. *Proc. Amer. Soc. Hort. Sci.* **59**:61–64.

Nason, A., and W. D. McElroy, 1963. Modes of action of essential mineral elements, in "Plant physiology," vol. 3, Inorganic nutrition of plants, chap. 4. Academic, New York, 811 pp.

Southwick, L., 1943. Magnesium deficiency in Massachusetts apple orchards. *Proc. Amer. Soc. Hort. Sci.* **42**:85–94.

(Iron)

Stewart, I., and C. D. Leonard, 1952. Chelates as sources of iron for plants growing the field. *Science* **116**:564–566.

(Manganese)

Lewis, A. H., 1939. Manganese deficiencies in crops. I. Spraying pea crops with solutions of manganese salts to eliminate marsh spot. *Emp. J. Exp. Agr.* **7**:150–154.

Reuther, W., and R. D. Dukey, 1937. A preliminary report on frenching of tung trees. *Bull. Fla. Agr. Exp. Sta.* no. 318, 21 pp.

Samuel, G., and C. S. Piper, 1929. Manganese as an essential element for plant growth. *Ann. Appl. Biol.* **16**:493–524.

Wallace, T., and J. O. Jones, 1943. The control of manganese deficiency in fruit trees. *Ann. Rep. Bristol Univ. Agr. Hort. Res. Sta.* (1942), pp. 19–23.

(Zinc)

Alben, A. O., and H. M. Boggs, 1936. Zinc content of soils in relation to pecan rosette. *Soil Sci.* **41**:329–332.

Krantz, B. A., and A. L. Brown, 1961. Zinc fertilization of field and vegetable crops in California. *Agr. Chem. W.* **4**:11:5–6.

Lingle, J. C., and D. M. Holmberg, 1957. Response of sweet corn to foliar and soil zinc applications on a zinc deficient soil. *Proc. Amer. Soc. Hort. Sci.* **70**:308–315.

Lyman, C., and L. A. Dean, 1942. Zinc deficiency of pineapples in relation to soil and plant composition. *Soil Sci.* **54**:315–324.

Skoog, F., 1940. Relationships between zinc and auxin in growth of higher plants. *Amer. J. Bot.* **27**:939–951.

Viets, F. G., Jr., 1961. Zinc deficiency of field and vegetable crops in the west. U.S.D.A. Leaflet no. 495, 8 pp.

(Boron)

Atkinson, J. D., 1948. Cracked stem of celery (*Apium graveolens*). *N. Z. Sci. Tech.* **A29**:261–264.

Dearborn, C. H., 1942. Boron nutrition of cauliflower in relation to browning. *Bull. Cornell Univ. Agr. Exp. Sta.* no. 778.

Dennis, R. W. G., and O. G. O'Brien, 1937. Boron in agriculture. *W. Scot. Agr. Coll. Res. Bull.* 5.

Gauch, H. G., and W. M. Duggar, Jr., 1953. The role of boron in the translocation of sucrose. *Plant Physiol.* **29**:457–466.

Grizzard, A. L., and E. M. Matthews, 1942. The effect of boron on seed production of alfalfa. *J. Amer. Soc. Agron.* **34**:365–368.

Jamelainen, E. A., 1936. The effect of boron on the occurrence of cork disease in apple. *State. Agr. Exp. Act. Helsinki*, Publ.˙m. 89.

Lorenz, A., 1942. Internal breakdown of table beets. *N.Y. Agr. Exp. Sta. Memo.* 246.

Naftel, J. A., 1937. Soil-liming investigations. B. The relation of boron deficiency to over-liming injury. *J. Amer. Soc. Agron.* **29**:761–771.

Nusbaum, C. J., 1946. Internal brown spot, a boron deficiency disease of sweet potato. *Phytopathol.* **36**:164–167.

Palser, B. F., and W. J. McIlrath, 1956. Responses of tomato, turnip and cotton to variations in boron nutrition. II. Anatomical responses. *Bot. Gaz.* **118**:53–71.

Reed, H. S., 1947. A physiological study of boron deficiency. *Hilgardia* **17**:337–411.

Reeve, E., and J. W. Shive, 1944. Potassium-boron and calcium-boron relationships in plant nutrition. *Soil Sci.* **57**:1–14.

Russel, D. A., 1957. Boron and soil fertility, in U.S.D.A. Yearbook of Agr. Soils, pp. 121–128.

Scott, L. E., H. Earl Thomas, and H. E. Thomas, 1943. Boron deficiency in the olive. *Phytopathol.* **33**:933–942.

Skok, J., 1958. The role of boron in the plant cell, in C. A. Lamb, O. G. Bentley, and J. M. Beatic (eds.), "Trace elements," pp. 227–243. Academic, New York.

Whittington, W. J., 1959. The role of boron in plant growth. II. The effect on growth of the radicle. *J. Exp. Bot.* **10**:93–103.

Woodbridge, C. G., 1937. The boron content of apple tissues as related to drought spot and corky core. *Sci. Agr.* **18**:41–48.

(Copper)

Anderson, F. G., 1932. Chlorosis of deciduous fruit trees due to a copper deficiency. *J. Pomol.* **10**:130–146.

Dickey, R. D., et al., 1948. Copper deficiency of tung in Florida. *Fla. Agr. Exp. Sta. Bull.* 447.

Floyd, B. F., 1917. Dieback, or exanthema of citrus trees. *Fla. Agr. Exp. Sta. Bull.* 140, pp. 1–31.

Harris, H. C., 1947. A nutritional disease of oats apparently due to the lack of copper. *Science* **106**:398.

Jones, J. O., and W. Dermott, 1952. Copper deficiency in pears. *J. Minn. Agric. (G. Br.)* **59**:35–37.

Millikan, C. R., 1944. Symptoms of copper deficiency in flax. *Proc. Roy. Soc. Victoria* **56**:113–117.

Oserkowski, J., and H. E. Thomas, 1938. Exanthema in pear and copper deficiency. *Plant Physiol.* **13**:451–467.

Reuther, W., 1957. Copper and soil fertility, in U.S.D.A. Yearbook of Agr. Soils.

Riceman, D. S., and A. J. Anderson, 1943. The symptoms and effects of copper deficiency in cereals and pasture plants in South Australia. *J. Dept. Agr. S. Aust.* **47**:64–72.

(Molybdenum)

Anderson, A. J., 1956. Molybdenum as a fertilizer. *Advan. Agron.* **8**:163–202.

Bear, F. E. (ed.), 1956. Symposium on molybdenum. *Soil Sci.* **81**:163–258.

Evans, H. J., and N. S. Hall, 1955. Association of molybdenum with nitrate reductase from soybean leaves. *Science* **22**:922–923.

Hewitt, E. J., and E. W. Bolle-Jones, 1952. Molybdenum as a plant nutrient. I. The influence of molybdenum on the growth of some brassica crops in sand cultures. *J. Hort. Sci.* **27**:257–265.

Stout, P. R., and C. M. Johnson, 1956. Molybdenum in horticultural and field crops. *Soil Sci.* **81**:183–197.

———— and ————, 1957. Trace elements, in U.S.D.A. Yearbook of Agr. Soils, pp. 139–197.

CHAPTER FOURTEEN

MINERAL TOXICITY

The mineral elements in the soil, whether required for normal nutrition or not, are all absorbed by the plant to some extent. Every plant requires a specific, optimum amount of the essential chemicals, but if a surplus of any exists, the plant may absorb and accumulate excessive toxic amounts. The ratio between many of these elements is also vital. Too much, as well as too little, of many mineral elements will upset this ratio and cause abnormal plant development. While not appearing with the frequency and consistency of deficiency diseases, mineral toxicities are still common, widespread, and deserving of consideration.

Tolerance to elements varies with the nutrient requirements of the species, its inherent tolerance and ability to absorb and accumulate different ions, and the ratio of different elements in the soil to each other. While nutrient requirements are largely inherited, nutrient absorption and accumulation depend also on such physical factors of the soil as structure and acidity. Such elements as aluminum are highly soluble and available in acid soils but bound tightly and unavailable in alkaline soils. Solubility and availability of the element, as well as the initial amounts of the elements present in the soil, are the most important factors influencing absorption.

The relative amounts of different elements also influence their toxicity. Excesses of certain mineral nutrients induce deficiencies of others: Excess nitrogen can lead to a magnesium or calcium deficiency; excess nitrogen or phosphorus produces potassium deficiency; excess potassium causes magnesium or calcium deficiency; and excess magnesium or sodium may induce calcium deficiency. Other elements, particularly the trace elements and alien metal ions, can induce deficiency or be directly toxic. Excess chromium, cobalt, copper, manganese, nickel, or zinc may induce iron deficiency in addition to being directly toxic.

Excessive concentrations, particularly of such salts as calcium, magnesium, and sodium, can alter the osmotic pressure of the soil solution to a degree which is toxic to plants and can cause growth suppression and leaf burning. Excess quantities of other elements may be directly toxic to the protoplast or surface membranes and kill the tissues. The most common effect of excesses, though, is in modifying the growth and fruit characteristics or in predisposing the plant to other environmental stresses and plant disease.

14-1 ESSENTIAL AND NONESSENTIAL ELEMENTS

Nitrogen

Up to a point, plant growth increases with increased nitrogen. Beyond this, nitrogen is not only wasted but detrimental. While commonly deficient in native soils, it is often used in excess in both agricultural plantings and home gardens. The farmer or homeowner, thinking that if a little is good, more is better, is prone to overfertilizing. An extreme example of this, known to everyone, is the "burning" of lawns on which the nitrogen application was calculated incorrectly, where the fertilizer spreader overlapped, or fertilizer was accidentally spilled. Such incidents of leaf necrosis are easily recognized and diagnosed. Less readily identified are symptoms arising from the application of slightly less, but still excessive quantities of nitrogen. Studies in Washington showed that approximately 25 percent of the apple orchards in that state were overfertilized even though there were no striking toxicity symptoms. Woodbridge et al. (1960) report as follows: "Tree growth increases linearly with increases in leaf nitrogen until the latter reaches 2.2 percent. If the leaf nitrogen is above this amount, there is no increase in growth; if it exceeds 2.4 percent the fruit usually lacks color and its storage life is adversely affected." Size of the root system was also decreased significantly at the higher nitrogen levels.

High nitrogen can be especially harmful when large amounts are applied to fruit crops such as peach and cherry shortly before harvest. Cells enlarge and fruit size increases rapidly, but unfortunately this rapid enlargement is at the expense of both fruit quality and shipping quality. Maturity is delayed, fruit color poor, and the thin-walled cells highly sensitive to bruising and breakdown. If shipped, fruits are readily invaded and damaged by biotic pathogens which render them worthless before reaching their destination. If excessive amounts of nitrogen are applied earlier in the season, blossom and fruit retention may be impaired.

The earliest indication of a slight nitrogen excess is the lush, deep green foliage color. Higher concentrations cause chlorosis, which would remind one of nitrogen deficiency except for the more rank, "soft" growth and brighter yellow color. Leaf shedding is also encouraged. Necrotic spots, irregular in outline, may subsequently develop. When concentrations are still greater, possibly ten times what would be regarded as ample, leaf burning, accompanied by chlorosis, growth retardation, and sometimes gummosis and dieback, can be expected.

Nitrogen levels must be extremely high in plant tissues, in excess of 4 percent of the dry weight, before causing direct tissue damage. More often an excess disturbs the balance or ratio of nitrogen to other elements such as potassium, magnesium, or calcium, inducing a deficiency of these elements of promoting an abnormally "soft" type of growth.

Too much nitrogen is particularly deleterious when plants are grown for their flowers or fruit. Nitrogen encourages vegetative growth, not reproduction. Overfertilized roses and camellias develop rank new growth and large, lustrous, deep green leaves, but few flowers.

The vegetative growth habit induced is equally undesirable in delaying maturity and prolonging shoot and leaf growth, so that shoots of normally deciduous plants remain lush well into fall and winter. The rank growth is low in sugar and other carbohydrates and therefore highly sensitive to low- or high-temperature injury and drought. The active, rapid growth produces thin-celled tissues which are readily penetrated by biotic pathogens and provide an excellent substrate for the further growth of disease-causing organisms.

The accelerated growth and long internodes induced by over-fertilization can also be harmful to cereal crops. As the head fills, the weak stem is unable to support the added weight, and the plant collapses. Such displacement of the plant from the vertical position, known as lodging, increases linearly with the nitrogen level in the crops.

High nitrogen levels are more likely to cause plant disorders indirectly than directly. Such is the case with *tip burn of sugarbeet*, which consists of a necrosis or burning of the tips of the outer leaves. Tip burn develops in plants growing in fertile, nitrogen-rich soil when the light intensity is low. It does not develop at high light intensity even when the nitrogen level is extremely high or the outer leaves are receiving full sunlight. Under low light conditions, toxic, presumably nitrogenous materials considered to be responsible for the tip burn accumulate in the root. Affected beets recover completely when high light intensity is provided or the nitrogen level is reduced. *Bud blasting*, or dropping, of Croft lilies is also indirectly caused by nitrogen excesses; its incidence has been demonstrated to increase an average of 37 percent at high nitrogen levels (Bald, Korfranek, and Lunt, 1955).

Potassium

Excessive, toxic concentrations of potassium in the soil are rare but can arise from prolonged heavy potassium or nitrogen fertilization. The high potassium levels are not directly toxic; rather, the principal effect seems to be to induce deficiency of other ions such as calcium, magnesium, or iron; symptoms of potassium excess can resemble a deficiency of these elements. The yellow-orange, blotchy chlorosis and decreased growth characterizing both high potassium and low magnesium have been induced on McIntosh apples following three or more years of heavy potassium fertilization. Chemical analysis showed that leaves were high in foliar potassium and low in magnesium. New shoot growth was low in both magnesium and calcium (Boynton and Burrell, 1944).

Excess potassium also caused pale yellow chlorosis with contrasting green veins characterizing iron deficiency. The actual damage from such a potassium surplus, then, is due to an iron deficiency.

Since potassium is an alkali much like sodium, high concentrations in excess of roughly 3 percent in the leaves might be expected to have similarly deleterious effects. Potassium may act with sodium or substitute for it and thereby unbalance the ratio of sodium to calcium. If the ratio of sodium (or potassium) to calcium is too high, a calcium deficiency is induced. High potassium also impairs calcium absorption, inhibiting plant development and causing calcium deficiency.

Sodium and Calcium

Excessive amounts of sodium or calcium can damage plants directly, but damage is more often related to the saline and/or alkaline characteristics that these elements impart to the soil. While they often occur together, salinity and alkalinity reflect distinct situations and are not at all the same.

A soil is saline when its total soluble salt content is excessive; that is, when there are enough salts to affect plant growth adversely. Growth of sensitive plants becomes impaired when the salt content of the soil exceeds about 0.1 percent. More precisely, a soil is said to be saline when the solution extracted from a saturated soil paste has an electrical conductivity (EC) value of 4 mmho per centimeter of soil extract.* Below 2 mmho, the effects on even sensitive plants are negligible; above 16, only a very few salt-tolerant plants remain productive. Soils may be naturally high in salts when the parent material has weathered and rains or irrigation waters are inadequate to leach them out of the soil. Various salts, of which sodium, calcium, and magnesium are most common, contribute to salinity. High levels of fertilization also contribute to salt accumulation and can be significant in agricultural situations.

Alkali soils contain excessive amounts of absorbed sodium but are not necessarily high in total salts. Sodium is said to be excessive when

* When a molecule of sodium chloride, calcium chloride, or other salt, goes into solution, it splits into electrically charged particles called ions, e.g., a positive sodium ion and a negative chloride ion. The ions carry an electric current, so that the more ions that are present, the greater will be the charge. Thus, the total amounts of salts, or salinity, of a solution can be measured by its electrical conductivity, and conductivity can be determined from the saturation extract.

the exchangeable sodium percentage exceeds 15 percent.* Sodium is often harmful when the exchangeable sodium is as low as 5 percent. It is the high sodium content, then, that is the damaging entity of alkali soils. In addition to being high in sodium, the pH of alkali soils is generally above 8.5. Such a high pH almost invariably indicates not only high sodium but the presence of alkaline earth carbonates (lime); but conductivity of the saturation extract of a solely alkali soil is usually less than 4 mmho/cm.

Soils may be saline, alkaline, or both. When both soluble salts and exchangeable sodium are high, the soil is referred to as saline-alkali. The soil is alkali only when the sodium is excessive, saline when total salts are excessive.

Salt excesses are reported to lower the productivity and value of an estimated one-fourth of the irrigated agricultural land in the United States. The harmful effect of saline soils is primarily in reducing the rate at which plants absorb water. The excessive amount of mineral ions of a saline soil increases the osmotic pressure to a degree where a tension is created in the soil solution against which the plant cannot absorb water. In essence, the soil is holding on to the water with a greater force than the plant can exert. The plant is unable to absorb enough water to function normally, and a water stress condition develops.

As salinity levels increase, growth of intolerant plants declines, and yields are reduced. Growth suppression is sometimes accompanied by leaf injury. Leaves become smaller and deeper blue-green than normal, and leaf tips or margins become bleached, tan, or brownish in proportion to the degree of salinity. Bronzing and early defoliation may also be prominent. Leaf injury may be the most prominent symptom of salinity but is not nearly as important to yield reductions as the growth suppression.

Leon Bernstein (1964), working with the Soil and Water Conservation Research Division of the U.S. Department of Agriculture has reported the EC at which yields of field, forage, and vegetable crops were reduced 10, 25, and 50 percent. As a group, vegetable crops were most sensitive, and yields often were suppressed 10 percent even when conductivity was just below 2 mmho/cm. Carrot, onion, and bean yields were reduced 50 percent at 4 mmho/cm.

* Soils have a rather definite capacity to absorb and exchange cations (positively charged particles such as sodium, calcium, etc.); the percentage of this capacity that is taken up by sodium is referred to as the exchangeable sodium percentage.

Apricot, apple, almond, plum, and pear trees have been shown to tolerate as much as 0.5 percent salts before the growth was adversely affected (Oganesjan, 1953). But so many factors influence salt tolerance that the predisposing factors must be stipulated. Many plants thrive even when irrigated with sea water if conditions are otherwise completely favorable (Boyko, 1967). Favorable conditions demand an extremely sandy soil in which salts will not accumulate but which will allow their return to the sea in surface or deeper waters.

While excessive amounts of almost any salts are harmful, the composition of the salts is still more important. Such salts as sodium can be toxic in themselves even when the soils are not saline.

High-sodium, alkali soils are responsible for such diseases as *white tip* on grains common in many parts of the West. Affected leaf tips turn white or greenish white; beginning from June to August, the sheath twists, heads emerge imperfectly, and seed may be deformed. Plants may be dwarfed and head formation prevented. It is believed that the alkali prevents the plants from obtaining sufficient iron and possibly other elements. The condition is partly corrected by acidifying the soil with sulfur or sulfuric acid.

Other disorders associated with high sodium include *tip burn of almond*, described by Lilleland et al. in 1945, and *sodium scorch* on avocado leaves, described by Ayers in 1950.

Leaves absorb sodium or chlorine rapidly both from the soil and through the leaves; consequently, water sprayed on vegetation may be highly toxic. The more rapidly a plant absorbs these chemicals, the more damage can be expected. Toxic concentrations of both elements may accumulate in sprinkle-irrigated trees if the salt content is high. Characteristic leaf-injury symptoms are most frequent on fruit and nut crops such as berries, grapes, stone fruits, citrus, avocado, pecan, and woody, ornamental species. Marginal foliar bleaching and necrosis appear when the leaves accumulate more than 0.25 percent of sodium or 0.5 percent of chlorine on a dry weight basis. The severity of injury increases in proportion to the level of sodium or chlorine.

Accumulation of sodium in avocado leaves can be mistaken for chloride burn even though the symptoms are distinct. Injury from sodium begins with the appearance of necrotic spots within or along the leaf tip. These enlarge to form large necrotic intercostal lesions delimited by the secondary veins. Often sodium burn exists in conjunction with chloride burn. When spot burning appears, the foliar sodium content usually will be above 0.4 percent.

Some species, notably citrus and certain shrubs, exhibit leaf bronzing and early defoliation rather than leaf burning. Killing of leaf tissues directly limits growth and production in proportion to the necrosis, but if plants are sensitive to salinity, the burning effect may be negligible compared with the effect on plant metabolism.

The ratio of sodium to calcium may influence the degree of toxicity as much as the sodium concentrations alone. This is exemplified by the *red root disease* of chrysanthemum (Raabe et al., 1966). Chrysanthemums irrigated with high-sodium water showed reddening of portions of the roots, loss of root hairs and small rootlets, and death of the root tips beyond the reddened sectors. The only above-ground symptom was the cessation of growth. While the disease can be prevented by reducing the sodium concentration, it can just as easily be corrected by raising the calcium level so that the ratio of sodium to calcium is decreased.

Sodium, calcium, and chlorine may accumulate in toxic concentrations when sodium chloride or calcium chloride is used on icy highways in the winter to improve traction and help melt the ice, or on dusty roads in the summer to settle dust. The chemicals are applied to icy driveways and on paths around homes as well as to state and county roads. When applications are excessive, or accumulated in impounded runoff waters, lawns, shrubs, and trees bordering treated surfaces are damaged, particularly on the side of the plant receiving most of the runoff. Severe leaf scorch and decline has been described on trees growing along calcium chloride–treated highways (Holmes, 1961; Lacasse, 1964). The *maple decline* disease has also been blamed on salt toxicity. Symptoms consist of prominent marginal scorch, leaf cupping, and pale, stunted new growth. Affected maples may die out in two to six years, while understory species such as wild cherry continue to grow profusely. Both sodium chloride and calcium chloride are known to be harmful whether applied in the soil or sprayed on the plant, but experimental applications proved sodium chloride to be five to ten times more toxic. Nevertheless, sodium chloride is generally used because of its lower cost and ease of handling.

Excess sodium chloride or calcium chloride is not damaging under all conditions, and maple decline has been attributed to moisture stress alone (Banfield, 1967). The disease is often most severe on trees in the driest sites and may occur in the absence of excess salts. Even when the amount of salt is high, the timing of applications before or after plowing, quality and drainage of the soil, slope of the terrain, dates of salting, depth and duration of freezing in the soil, depth of plowed snow piles, amount of runoff from melted snow before the ground thaws, and the inherent susceptibility of the individual tree all contribute to the differences between damage and tolerance.

Sodium chloride may also accumulate in areas close to the sea coast where salt water sprays prevail and influence the species which can grow successfully. In the East, sensitive species such as sweet gum and other hardwoods are often excluded from such areas, leaving the habitat for more tolerant species.

Chlorine

While an overabundance of chlorine is often found in association with sodium or calcium, toxic concentrations of chlorine alone may exist in soil or irrigation water in the absence of sodium or calcium excesses. Toxicity symptoms are much like those described for salt toxicity; much of the damage ascribed to salts may actually be due to chloride alone. In a general way, symptoms consist of chlorosis, necrosis, and decline.

Avocado trees are among the species most sensitive to chlorine and provide a good example of the response of plants to high chlorides (Kadman, 1963). Chloride causes a disease of avocado known as *tip burn* which has been observed in southern California for many years. It occurs when the irrigation water is high in chlorine and is particularly noticeable in years of lower-than-normal rainfall. The first expression of toxicity is the appearance of chlorosis, and soon necrosis, of the leaf tip and margins. If foliar concentrations exceed about 1 percent, fruit quality is impaired and defoliation and tree mortality increased. Trees grown at varying chlorine concentrations showed yield reductions from 8 to 60 percent due both to a decreased functional leaf area and early leaf abscission. Production losses and necrosis both tended to be correlated with chloride content of the leaves (Hayward et al., 1946).

Growth suppression has been demonstrated on tomato and corn. It may occur at chlorine levels below those causing chlorosis or necrosis, but, more commonly, growth reduction is proportional to the amount of leaf burning.

Chlorine has also been reported to contribute to tree decline. It is considered to be the most toxic component of the calcium chloride or sodium chloride applied to icy roads. Continued applications of these chemicals are regarded to be toxic if the foliar chlorine content exceeds 1 percent (Holmes, 1966). Chlorine injury was simulated by spraying it at varying concentrations on test trees. The symptoms produced consisted of marginal necrosis and small angular to circular brown spots. Chlorine concentrations of 0.2 percent were toxic; at higher concentrations leaves turned pale green rather than brown, becoming dry and brittle and shedding prematurely.

Damage from chlorine is most severe when temperatures are high and evaporation rapid. Under such conditions uptake and accumulation is greatest, and chlorine soon reaches toxic concentrations. The concentration of chloride in the leaf required to cause necrosis varies with the species, but a broad threshold can be given as from 0.5 percent to 1 percent of the dry weight of the leaf.

Manganese

Most of the manganese in soils is held tightly in an insoluble form which is unavailable to the plant. But when the soil pH falls below 5.5, manganese becomes increasingly soluble and available in toxic concentrations.

The degree of toxicity and injury depends on the inherent capacity of a species to absorb or exclude manganese. The ability of such plants as oats and strawberries to grow in high-manganese soils is attributed to their low intake or selective exclusion of manganese and their decreased capacity for translocating manganese from the roots to the shoots.

The symptoms of manganese toxicity differ from species to species, but in a general way consist of a "frenching" or crinkling of the leaf margin and restriction of marginal growth, which causes a leaf cupping; often the margin, which accumulates the highest concentrations of manganese, becomes extremely chlorotic to white. In cauliflower and kale, dark brown to purplish necrotic spots develop in the chlorotic areas. In potato, early symptoms consist of profuse minute, black, necrotic spots which develop along the petioles, on the underside of leaf veins, and along the stems. These spots gradually coalesce, forming a stem streak necrosis.

High manganese also causes internal stem necrosis on plants as diverse as apple and potato (Berg and Clulo, 1946). On red Delicious apples, the toxicity symptoms have been described as *internal bark necrosis;* on potato, the disease is called *stem streak necrosis* (Berger and Gerloff, 1947). In both cases, dark brown, elongated, pitted areas appear near the base of the stem and petioles and extend into the pith. Chlorosis subsequently develops between the veins; as symptoms progress, chlorosis intensifies, and leaves gradually become pale yellow and brittle, dry up, and fall. In severe cases, small necrotic irregular flecks appear in the chlorotic areas between the veins nearest the midrib. Finally the terminal bud dies, followed by the premature death of the entire plant.

Crinkle leaf of cotton was among the earliest-named diseases attributed to manganese deficiency (Neal, 1937). Affected young leaves were variously distorted and chlorotic, with necrotic lesions developing along and between the veins. The cotton fibers were weak and worthless. Crinkle, as well as other manganese toxicity diseases, can be corrected by reducing the soil acidity by applying calcium carbonates or related materials, thus decreasing the solubility and availability of manganese.

Zinc

Zinc toxicity is not common, but toxic concentrations may exist in some acid soils in the neighborhood of large deposits of zinc ores or near zinc smelters.

Symptoms on soybeans are characteristic (Earley, 1943). The base of the midrib becomes red, and leaves begin to curl under; the younger leaflets become chlorotic, reddish pigmentation of the leaf intensifies, and the shoot apex dies.

Symptoms generally resemble those of manganese deficiency. Zinc apparently is closely enough related to manganese to replace it physically in essential enzymes, but not closely enough related to replace it chemically. Zinc only interferes with its activity by blocking enzyme activity and causing a shortage of manganese.

Boron

Boron toxicity represents an important agricultural problem in many geographic areas. Boron is naturally high in some soils and can accumulate in others when the concentrations in the irrigation water are excessive. In California, excesses in fruit crops such as almond, apricot, cherry, and peach may cause shoots to start growth in the spring, only to die back a few weeks later.

Leaves are less often damaged, but when symptoms do appear, injury is generally associated with the vein endings. In plants with parallel venation, such as corn, grasses, and lilies, tip burn (often with necrotic blotches near the tip) is most prominent. In net-veined species, including geranium, cotton, and cantaloupe, marginal necrosis is characteristic. In species including citrus, gerbera, and aster, spotted, blotchy chlorosis between the secondary veins is a prominent early symptom.

Marginal chlorosis, and sometimes necrosis, downward curling, poor leaf lobation, and the appearance of brown, interveinal flecks about 1 mm in diameter toward the margin are also characteristic.

Symptoms of boron toxicity on citrus first consist of a yellowing along the margins which extends between the veins of older leaves. This is followed by necrosis at the tip and margin and abnormally heavy defoliation in winter and early spring.

Walnut leaves become necrotic at the leaflet tips and margins, particularly in August and September, but little chlorosis appears (Kelley and Brown, 1928).

Excess boron may inhibit flower development, especially when calcium is ample, but the effect of boron toxicity on fruit production is largely indirect and due to the leaf tissue damage. Since necrosis is usually slight, yield reductions also are largely insignificant. However, in some instances fruits are directly affected. Symptoms on peach fruits consist of dark brown woody lesions which extend almost into the pit. Injury on apricot fruits consists of dark, circular spots about $\frac{1}{4}$ in. in diameter.

Boron is considered excessive for the most sensitive species at concentrations above 0.5 ppm in water or over 190 ppm in the leaf tissues, but tremendous variation in sensitivity exists, due largely to differences in the rate of boron accumulation.

Copper

The toxic nature of copper has been recognized for many years, and the chemical has been utilized to kill and control numerous fungus and algal pests. Excessive amounts of copper are equally harmful to higher plants, suppressing development of fibrous roots and reducing yields. Whenever the concentration of copper exceeds about 0.5 ppm in the water, plant growth is reduced, and only slightly higher concentrations cause chlorosis resembling iron deficiency. The mechanism of toxicity is one of interference with normal metabolic reactions, primarily the blocking of specific enzymatic reactions which require iron.

Toxic copper concentrations have arisen largely where soils were initially deficient in it and had copper applied annually to correct the deficiency and improve growth and production. Applications of as little as 10 lb/acre of copper sulfate have retarded plant growth on sandy soils. Where copper fungicide sprays have been used for prolonged periods, concentrations of over 800 ppm have been reported.

Aluminum

Toxic concentrations of aluminum in the soil occur naturally in areas of high rainfall; they may also be induced deliberately or inadvertently by man's use of soil amendments or fertilizers such as sulfur, aluminum sulfate, ferric sulfate, or ammonium sulfate. Aluminum occurs in different forms, depending on the soil acidity (pH). Some aluminum can be absorbed without deleteriously affecting plants, and it may even be beneficial, but excessive quantities may accumulate in acid soils. Aluminum may be harmful when present in a soluble form in concentrations above 10 ppm. Such concentrations are unlikely unless some highly soluble acidifying fertilizer salt such as ammonium sulfate is used on already acid soils. As the pH approaches 5.0, aluminum becomes increasingly soluble, and toxicity becomes increasingly important.

When nutrient cations such as calcium, magnesium, and potassium are wanting, as is common in acid soils, water-soluble aluminum concentrations as low as 1 to 2 ppm may inhibit growth of rice roots (Cate and Sukhai, 1964). Higher concentrations prevent root growth and cause a characteristic brown mottling of the leaves, especially at the tip and along the edges. Symptoms on other cereals, e.g. barley, consist of chlorosis of the older leaves, limited and distorted root growth, and root discoloration.

Toxicity symptoms closely resemble those of calcium deficiency. This would be expected, since aluminum reduces both the adsorption and accumulation of calcium and restricts transport of calcium to the shoots. Further, aluminum inhibits cell elongation and division (Clarkson, 1965). Studies with onion showed that root elongation was almost completely suppressed following treatment with 10^{-3} and 10^{-4} M aluminum sulfate in solution.

Nickel

Nickel is toxic to plants at relatively low concentrations of about 40 ppm (Hunter and Vergano, 1952). While the total nickel content of agricultural soils mostly ranges between 10 to 40 ppm, it may be substantially higher in soils derived from serpentine rock. Symptoms produced by nickel toxicity closely resemble those of manganese deficiency. Leaves show interveinal and marginal chlorosis with some necrotic spotting.

Beryllium

Beryllium may inhibit plant growth significantly at concentrations of 3 to 5 ppm and is considered toxic when its concentration in water exceeds 1 ppm (Romney, et al., 1962). Roots turned brown within five days after exposure and failed to resume normal growth when removed from the beryllium solution. Treated plants flowered abnormally early.

The role of excessive beryllium was to decrease the magnesium content in the roots and stem. Beryllium also caused a decrease of calcium in roots, leaves, stem, and fruit, and a decrease of phosphorus in the roots. Fundamentally, beryllium is thought to inhibit the normal function of the plant phosphotase enzyme systems.

Lithium

Lithium is present in some irrigation waters in concentrations over 0.1 ppm, which can impair plant growth and cause leaf chlorosis and burning (Bingham et al., 1964). Toxicity symptoms of lithium, like those resulting from excessive amounts of many other elements, are not distinctive. In a general way, the first response of broad-leaved plants is depressed growth and general marginal leaf necrosis followed by interveinal chlorosis and leaf abscission.

Lithium toxicity symptoms are correlated with the accumulation of lithium in the petiole and leaf tissues of the plant, and when foliage concentrations reach 100 ppm, plant damage can generally be expected.

At least one specific disease has been attributed to lithium toxicity, namely, *leaf scorch of easter lily* (Furuta et al., 1963). Before lithium was established as the causal agent, the disease was attributed to such environmental factors as adverse temperature relations and light intensity. These factors influence development of leaf scorch but do not cause it. To determine the role of lithium, studies were conducted in which lithium was sprayed on leaves at concentrations similar to those occurring in the irrigation water. Soon after treatment, older lily plants developed a brownish cast on the top half of the leaf blades. New growth was light green to whitish, tips of older leaves gradually died, and necrotic flecks appeared on newly developed leaves. The margin between healthy and dead tissues was irregular and indistinct. The lithium content of affected foliage was 156 ppm.

Frank Bingham and his coworkers at the University of California (1964) list the sensitivity of plants to lithium in order of increasing tolerance as avocado, soybean, sour orange, grape, tomato, red kidney bean, cotton, red beet, rhodes grass and sweet corn.

14-2 METABOLIC EXUDATES

Guttation

Water containing copious amounts of dissolved solutes is exuded through the hydathodes along the leaf margin. Hydathodes are similar to stomata except that no guard cells are available to adjust the pore size or regulate water movement. These masses of small, thin-walled cells occur principally between the epidermis and the bundle cells of the tracheids.

Water moving rapidly out of the hydathodes carries sufficient salts out of the leaf so that when the water evaporates, the salts deposited on the surface may be present in concentrations lethal to the adjacent tissues.

Tip burn of potato, in which the tips of the leaflets become necrotic, is attributed to the accumulation of toxic quantities of salts around the hydathodes. The first sign of salt damage to the cells is chlorosis, followed by necrosis. As the lesions coalesce, larger bands of dead tissue and tip burning appear. Injury from intercostal, or laminar, guttation is more frequent than from marginal guttation and may affect several square centimeters or even the entire blade. A thin film of "cell sap" or exudate is formed over the leaf as the individual drops coalesce. As the liquid evaporates, crystalline lumps of salt remain which can be seen as conspicuous deposits on the leaves. Plants are rarely killed, but development may be impaired and fruiting delayed. Identical lesions have been induced artificially by applying drops of 4 percent sodium phosphate and other phosphorus salts to the leaf.

Guttated exudates are not only inorganic—as much as half of the exuded solids may be organic. In one incident (Curtis, 1944), under cloudless, cool nights, grass guttated profusely, and lawns took on a whitish cast from the exudate. Dr. Curtis postulated that guttation provided the plant with an excretory system to eliminate not only excesses of water but various organic and inorganic chemicals.

Metabolites

Metabolic products secreted by the roots of certain plant species may be toxic to other species. Toxic substances inhibitory to growth are contained in the roots of many plant species including peach, brome grass, locust, and walnut. Wilting of potato and other plants growing around walnut trees has been observed since at least 1921 (Cook). Antagonistic excretions from walnut roots cause wilting and death of such plants as alfalfa, tomato, and potato (Massey, 1925). The toxin does not diffuse far from the source but is localized in the immediate vicinity of the walnut roots.

Studies by Z. A. Patrick (1963) in the Salinas Valley of California revealed that phytotoxic substances were excreted from residues of barley, rye, wheat, sudan grass, vetch, broccoli, and broad bean which had been decomposing ten to twenty-five days. Seedlings growing in soils which contained these residues were damaged. Injury consisted of discoloration and death of the apical meristem of the root, which prevented water uptake and led to wilting of the above-ground portions. Injury to roots of lettuce and spinach seedlings was confined mainly to those parts in direct contact with, or in the immediate vicinity of, decomposing plant fragments in the soil.

Carley and Watson (1967) found that extracts of twenty-three crop residues retarded root development. Materials contained in aqueous extracts from barley, bean, alsike clover, carrot, green soybean, sagebrush, onion, red clover, pea, sugarbeet, and potato residues were selectively toxic to clover, radish, and wheat seedlings used as assay plants. The extracts killed the root promeristem and cells in the zone of cell elongation. Secondary roots generally proliferated above the necrotic areas of the root tips. Secondary symptoms caused by the phytotoxins consisted of severe stunting and chlorosis of a manganese-deficiency type.

While the toxicity is important in itself, it has a still greater damage potential in predisposing the plants to infection by root parasites. The weak, partly killed, necrotic roots provide an ideal infection court for disease-causing fungi (Toussoun and Patrick, 1963). In natural situations, plant secretions appear to be helpful in preventing other plants from invading an area and growing in close proximity, where they might compete for water, light, and nutrients. In desert situations, sage and many other species utilize this mechanism to avoid overpopulation.

BIBLIOGRAPHY

Ayers, A. D., 1950. Salt tolerance of avocado trees grown in culture solution. *Calif. Avocado. Soc. Yearbook*, pp. 139–148.

Bald, J. G., A. M. Kofranek, and O. R. Lunt, 1955. Leaf scorch and rhizoctonia on croft lilies. *Phytopathol.* 45:156–162.

Banfield, W. M., 1967. The significance of water deficiency in the etiology of maple decline. *Phytopathol.* 57:338.

Berg, A., and G. Clulo, 1946. The relation of manganese to internal bark necrosis of apple. *Science* 104:265–266.

Berger, K. C., and G. C. Gerloff, 1947. Stem streak necrosis of potatoes in relation to soil acidity. *Amer. Potato J.* 24:156–162.

Bernstein, L., 1964. Salt tolerance of plants to lithium. *Soil Sci.* 98:4–8.

Bingham, G. R., G. R. Bradford and A. L. Page, 1964. Toxicity of lithium to plants. *Calif. Agr.* 18:6–7.

Boyko, H., 1967. Salt-water agriculture. *Sci. Amer.* 216:89–96.

Boynton, D., and A. B. Burrell, 1944. Potassium-induced magnesium deficiency in the McIntosh apple tree. *Soil Sci.* 58:441–454.

Carley, H. E., and R. D. Watson, 1967. Plant phytotoxins as possible predisposing agents to root rots. *Phytopathol.* 57:401–404.

Cate, R. B., and A. P. Sukhai, 1964. A study of aluminum in rice soils. *Soil Sci.* 98:85–93.

Clarkson, D. T., 1965. The effect of aluminum and some other trivalent metal cations on cell division in the root apices of *Allium cepa*. *Ann. Bot.* (*London*) 29:310–315.

Cook, M. T., 1921. Wilting caused by walnut trees. *Phytopathol.* 11:346.

Curtis, L. C., 1944. The exudation of glutamine from lawn grass. *Plant. Physiol.* 19:1–5.

Earley, E. B., 1943. Minor element studies with soybeans. I. Varietal reaction to concentrations of zinc in excess of nutritional requirement. *J. Amer. Soc. Agron.* 35:1012–1023.

Eaton, F. M., R. D. McCallum, and M. S. Mayhew, 1941. Quality of irrigation water with special reference to boron content and its effect on apricots and prunes. *U.S.D.A. Tech. Bull.* 746, 59 pp.

Furuta, T., W. C. Martin, and F. Perry, 1963. Lithium toxicity as a cause of leaf scorch on easter lily. *Proc. Amer. Soc. Hort. Sci.* 83:803–807.

Gerloff, G. C., 1963. Comparative mineral nutrition of plants. *Ann. Rev. Plant Physiol.* **14**:107–124.

Hayward, H. E., E. M. Long, and R. Uhvits, 1946. The effect of chlorine and sulfate salts on the growth and development of the Elberta peach on Shalil and Lovell root stocks. *U.S.D.A. Tech. Bull.* 922, 48 pp.

Holmes, F. W., 1961. Salt injury to trees. *Phytopathol.* **51**:712–718.

———, 1966. Salt injury to trees. II. Sodium and chloride in roadside sugar maples in Massachusetts. *Phytopathol.* **56**:6, 633–636.

Hunter, J. G., and O. Vergano, 1952. Nickel toxicity in plants. *Ann. Appl. Biol.* **39**:279–284.

Kadman, A., 1963. Uptake and accumulation of chloride in avocado leaves and the tolerance of avocado seedlings under saline conditions. *Proc. Amer. Soc. Hort. Sci.* **83**:280–286.

Kelley, W. P., and S. M. Brown, 1928. Boron in the soils and irrigation waters of southern California and its relation to citrus and walnut culture. *Hilgardia* **3**:445–458.

Lacasse, N. L., and A. E. Rich, 1964. Maple decline in New Hampshire. *Phytopathol.* **54**:9, 1071–1075.

Lilleland, O., J. G. Brown, and C. Swanson, 1945. Research shows excess sodium may cause leaf tip burn. *Almond Facts* **9**:1, 5.

Massey, A. B., 1925. Antagonism of the walnuts (*Juglans nigra* L. and *J. cinerea* L.) in certain plant associations. *Phytopathol.* **15**:773–784.

Neal, D. C., 1937. Crinkle leaf, a new disease of cotton in Louisiana. *Phytopathol.* **27**:1181–1175.

Oganesjan, A. P., 1953. Salt tolerance of some fruit crops. *Bot. Z.* **38**:744–51 (Hort. Abstr. 1954, no. 2412).

Patrick, Z. A., 1963. Phytotoxic substances in arable soils associated with decomposition of plant residues. *Phytopathol.* **53**:152–161.

Raabe, R. D., J. Vlamis, J. H. Hulimann, and J. Quick, 1966. Sodium injury to cuttings of chrysanthemums. *Calif. Agr.* **20**:11–13.

Romney, E. M., J. D. Childness, and G. V. Alexander, 1962. Beryllium and the growth of bush beans. *Science* **135**:786–787.

Toussoun, T. A., and Z. A. Patrick, 1963. Effect of phytotoxic substances from decomposing plant residues on root rot of bean. *Phytopathol.* **53**:265–270.

Woodbridge, C. G., N. R. Benson, and L. P. Batjer, 1960. Nutrition of fruit trees in the semi-arid regions of the Pacific Northwest in W. Reuther (ed.), "Plant analysis and fertilizer problems." *Am. Inst. Biol. Sci. Publ.* 8, Washington, D.C., 454 pp.

SELECTED REFERENCES

Aldrich, D. G., A. P. Vanselow, and G. R. Bradford, 1951. Lithium toxicity in citrus. *Soil Sci.* **71**:291–295.

Bonner, J., 1950. The role of toxic substances in the interaction of higher plants. *Bot. Rev.* **16**:51–65.

Borner, H., 1960. Liberation of organic substances from higher plants and their role in soil sickness problem. *Bot. Rev.* **26**:393–424.

Bradford, G. R., 1963. Lithium survey of California water resources. *Soil Sci.* **96**:77–81.

Brian, P. W., 1957. The ecological significance of antibiotic production, in "Microbiological ecology," pp. 166–188. 7th Symposium Soc. Gaz. Microbiol, Cambridge University Press, London.

Curtis, L. C., 1943. Deleterious effect of guttated fluid on foliage. *Amer. J. Bot.* **30**:778–781.

Eaton, F. M., 1944. Deficiency, toxicity and accumulation of boron in plants. *J. Agr. Res.* **69**:237–279.

Edlin, H. L., 1957. Salt burn following a summer gale in southeastern England. *Quart. J. Forest.* **51**:46–50.

Forster, W. A., 1953–54. Toxic effects of heavy metals on crop plants. Doctoral dissertation, Univ. of Bristol, England.

Hewitt, E. J., 1953. Metal interrelationships in plant nutrition. I. Effects of some metal toxicities on sugar beet, tomato, oat, potato and narrow stem kale grown in sand culture. *J. Exp. Bot.* **4**:59–64.

———, 1963. The essential nutrient elements: requirement and interactions in plants, in F. C. Steward (ed.), "Plant physiology," vol. 3, chap. 2, pp. 137–360. Academic, New York, 811 pp.

Ivanoff, S. S., 1963. Guttation injuries of plants. *Bot. Rev.* **29**:202–242.

Jacobsen, H. B. M., and T. R. Swanback, 1929. Manganese toxicity in tobacco. *Science* **70**:283–284.

Johnson, R. E., and W. A. Jackson, 1964. Calcium uptake and transport by wheat seedlings affected by aluminum. *Proc. Soil Sci. Soc. Amer.* **28**:381–386.

Jones, W. W., J. P. Martin, and W. P. Bitters, 1957. Influence of exchangeable sodium and potassium in the soil on the growth and composition of young lemon trees on different root stocks. *Proc. Amer. Soc. Hort. Sci.* **69**:289–296.

Kurauchi, I., 1956. Salt spray damage to the coastal forests. *Jap. J. Ecol.* **5**:213–17.

Lindner, R. C., 1943. Arsenic injury of peach trees. *Proc. Amer. Soc. Hort. Sci.* **42**:275–279.

Little, S. J., J. J. Mohr, and L. L. Spicer, 1958. Salt-water storm damage to Loblolly pine forest. *J. Forest.* **56**:27–28.

Ljones, B. K., 1954. Refsdal klorskade pa jordbaer (Chlorine injury to strawberries). *Frukt og Baer* **7**:72–80. (*Hort. Abstr.* **24**:2411, 1954).

Loneragan, J. F., M. D. Carroll, and K. Snowball, 1966. Phosphorus toxicity in cereal crops. *Aust. Inst. Agr. Sci. J.* **32**:221–223.

Lutman, B. J., 1922. The relation of the water pores and stomata of the potato leaf to the early stages and advance of tip burn. *Phytopathol.* **12**:305–333.

McIlrath, W. J., and B. F. Palsen, 1956. Responses of tomato, turnip and cotton to variation in boron nutrition. I. Physiological responses. *Bot. Gaz.* **118**:43–52.

McLean, F. T., and B. E. Gilbert, 1927. The relative aluminum tolerance of crop plants. *Soil Sci.* **24**:163–175.

Millikan, C. R., 1947. Effects of molybdenum on the severity of toxicity symptoms in flax induced by an excess of either manganese, zinc, copper, nickel, or cobalt in the nutrient solution. *J. Aust. Inst. Agr. Sci.* **13**:180.

Monk, R. W., and H. H. Wieke, 1961. Salt tolerance and protoplasmic salt hardiness of various woody and herbaceous ornamental plants. *Plant Physiol.* **36**:478–482.

Nicholas, D. J. D., 1951. Some effects of metals in excess on crop plants grown in soil cultures. I. Effects of copper, zinc, lead, cobalt, nickel and manganese on tomato grown in an acid soil. *Ann. Rep. Long Ashton Res. Sta.* 1950, 96.

―――― and W. D. E. Thomas, 1953. Some effects of heavy metals on plants grown in soil culture. I. The effect cobalt on fertilizer and soil phosphate uptake and the iron and cobalt status of tomato. *Plant and Soil* **5**:67–80.

―――― and ――――, 1954. Some effects of heavy metals on plants grown in soil culture. II. The effect of nickel on fertilizer and soil phosphate uptakes and the iron and nickel status of tomato. *Plant and Soil* **5**:182–183.

――――, 1961. Minor mineral nutrients. *Ann. Rev. Plant Physiol.* **12**:63–90.

Reinking, O. A., 1946. Tomato wilt caused by black walnut. *Farm Res. N.Y.S. Sta.* **12**:2–3.

Reuther, W. T., T. W. Embleton, and W. W. Jones, 1968. Mineral nutrition of tree crops. *Ann. Rev. Plant Physiol.* **9**:175–206.

Richards, L. A. (ed.), 1954. Diagnosis and improvement of saline and alkali soils. *U.S.D.A. Handb.* 60, 160 pp.

Richardson, S. D., 1955. Effects of sea-water flooding on tree growth in the Netherland. *Quart. J. Forest.* **49**:22–28.

Rossiter, R. C., 1952. The influence of soil type on phosphate toxicity in subterranean clover (*Trifolium subterraneae* L.) *Aust. J. Agr. Res.* **6**:1–8.

Smith, P. E., 1956. Effects of high levels of copper, zinc and manganese on tree growth fruiting. *Proc. Amer. Soc. Hort. Sci.* **67**:202–209.

Stiles, W., 1961. The effect on plants of trace-elements in excess, in "Trace elements in plants and animals," 3d ed., pp. 106–113. Cambridge, London, 249 pp.

Strong, F. C., 1944. A study of calcium chloride injury to roadside trees. *Quart. Bull. Mich. Agr. Exp. Sta.* **27**:209–24 (R.A.M. 1945, p. 294).

Taubenhaus, J. J., and W. N. Exekiel, 1934. Alkali scorch of Bermuda onions. *Amer. J. Bot.* **21**:69–71.

Traaen, A. E., 1950. Injury to Norway spruce caused by calcium chloride used against dust on roads. *Int. Bot. Congr. Proc.* **7**:185–186.

Vlamis, J., and D. E. Williams, 1962. Liming reduced aluminum and manganese toxicity in acid soils. *Calif. Agr.* **16**:6–7.

Wallace, R. H., 1949. Salt spray damage from recent New England hurricane. *Proc. Nat. Shade Tree Conf.* **15**:112–119.

Wallace, T. (ed.), 1950. "Trace elements in plant physiology." Chronica Botanica, Waltham, Mass., 144 pp.

PART FOUR

POLLUTION IN
THE ENVIRONMENT

CHAPTER FIFTEEN

SULFUR DIOXIDE

Sulfur dioxide emanating from home fires and furnaces, smelters and other industries, refineries, and power plants has destroyed vegetation, impaired crop production, and threatened human health in every industrial country in the world. Small quantities of sulfur oxides are normally present in the atmosphere from biological oxidation of sulfides, but excessive and often harmful quantities have been released by man's activities since he first began to refine sulfide-containing iron and copper ores over two thousand years ago. When man aggregated into large cities and used open coal fires for cooking and warmth, sulfur dioxide pollution posed an added threat. The situation was further intensified with the industrial expansion of the nineteenth century, when pollutants near smelters destroyed vegetation over the surrounding farms and countryside.

The chemical responsible for the damage remained unidentified for many years. The toxic nature of sulfur dioxide was not established in Germany until the 1890s; not until 1907 was this chemical accepted as the toxic component of smelter smoke in the United States, and only then were methods sought to reduce emissions.

15-1 SULFUR DIOXIDE SOURCES

Smelters, historically, have contributed the greatest amounts of sulfur oxides to the atmosphere; more significantly, their emissions have done much to call air-pollution hazards to the public's attention. Ores of copper, zinc, lead, nickel, and iron often form as sulfides in minerals which may contain as much as 10 percent or more sulfur. This waste product combines with oxygen in the air at the higher temperature of the smelter to produce principally sulfur dioxide, which is released into the atmosphere.

When Dr. Morris Katz summarized the total sulfur dioxide emissions on a worldwide basis in 1961, he estimated that 11 to 12 million tons were still being released from copper smelters and 3.5 to 4 million tons from lead and zinc smelters. But continually improving recovery methods, their economic feasibility, and public pressure are reducing the role of smelters as major contributors to sulfur oxide pollution. In 1965, Rohrman and Ludwig (1965) estimated that smelting of ores contributed only 7.4 percent of the total sulfur dioxide emission. Other sources of sulfur wastes, some of which have been present much longer than smelters, persist, and their relative importance is increasing.

Coal burning was recognized as a major contributor to air pollution hundreds of years before the Industrial Revolution, and it presently accounts for some 60 percent of the sulfur dioxide pollution (Rohrman and Ludwig, 1965). Some 41 percent of the coal is used in generating electrical power; the remaining 19 percent goes into space heating and similar combustion processes. Domestic coal burning, largely for heating, further increases the sulfur dioxide pollution of the atmosphere; more sulfur oxides originate from refining and combustion of petroleum products, which Rohrman and Ludwig calculate to comprise some 20.7 percent of the total sulfur dioxide emissions. Lesser amounts come from coke processing and incineration.

15-2 DISPERSION OF SULFUR OXIDES

Once sulfur oxides are released into the atmosphere, what is their dispensation? Gaseous contaminants are absorbed by large bodies of water or by precipitation and are diluted by anything with which they come into contact—land and water surfaces, rain clouds, rain, dew, or snow. Sulfur dioxide readily goes into solution in water (Terraglio and Manganelli, 1967). Over the concentration range of 0.81 to 8.73 mg SO_2/m^3, the rate of solution is a function of the sulfur dioxide concentration; the higher the concentration, the more rapidly is saturation reached. This suggests that rainwater, with constantly renewed surfaces, can be very effective in removing sulfur dioxide from the atmosphere. More is removed by impact against buildings, vegetation, or the ground.

If not absorbed on surfaces, sulfur dioxide is soon oxidized to the less toxic SO_3 and finally to sulfuric acid mist. Oxidation of SO_2 to H_2SO_4 is most rapid when wind speeds are low and oxidants high (Thomas, 1961). Sulfuric acid aerosols are neutralized by reacting with numerous compounds in the atmosphere or on contact surfaces such as ammonia and calcium sulfate. A substantial portion of the sulfur dioxide is directly neutralized by ammonia, calcite dust, or one of many other alkalis, and rapidly oxidized to sulfate. These salts are ultimately removed by precipitation or settle out by gravity. Sulfuric acid mists are also quick to form condensation nuclei which become sufficiently large to be removed by gravitational settling.

In regions where precipitation is sparse, sulfur dioxide is longer-lived and settles out slowly together with other aerosols and dusts. Once it falls upon the vegetative canopy, sulfur dioxide may be absorbed directly by the plants. Large quantities are absorbed by the external plant surfaces, which act as filters screening out much of the sulfur dioxide in the lower atmosphere. Fried (1949) has demonstrated that alfalfa plants and lemon trees can absorb sulfur dioxide directly through the leaves and utilize it; but most of the sulfur probably settles on the foliage and is ultimately washed to the ground. Sulfur dioxide, then, does not remain in the air indefinitely; its persistence depends on the amount of reactive and surfaces available and rarely persists more than forty-three days (Junge and Werby, 1958).

15-3 CONCENTRATION REACHING THE PLANT

The concentration of sulfur dioxide to which the plant is exposed depends first on the concentration at the source and second on the degree of dilution in transit. The extent of dilution is determined in part by the density of vegetation and other surfaces but primarily by the length of time between emission and deposition. In the case of single sources of pollution such as power plants or smelters, this period is extended by use of tall smokestacks which release the pollutants several hundred feet high so they are dispersed over great distances and diluted correspondingly.

Meteorological conditions have the greatest influence on the dispersal and removal of air pollutants. Winds or vertical turbulence are necessary to dilute the pollutants; without these, they accumulate in concentrations which may easily become toxic. Other meteorological parameters, such as rainfall, the turbulent structure of the wind, and low-level temperature and wind gradients, also influence the extent to which pollutants are diluted.

15-4 FOLIAR ACCUMULATION AND CONCENTRATION

Sulfur dioxide enters the plant principally through the leaf stomata and passes into the intercellular spaces of the mesophyll, where it contacts, and is absorbed on, the wet cell walls (Thomas et al., 1950). Here it combines with water to form sulfurous acid and sulfate. If stomates were the only means of entry, or if sensitivity were solely a function of sulfur dioxide concentrations, a correlation might be expected betweeen the size or number of stomates and the plant's sensitivity to sulfur dioxide. The greater the number of stomates, the more sulfur dioxide would be expected to enter. But no correlation has been disclosed between susceptibility and the number of stomata per unit area. Rather than implying that stomates are not important routes of sulfur dioxide access, this indicates that inherent, physiological factors are still more important in determining susceptibility. Once within the plant, sulfur has been found to be distributed rather uniformly throughout the leaves according to the normal distribution of sulfur-bearing proteins. The total sulfur content of the leaf, then, might be expected to provide a measure of the degree of leaf contamination and an index of the amount of air pollution. But such is not always the case. While leaves exposed to sulfur dioxide may contain several percent more sulfur than normal for a given species in a given location, the relation of sulfur dioxide exposure to sulfur content is not quantitative. Similarly, no quantitative relation exists between the total sulfur content and the degree of injury. Sulfur content of leaves is useful mainly to obtain a general idea of concentration increases once the normal levels are known and to define the scope of the area which has been exposed to sulfur dioxide.

15-5 PLANT RESPONSES

Mechanism of Injury

Ever since the toxic nature of sulfur dioxide was first recognized, workers have speculated as to the mechanism by which plants are injured. In 1903, Haselhoff and Lindau postulated that sulfur dioxide combined with aldehydes and sugars, forming secondary products which, when they slowly decomposed, released sulfurous or sulfuric acid into the plant cells, causing injury. A drawback to this hypothesis, which tends to refute it, is that leaves are most sensitive to sulfur dioxide in the morning, when the sugar content is low.

In 1929, Noack suggested that sulfur dioxide inactivated the iron in the chloroplasts, interfering with its catalytic properties. This interference was believed to promote secondary processes which broke down the chlorophyll and killed the cells.

Dorries (1932) later proposed that sulfur dioxide caused a local acidity which split magnesium from the chlorophyll molecule, converting it to phaeophytin. Phaeophytin then accumulated at the expense of chlorophyll, causing chlorosis and reduced photosynthesis.

The potential toxicity of excess sulfate has also been considered. Although sulfates ordinarily are not in themselves considered toxic, they can be harmful if present in excessive quantities. High sulfate may limit calcium utilization and induce a deficiency of this element, interfere with ion absorption, or otherwise disrupt the nutrient balance.

Bleasdale (1952) attributed the toxic nature of sulfur dioxide to its reducing property, postulating that there is normally an equilibrium between sulfhydryl groups and more oxidized sulfur compounds, such as sulfites. Any imbalance in this equilibrium, as might be caused by an excess of oxidized sulfur compounds, would upset sulfur utilization and protein synthesis. Sulfur dioxide would reduce the sulfites to sulfhydryl, causing an accumulation of sulfhydryl and an imbalance in the ratio of sulfhydryls to oxidized sulfur compounds. The tolerance of plants to sulfur dioxide may be a function of the stability of sulfhydryl. Although not yet proved, this concept appears to be the most plausible.

Cellular Effects

When plant tissues accumulate sulfur dioxide faster than it can be oxidized and assimilated, a threshold of accumulation is exceeded, and phytotoxic concentrations presumably arise in the intercellular spaces of the leaf, causing cell injury (Thomas, 1961).

Detailed anatomical studies of sulfur dioxide injury on alfalfa were described by Katz and Ledgham in 1939. Cells in the palisade tissue began to shrink and collapse, and the entire leaf shrank in thickness before any external symptoms appeared.

Shrinkage in the absence of chlorosis and necrosis most commonly occurs following prolonged exposure to low concentrations of sulfur dioxide. External, macroscopic symptoms arise as the chloroplasts disintegrate, chlorophyll becomes dispersed in the cytoplasm, and the protoplast of the spongy and palisade parenchyma becomes plasmolyzed.

Studies of sulfur dioxide–injured bean cells and tissues conducted by Richard Solberg at Washington State University (Solberg and Adams, 1956) showed that the spongy mesophyll cells closest to the lower epidermis were the first to be damaged. Chloroplasts disintegrated, cells plasmolyzed, and the protoplast collapsed. Cell walls become distorted and "crinkled" but retained their connection with each other. The lower epidermal cells collapsed at nearly the same time. The palisade tissue and upper epidermis broke down next. Vascular tissues, especially xylary elements, were most tolerant and retained their form even when surrounding tissues were completely collapsed. Phloem cells were distorted and collapsed only when leaf injury was severe.

Macroscopic Effects

When a sufficient number of cells have been plasmolyzed, the affected tissues collapse and dry out, revealing a pattern of macroscopically visible, acute or chronic symptoms. Acute injury, caused by high concentrations of sulfur dioxide, is characterized by the rapid disappearance of chlorophyll, breakdown of cells, and development of necrosis. Chronic injury is slower and results from the gradual breakdown of chlorophyll and development of chlorosis without any cellular collapse. It involves reduced metabolic activity, decreased photosynthesis, and general growth suppression. There is no sharp line of demarcation between acute and chronic injury; the terms represent more a degree of injury than a type and reflect the sulfur dioxide concentrations present and the metabolic activity and susceptibility of the plant.

The basic cellular responses to sulfur dioxide are the same for all species, but because of variations in their anatomy different species show variations in symptoms. Because of their different structural characteristics, the greatest difference occurs between needle-leaved and broad-leaved species. Symptoms, relative sensitivity, and the sulfur dioxide concentrations causing toxicity are discussed separately for these groups.

NEEDLE-LEAVED SPECIES   The most pronounced symptoms of sulfur dioxide injury on needle-leaved evergreens are the reddish-brown discoloration of leaves, shrinkage of tissues, and early defoliation, which give the tree a thin, sparse-foliaged, weak appearance with correspondingly reduced growth. Damaged pine needles may drop after one to three years rather than clinging for the normal five or more years; Douglas fir needles may be shed within a few days or weeks of an injurious exposure to sulfur dioxide (Katz et al., 1939).

When plants are subjected to toxic concentrations of sulfur dioxide during the summer months, necrosis most often begins at the tip of the needle and progresses toward the base. The entire length may be affected, or only a portion; sometimes only a limited area at the base, middle, or tip of the leaf is marked. Thus, necrosis may appear as tip burn, banding, or basal burn and may develop within a few days of exposure. When the sulfur dioxide concentration is low, and close to the threshold of injury, chloroplasts and cells are damaged but not killed. Injury then appears only as a chronic chlorosis or yellowing of the affected portions of the leaf. Necrosis and chlorosis often develop on scattered needles in a cluster rather than on all the needles of a shoot.

The plant may respond differently during the winter months. Winter exposures are more apt to cause a general chlorosis followed by definite yellowish-brown banding of the most severely injured portions. Gradually, over a period of one or two months, the intervening green areas fade to a general yellow, brown, and ultimately reddish-brown discoloration. Because of the genetic variation, symptoms may vary somewhat among species and even among different individuals of a single species.

Needle damage and defoliation are usually most pronounced on the current year's needles at the ends of limbs. Older leaves on parts of the limbs closest to the trunk, having a lower metabolic acitivity, are most likely to be spared. If the tree recovers, numerous adventitious buds develop which give rise to new growth on the main stem and inner portion of the branches.

The sulfur dioxide levels required to be injurious depend on the sensitivity and metabolic activity of the plant, the time of year, and the rate of sulfur dioxide absorption. Katz (1952) found that injury occurs whenever the sulfur dioxide is accumulated faster than it can be assimilated. Large amounts could be absorbed without injury if absorption were slow; but if accumulated at over 10 ppm per day by dry weight of leaf tissue, sulfur dioxide was highly destructive. Damage is generally least during the winter months when the leaves are relatively dormant, since gas exchange and sulfur dioxide accumulation is minimal. Plants are most sensitive during April and May when leaves are developing. For instance, fumigation of sensitive conifers at 2.0 ppm for 69 hours in February failed to cause injury, while a 35-hour fumigation in March destroyed 75 percent of the foliage.

In spring, species such as larch are damaged by sulfur dioxide concentrations of only 0.3 ppm for 8 hours. Slight markings were produced on one-year-old pine and two-year-old Douglas fir seedlings by 0.29 ppm for 44 hours and 0.78 for 8 hours. Charles Berry (1967, personal communication) has found some clones of Eastern white pine to be even more sensitive; these were injured by only 0.1 ppm for 8 hours when the humidity was high and the nitrogen nutrition low (Fig. 15.1).

Different species vary greatly in their susceptibility to sulfur dioxide. The order of susceptibility is given by Katz (1939) for plants studied in British Columbia. Beginning with the most sensitive species, Katz ranks the susceptibility of native conifers as western larch, Douglas fir, yellow pine, Engelmann spruce, white pine, hemlock, lodgepole pine, silver fir, white fir, and red cedar. Junipers, with their small, cutinous, awl-like leaves, are among the most tolerant species.

BROAD-LEAVED SPECIES   The wide range of chlorotic and necrotic markings caused by sulfur dioxide on dicotyledonous plants can perhaps best be illustrated by the response of alfalfa. Alfalfa is among the most sensitive and important species affected and has been studied

FIGURE 15.1 *Needle-tip necrosis of Eastern white pine. Plants were fumigated by Dr. Charles Berry at 0.25 ppm.*

intensively. The markings may be either acute or chronic; if acute, the injury may be marginal, intercostal, or veinal.

Acute injury occurs when cells are killed. Cells in tissues accumulating the greatest amount of sulfur dioxide lose their capacity to retain water; the cell sap diffuses through the intercellular spaces, giving the area first a water-soaked, dull, gray-green appearance. The flaccid area dries out, leaving bleached, light tan to ivory, necrotic zones extending through the leaf (Figs. 15.2, 15.3).

The mildest acute symptoms, usually produced following long exposures to sulfur dioxide concentrations of 0.3 to 0.5 ppm, consist of a narrow border of necrotic tissue along the margins, or at the tip, of the leaflet. Necrosis may be sharply delimited but characteristically extends irregularly toward the midrib between the veins.

At concentrations above 0.5 ppm, or under conditions of high

FIGURE 15.2 *Acute marginal and intercostal necrosis on alfalfa from sulfur dioxide.*

light intensity and humidity, necrosis tends to be intercostal. Lesions may extend inward from the margin or be limited to irregular areas over the leaflets and generally confined between the larger veins. Affected areas may be of any size or shape, ranging from minute flecks through involvement of most of the leaf area. The intercostal lesions, rather than being sharply delimited, may have a diffuse chlorotic border. Chlorosis may range from an almost normal green color to pale yellow on the same leaf and be more intense on either the upper or lower leaf surface.

On other occasions, necrosis is restricted to tissues along and over the veins, producing a "Christmas tree" pattern. This veinal type of necrosis may be confined to the veins or overlap them so that irregular portions of the interveinal tissue are also affected.

When the chlorophyll is only partially destroyed, and the protoplast survives, the cells remain partially functional. This injury is considered to be chronic and results from prolonged exposures to low concentrations of sulfur dioxide in the order of 0.1 to 0.3 ppm. It is best characterized by irregular, blotchy chlorosis developing over the same tissues as acute injury. In its mildest form the chlorosis is transient, and tissues return to normal in a day or two; in its severest form, chlorophyll is completely lost, and the leaf becomes yellow-green or brown. Sometimes, then, there is no clear demarcation between acute and chronic injury, and the two often occur together.

The response of monocotyledonous plants to sulfur dioxide can be represented by a cereal crop such as barley which is highly sensitive and readily damaged by concentrations of 0.3 to 0.5 ppm sulfur dioxide. Injury first appears as a diffuse gray-green discoloration of the leaf tip. Chloroplasts break down, and chlorophyll diffuses into cytoplasm. If exposed to sunlight, affected tissues become desiccated and flaccid, rapidly shrink, and become bleached. Tip and marginal necrosis is often accompanied by spotted, interveinal flecks or lesions between the midrib and margins. While the most sensitive part of young leaves is usually the tip, the tissues at the bend of older, bent-over leaves most directly exposed to the sun are equally sensitive.

FIGURE 15.3 *Sulfur dioxide injury on alfalfa, showing the variation in the field.*

At still lower concentrations below about 0.3 ppm, chlorosis rather than necrosis is most prominent.

Symptoms on other broad-leaved species are similarly characterized by intercostal chlorosis and necrosis (Fig. 15.4). The earliest necrosis to develop is usually pale brown in color; this intensifies, finally becoming dark brown, reddish-brown or even reddish-purple. Leaves on plants such as pear and aspen, which tend to blacken in response to almost any pathogen, turn black in response to sulfur dioxide. The necrotic areas consist of numerous small, interveinal spots with a definite margin. These coalesce, forming elongated lesions between the main veins. Necrosis may also start at the margin and extend inward between the veins. In rare instances, necrosis is most prominent near the midrib and develops outward.

Sulfur dioxide damage around major smelters has provided botanists with an excellent opportunity to study the relative sensitivity of native species. Broad-leaved trees and shrubs are generally more tolerant than conifers and continue to grow in areas where the conifers have been killed out. Katz (1939) ranked birch, bitter cherry, aspen, ninebark, ocean spray, and service berry as the most sensitive species. All are sensitive to sulfur dioxide and can serve as useful indicator plants in natural areas.

FIGURE 15.4 *Sulfur dioxide–induced necrosis of red clover leaves.*

The most extensive studies to determine the sensitivity of species to sulfur dioxide were conducted by P. J. O'Gara at the American Smelting and Refining Agricultural Station, who examined the response of over 300 plant species and varieties (Thomas et al., 1950). Sensitivity of the more common of these species, and additional native plants, are presented in Table 15.1. The "O'Gara" factor represents the relative sensitivity compared with alfalfa, the most sensitive species, which is given a value of 1.0. The higher numbers represent increasing resistance.

TABLE 15.1. *Relative sensitivity of native and cultivated plants to sulfur dioxide.* (*A low number indicates high sensitivity.*)

Sensitive		Intermediate		Resistant	
Alfalfa	1.0	Yellow pine†	1.6	Gladiolus	1.1—4.0
Barley	1.0	Dandelion	1.6	Sweet cherry	2.6
Endive	1.0	Sugarbeet	1.6	Purslane	2.6
Cotton	1.0	Aster	1.6	Rose	2.8—4.3
Gaura	1.0	Tomato	1.3—1.7	Sumac	2.8
Cheatgrass	1.0	Lambs' quarter	1.8	Shepherds' purse	3.0
Mallow	1.1	Apple	1.8	Maple	3.3
Ragweed	1.1	Catalpa	1.9	Box elder	3.3
Rhubarb	1.1	Sweet clover	1.9	Virginia creeper	3.8
Radish	1.2	Cabbage	2.0	Onion	3.8
Lettuce	1.2	Marigold	2.1	Lilac	4.0
Zinnia	1.2	Pea	2.1	Corn	4.0
Spinach	1.2	Linden	2.3	Cucumber	4.2
Bean	1.1—1.5	Douglas fir	2.3	Salt grass	4.6
Curly dock	1.2	Peach	2.3	Chrysanthemum	5.3—7.3
Table beet	1.3	Apricot	2.3	Citrus	6.5—6.9
Buckwheat	1.3	Cocklebur	2.3	Arborvitae	7.8
Plantain	1.3	Elm	2.4	Currant blossoms	12.0
Sunflower	1.3—1.4	Iris	2.4	Live oak	14.0
Clover	1.4	Poplar	2.5	Apple Blossoms	25.0
Rye	1.4	Yellow pine	2.4—4.7	Apple buds	87.0
Carrot	1.5				
Wheat	1.5				
Larch	1.5				

* Adapted from Thomas et al., 1950.
† Year-old seedlings in May, 1.6; in August, 2.4–4.7.

Growth and Physiological Effects

Any time plant tissues are injured, whatever the cause, growth and production are reduced at least in proportion to the degree of tissue destruction. Some of the most intensive, quantitative research to determine the relation of reduced leaf area to production was conducted by M. D. Thomas and George R. Hill at the American Smelting and Refining Company Research Station in the Salt Lake Valley (Hill and Thomas, 1933). In one of a series of studies, alfalfa plots were fumigated with sulfur dioxide one, two, or three times during the growing season. Ten to fourteen days were allowed to lapse between fumigations and between the final fumigation and harvest. The amount of leaf destruction caused by the sulfur dioxide was determined by counting the injured and uninjured leaves on representative samples and estimating the average areas of injury on the marked leaves. These data provided a leaf destruction value. Crops were harvested at the normal time, and the dry-weight yield was determined and compared with that of the control plants. Correlation coefficients calculated between the amount of leaf destruction and yields showed that yield loss was directly proportional to the amount of leaf necrosis. Similar results were found by the National Research Council of Canada in 1939.

Studies of a similar nature with barley and wheat in Utah, California, and Arizona showed the yields of these crops were less affected by leaf destruction than alfalfa. Since the fruit of cereal crops, not the leaf area, is the primary consideration, the stage of plant development when leaf destruction occurs becomes critical. Leaf destruction during tillering is relatively unimportant since the plant simply develops more tillers, or the undamaged portions of the leaves enlarge sufficiently to compensate for the damage. Injury becomes increasingly important after the shoots form, especially during the bloom period. The importance of leaf injury again diminishes after the dough stage, since by then the plants are beginning to dry up normally. At no stage of development is the yield less than proportional to leaf damage.

In studies with wheat, Brisley and Jones (1950) calculated the actual square centimeters of leaf tissue and the area which was rendered necrotic or chlorotic by sulfur dioxide. The leaf area destroyed was compared with the yield. The loss generally ranged from 0.26 to 0.62 percent crop reduction for each 1 percent of foliage damage.

The effect of sulfur dioxide on growth of conifers was studied by Lathe and McCallum (1939) using the increment diameters of yellow pine and Douglas fir as the criteria. During a four-year period, more than 175,000 measurements were made from borings in 10,043 trees extending nearly 100 miles from the smelters at Trail, British Columbia. Growth was markedly reduced in all areas where leaf burning and defoliation occurred, but not beyond this area.

Controlled fumigation studies later reported by Katz (1952) confirmed that sulfur dioxide did not affect growth of conifers in the absence of necrosis. He also found that even severely injured trees gradually recovered. The defoliation and needle burning, however, caused a retardation of growth up to three years after exposure even when less than 10 percent of the leaf area was necrotic.

The relation of growth suppression to necrosis was again demonstrated in the vicinity of Anaconda, Montana, smelters (Sheffer and Hedgecock, 1955). Growth of the alpine fir, Douglas fir, and lodgepole pine was most seriously retarded. Similarly, reduced growth from sulfur dioxide toxicity has been reported by Karlen and Tyden (1958) for conifers in Sweden (cited by Scurfield), and Linzon (1958) for Eastern white pine in Canada, but only when necrosis was present. Sheffer and Hedgecock (1955) reported that height growth of ponderosa pine was also limited whenever necrosis was present.

The most difficult question to resolve has been whether or not sulfur dioxide restricts growth, production, or metabolic activity in the absence of chlorosis or necrosis. In 1923, Julius Stoklasa postulated that toxic gases, at concentrations below those causing any visible symptoms, caused a reduction of photosynthesis, early senescence, an unthrifty appearance, reduced growth and yield, and increased susceptibility to disease and insects. This has come to be known as the "invisible injury" theory of plant damage and has been the subject of intensive research for many decades. Usage of this term has been opposed on the grounds that if injury were "invisible," a loss could not be measured. Possibly "physiological injury" would be more accurate, although all injury is basically physiologic.

Attempts to determine if such sublethal, "physiological" injury actually exists have provided excellent data concerning the hazards of sulfur dioxide. The results of studies at the American Smelting and Refining Company, summarized by Thomas (1956), are representative. Alfalfa plants were fumigated with sulfur dioxide at concentrations from 0.08 to over 8 ppm for varying periods of time, and carbon dioxide absorption and evolvement were determined as a measure of photosynthesis and respiration activity. Continuous fumigation with 0.08 to 0.22 ppm for over six weeks failed to reduce photosynthetic activity, although older leaves of treated plants became senescent earlier than leaves of control plants. Continued exposures to concentrations about 0.2 ppm had no effect the first ten days, but during the next two weeks at this concentration photosynthetic activity diminished gradually by about 10 percent. At 0.3 ppm, photosynthetic activity was reduced by twenty-five percent in the course of two weeks. Concentrations of 0.3 to 0.4 ppm for about four hours per day were tolerated for over a month without causing any significant reduction in photosynthesis, but at 0.45 ppm for four hours per day photosynthesis was reduced slightly but insignificantly during the period of fumigation. If the fumigations were discontinued before lesions appeared, photosynthesis returned to normal within one to two hours.

Concentrations of 0.6 ppm for four hours per day reduced photosynthetic activity twenty-five percent during fumigation. Six such fumigations during a two-week period reduced the total assimilation by 1.0 percent and the crop yield by 14 percent. However there was 2.5 percent leaf destruction.

Apparently, sulfur dioxide can cause some growth reduction in the absence of visible markings when a threshold concentration is exceeded. Thomas (1956) concludes: "Under these conditions it appears that sulfurous acid or sulfite can accumulate in the cells of the plants and inhibit photosynthesis without necessarily killing the cells. At subthreshold concentrations, the sulfite is oxidized to the nontoxic sulfate as rapidly as it is absorbed, so that inhibition of photosynthesis does not occur."

But the extent of impairment of growth and other metabolic processes by sublethal sulfur dioxide concentrations is far from resolved. Indications of sublethal physiological injury have appeared in some studies (Guderian and Stratman, 1968). Katz (1952) reported instances at Trail, British Columbia, where otherwise healthy-appearing trees in polluted areas prematurely lost their needles. Grayson (1956) suggested that sulfur dioxide, particularly when present in combination with fluorides, might reduce growth indirectly by accelerating leaf abscission. Linzon (1958) observed that mortality from unknown causes around Ontario smelters was three times higher in areas of high sulfur dioxide concentrations than in control areas. The reasons for this were not known.

Reproduction

Sulfur dioxide might be expected to impair reproduction in one or more of several ways: by direct injury to reproductive structures, damage to pollen, damage to the seed, reduced viability of seeds, or damage to seedlings. Some of these processes were reviewed by Katz (1939).

Pine pollen was fumigated with 10 ppm sulfur dioxide for six days without any effect on germination. Germination was prevented only when concentrations were raised to 200 ppm. The threshold was not determined, but apparently the levels necessary to impair pollen development were far higher than levels ever recorded in the field.

Observations of flowers exposed to high sulfur dioxide concentrations (Katz, 1939) revealed few markings, even when leaves were severely burned. Flowers which were fumigated also proved to be highly resistant to sulfur dioxide; it was concluded that reproduction was not adversely affected by injury to flowering organs.

Reproductive organs of conifers may be far more sensitive; in the area near Anaconda, Montana, cones of alpine fir, Douglas fir, and lodgepole pine trees which survived near the smelter were to a large extent abortive, and natural regeneration was limited (Hedgecock, 1912). However, germination showed no impairment from high sulfur dioxide exposures. Sulfur dioxide appears to have its greatest influence on reforestation when only very few parent trees are present and direct injury affects the sparse young seedlings.

Environmental Variables

The severity of injury produced is determined not only by the sulfur dioxide concentrations and duration of exposure but by many environmental factors, including light, moisture, temperature, and nutrient relations.

Since sulfur dioxide enters the leaf primarily through the stomata, anything helping to keep the stomata open would facilitate sulfur dioxide accumulation. Light is considered the most important factor influencing the stomatal aperture. Susceptibility to sulfur dioxide increases with light intensity up to full sunlight so long as the tissues do not dry out and there is no moisture stress. At least four times as much sulfur dioxide is required to produce injury at night compared with daytime.

Factors producing high leaf turgor and maximum photosynthesis also render plants more susceptible to sulfur dioxide injury. Plants grown in an ample supply of water are much more susceptible to injury than those growing under even a slight stress. When plants were slightly wilted, the stomata closed, and plants were able to tolerate increased concentrations of sulfur dioxide without being injured. Relative humidity, together with soil moisture, influences sensitivity. Generally, sensitivity to sulfur dioxide increases with increased relative humidity, and at 100 percent the time required for injury decreases markedly (Thomas, 1956).

Temperature is also important, and susceptibility is greatest at temperatures over 40°F. The far higher concentrations required to injure conifers during the winter months are due partly to the lower temperatures as well as the much-reduced gas exchange (Katz, 1939).

The age of plants and tissues further influences sensitivity. Young plants tend to be more resistant to sulfur dioxide, although there are notable exceptions. Seedlings of conifers are far more susceptible than mature plants or those in the flowering stage. Middle-aged leaves are most susceptible to acute and chronic markings, although the older leaves have the highest sulfur content and are first to become senescent.

Berry (1967) has shown the importance of nutrition. Eastern white pine trees growing under low nitrogen conditions were found to be far more sensitive to sulfur dioxide than those growing on better sites. Over 25 percent necrosis developed on 90 percent of the needles on infertilized plants, while plants provided with normal amounts of nitrogen remained unmarked.

Ecological Aspects

When sufficient numbers of plants in a natural population are damaged, the entire population and the associated vegetation may be destroyed. There are many instances where entire plant communities have been devastated by prolonged exposures to high concentrations of sulfur dioxide. The classic illustration of such destruction is the Copper Basin area of Tennessee, where some 7,000 acres of once-rich deciduous forest were completely denuded and another 17,000 acres replaced with grassland species following the destruction of the native forest species (Hedgecock, 1912). Gully erosion stripped away the soil, and even the climate was shown to have been altered (Hursh,

1948). In many instances, the topsoils are so severely eroded that the original type of vegetation cannot be supported and only a few highly tolerant introduced weeds predominate.

Timber and watershed have been equally decimated in areas surrounding Trail, British Columbia, in Ontario, Canada, and in Anaconda, Montana, where sulfur oxides again were present in concentrations lethal to dominant conifers (Scheffer and Hedgecock, 1955). Damage to timber near the smelters was obvious, but damage to understory species or to changes in plant populations beyond the area of mortality were not considered. No information is available on the effects of the pollutants on the bulk of the native species or on the plant community itself, although an obvious decrease in the number of different species present near the smelters is still apparent over fifty years after the initial damage.

One of the earliest instances of sulfur dioxide affecting plant populations was given by Toumey (1921), who reported that conifers were gradually disappearing along a narrow valley in Connecticut where coal consumption and output of copper, zinc, and other metals had increased rapidly from 1914 to 1918. Increment studies revealed that tree growth had been reduced 25 to 50 percent in this area. The weakened trees were unable to compete successfully and were gradually being replaced by more tolerant, broad-leaved species.

Another effort to learn what air pollutants were doing of the ecosystem was conducted by Gordon and Gorham (1963) in the vicinity of an Ontario iron-smelting plant where sulfur dioxide emissions exceeded 100,000 tons per year. Quadrats were laid out along a 36-mile transect, and all macrophytic ground flora species and their densities were listed for each quadrat. A striking decline in the variety of plants occurred within about 10 miles from the pollution source. The numbers of species per quadrat declined from 28 beyond 10 miles to 0 to 2 within 3 miles of the pollution source. Vegetation seemed "to have been peeled off in layers." The tree cover which dominated in more remote areas where injury was not obvious was almost wholly destroyed closer in and replaced by shrubs and other understory species. Closer yet, the understory consisted only of low-growing vegetation; even this was lacking closest to the sintering plant.

Population changes have also been demonstrated in controlled fumigation studies (Guderian, 1967). In mixed populations of forage crops, the more sensitive clover was visibly injured and gradually disappeared, to be replaced by more tolerant grasses.

BIBLIOGRAPHY

Bleasdale, J. K. A., 1952. Air pollution and plant growth, in "Smokeless air," report in *J. Park Admin.* **18**:300–301 (*Hort. Abstr.* 1954, no. 1084).

Brisley, H. R., and W. W. Jones, 1950. Sulfur dioxide fumigation of wheat with special reference to its effect on yield. *Plant Physiol.* **25**:666–681.

Dorries, W., 1932. Uber die Brauchbarkeit der spectroscopic Phaophytin-probe in der Rauchshaden-Diagnostik. *Z. Pflanzenkr. Gallenk.* **42L**:257–273.

Fried, M., 1949. The absorption of sulfur dioxide by plants as shown by the use of radioactive sulfur. *Proc. Soil Sci. Soc. Amer.* **13**:135–138.

Gordon, A. G., and E. Gorham, 1963. Ecological aspects of air pollution from an iron-sintering plant at Wawa, Ontario. *Can. J. Bot.* **41**:1063–1078.

Grayson, A. J., 1956. Effects of atmospheric pollution in forestry. *Nature (London)* **178**:719–721.

Guderian, R., 1967. Reaktionen von Pflanzengemeinschaften des Feldfutter-baues auf Schwefeldioxideinwirkungen. Schriftenreihe der Landesan-stalt für Immissions- und Bodennützungsschutz des Landes. *Naturwiss.* **4**:80–100.

Guderian, R., and H. Stratmann, 1968. "Freilandversuche zur Ermittlung von Schwefeldioxid wirkungen auf die vegetation." Teil 3: Grenzwerte schädlicher SO₂-Immissionen für Obst- und Forstkulturen sowie für landwirtschaftliche und gärtnerische Pflanzenrten. Köln und Oplanden, 114 pp.

Haselhoff, E., and G. Lindau, 1903. Die Beschädigung der Vegetation durch Rauch. Borntraeger, Leipzig.

Hedgecock, G. G., 1912. Winter killing and smelter injury in forests of Montana. *Torreya* **12**:25–30.

Hill, G. R., and M. D. Thomas, 1933. Influence of leaf destruction by sulfur dioxide and by clippings on yield of alfalfa. *Plant Physiol.* **8**:223–45.

Hursh, C. R., 1948. Local climate in the copper basin of Tennessee as modified by the removal of vegetation. *U.S.D.A. Circ.* 774, 38 pp.

Junge, C. E., and R. T. Werby, 1958. The concentration of chloride, sodium, potassium, calcium, and sulfate in rain water over the United States. *J. Meteorol.* **15**:417.

Katz, M., G. A. Ledgham, and A. W. McCallum, 1939. Symptoms of injury on forest and crop plants, in "Effect of Sulfur dioxide on vegetation." *Nat. Res. Counc. (Canada) Publ.* no. 815, 447 pp.

——— (ed.), 1939. "Effect of sulfur dioxide on vegetation." *Nat. Res. Counc. (Canada) Publ.* no. 815, 447 pp.

———, 1952. The effect of sulfur dioxide on conifers, in L. McCabe, "Air pollution," pp. 84–96. McGraw-Hill, New York.

———, 1961. Some aspects of the physical and chemical nature of air pollution, in "Air pollution," pp. 97–158. *W.H.O. (Geneva) Monogr.* no. 46.

Lathe, F. E., and A. W. McCallum, 1939. The effect of sulfur dioxide on the diameter increment of conifers, in M. Katz, "Effect of sulfur dioxide on vegetation," pp. 174–206. *Nat. Res. Counc. (Canada) Publ.* no. 815, 447 pp.

Linzon, S. N., 1958. The influence of smelter fumes on the growth of white pine in the Sudbury region. *Can. Dept. Agr. Forest. Biol. Div. Sci. Ser. (Toronto)*, 45 pp.

Noack, Kurt, 1929. Damage to vegetation from gases in smoke. *Z. angew. Chem.* **42**:123–126.

Rohrman, F. A., and J. H. Ludwig, 1965. Sources of sulfur dioxide pollution. *Chem. Eng. Prog.* **61**:59–63.

Scheffer, T. C., and G. G. Hedgecock, 1955. Injury to northwestern forest trees by sulfur dioxide from smelters. *U.S.D.A. Tech. Bull.* 1117, 49 pp.

Scurfield, G., 1960. Air pollution and tree growth. *Forest. Abstr.* **21**:3, 4.

Solberg, R. A., and D. F. Adams, 1956. Histological responses of some plant leaves to hydrogen fluoride and sulfur dioxide. *Amer. J. Bot.* **43**:775–760.

Stoklasa, Julius, 1923. "Die Beschädigungen der Vegetation durch Rauchgase und Fabriksexhalationen." Urban & Schwarzenberg, Berlin.

Terraglio, F. P., and R. M. Manganelli, 1967. The influence on the absorption of atmospheric SO_2 by soil. *Air Water Pollut. Int. J. (Oxford)* **10**:783–406.

Thomas, M. D., R. H. Hendricks, and G. R. Hill, 1950. The sulfur metabolism of plants. Effect of sulfur dioxide on vegetation. *Ind. Eng. Chem.* **42**:2231–2235.

———, 1956. The invisible injury theory of plant damage. *J. Air Pollut. Contr. Assoc.* **5**:205–206.

———, 1961. Effects of air pollution on plants, in "Air pollution," pp. 233–278. *W.H.O. (Geneva) Monogr.* no. 46.

Toumey, J. W., 1921. Damage to forest and other vegetation by smoke, ash, and fumes from manufacturing plants in Naugatuck Valley, Conn. *J. Forest.* **19**:263–373.

SELECTED REFERENCES

Berge, H., 1963. "Phytotoxische Immissionen." Paul Parey, Berlin, 100 pp.

Bleasdale, J. K. A., 1952. Air pollution and plant growth. Ph.D. thesis, Univ. of Manchester, England.

Hepting, G. H., 1964. Damage to forests from air pollution. *J. Forest.* **62**:630–634.

Linzon, S. N., 1966. Damage to Eastern white pine by sulfur dioxide, semimature tissue needle blight, and ozone. *Can. Dept. Agr.* **16**:3.

Setterstrom, C., and P. W. Zimmerman, 1939. Factors influencing susceptibility of plants to sulfur dioxide injury. *Contrib. Boyce Thompson Inst.* **10**:155–181.

Spaleny, J., M. Kutacek, and K. Oplistilova, 1965. On the metabolism of $S^{35}O_2$ in the leaves of cauliflower *Brassica oleracea* var. Botrytis L. *Int. J. Air Water Pollut.* (Pergamon Press) **9**:525–530.

Stern, A. C. (ed.), 1962. "Air pollution," vol. 2. Academic, New York, 586 pp.

Thomas, M. D., and G. R. Hill, Jr., 1935. Absorption of sulfur dioxide by alfalfa and its relation to leaf injury. *Plant Physiol.* **10**:291–307.

———— and ————, 1937. The continuous measurement of photosynthesis respiration and transpiration of alfalfa and wheat growing under field conditions. *Plant Physiol.* **12**:285–307.

———— and R. H. Hendricks, 1956. Effect of air pollution on plants, in P. L. Magill, F. R. Holden, and A. C. Ackley (eds.), "Air pollution handbook," chap. 9. McGraw-Hill, New York.

————, 1958. Assimilation of sulfur and physiology of essential S-compounds, in "Handbuch der Pflanzenphysiologie (Encyclopedia of Plant Physiology," vol. 9, pp. 37–63. Springer-Verlag, Berlin.

U.S. Public Health Service National Center for Air Pollution Control, 1967. Air quality criteria for sulfur oxides. P.H.S. Publ. no. 1619, 175 pp.

Zahn, R., 1963. Investigations of plant reactions to continuous and intermittent sulfur dioxide exposures. *Staub.* **23**:343–352.

CHAPTER SIXTEEN

FLUORIDE AND ITS EFFECTS ON PLANTS

16-1 FLUORIDE SOURCES

The hazards of fluoride to crops, livestock, and man have been recognized for at least a hundred years; but the inherent problems of toxicity have greatly intensified with the vast expansion of the aluminum industry and other major users of minerals high in fluoride since the Second World War. Fluoride is widespread in the earth's crust as a natural component of soil, rocks, and minerals such as cryolite, topaz, micas, and hornblends. When these materials are heated in the course of refining, toxic quantities of fluoride may be released into the atmosphere; more fluoride is emitted from a wide variety of industrial processes in which fluoride compounds are manufactured or utilized as catalysts or fluxes.

16-2 FLUORIDE CONCENTRATIONS IN THE ATMOSPHERE

The amount of fluoride which is released into the atmosphere may range from a few pounds to several hundred pounds per day depending on the volume of production and the amount of fluoride in the raw materials, fuel, or flux. Far more fluoride, measurable as thousands of pounds per day, was allowed to escape before the hazards were recognized and industries were equipped with scrubbers, filters, and electrostatic precipitators designed to remove fluoride from the waste emissions. But the increased number of sources in some areas more than offset the reduced emissions, and the total fluoride output may remain sufficiently high to be toxic despite control efforts.

While the atmospheric fluoride concentrations depend largely on stack emissions, other factors such as the degree of dispersion, which is a function of the stack height and air movement, are equally important. Industries located in sites having adequate air drainage contribute minimum pollution to local areas, while the unfortunate location of some industries in narrow valleys where air often becomes stagnant has intensified problems in these areas. The amount of precipitation present to cleanse the atmosphere, and the density of vegetation and other obstacles or structures which may absorb fluoride, also help determine the atmospheric fluoride concentrations.

Air pollutants have been measured in different ways, but the amount of fluoride in the air is best measured in micrograms per cubic meter ($\mu g/m^3$). This is roughly equivalent to the more commonly used, but less accurate, parts per billion (ppb).*

The atmospheric fluoride concentrations within a mile of some industrial sources exceed 100 ppb, but this is exceptional. Even before the installation of modern control equipment, air levels rarely exceeded 10 ppb. Now that emissions are largely controlled, concentrations rarely exceed the 1 to 5 ppb gaseous fluoride injurious to sensitive species.

16-3 FLUORIDE ACCUMULATION BY PLANTS

Entrance and Accumulation from the Air

Fluoride enters the plant primarily through the leaf stomata, passes into intercellular spaces, contacts the mesophyll, and is either directly absorbed into the cell or dissolved in water and transported through the vascular tissues to the leaf tips and margins where it accumulates.

* 1 $\mu g/m^3$ of fluoride is equal to 0.874 ppb fluoride by weight or to 1.33 ppb by volume of any gas containing 1 fluoride atom per molecule. Use of ppb might be preferable, except that values are reported on a volume basis by some workers and a weight basis by others, and the basis of expression is seldom stated.

Several studies have been conducted demonstrating the accumulation of fluoride at the leaf extremities. Zimmerman and Hitchcock (1956) showed that the tips of gladiolus leaves may contain 25 to 100 times as much fluoride as the basal section. Oats, cattails, onions, pine needles, and rushes show the same tendency; margins of fruit-tree leaves may be 2 to 10 times as high in fluoride as the section adjacent to the midrib. The tip half of corn leaves may contain 2 to 3 times as much fluoride as the basal half, and the margins 2 to 3 times more than the center.

Chang and Thompson (1966) at the University of California determined the distribution of fluoride within the cells of citrus leaves. Leaves high in fluoride were fractionated into the subcellular components, and the fluoride content of each fraction was determined. Chloroplasts were found to contain about 60 percent of the total fluoride and were presumed to be the major site of accumulation. Ledbetter and his associates (1959), using radioactive hydrogen fluoride to trace the distribution of fluoride within the cell, also found that chloroplasts contained a considerable portion of the fluoride. However, both studies showed that appreciable amounts of fluoride also accumulate in the cell wall, nuclei, and cytoplasm.

One problem with centrifugal separation is that most of the soluble fluorides continue to be washed from their initial site during centrifugation, so that much of the fluoride ends up in the final supernatant. It is not clear how much fluoride is initially associated with this protein fraction and how much is washed out from chloroplasts and other organelles. Fluoride is highly soluble and therefore would be expected to be widely distributed within the cell. Most may be in the epidermal cells or in the vacuole, where it would be inactive, but a sufficient amount obviously finds its way to the chloroplasts and other organelles to be highly toxic.

H. M. Benedict at the Stanford Research Institute (1962, 1964) studied the distribution of fluoride in different organs and tissues of representative herbaceous plant species which included orchard grass, alfalfa, spinach, and lettuce. Plants were fumigated to build up the foliar fluoride content to about 2,000 ppm. Top, center, and bottom portions of fumigated alfalfa plants were divided into separate samples. The older leaves, which had been exposed for the longest period, consistently contained the most fluoride. In one group of alfalfa plants following five weeks of fumigation with 0.8 $\mu g/m^3$ fluoride, the lower, middle, and upper leaves contained 324, 121, and 51 ppm fluoride, respectively. In the field, the oldest alfalfa leaves are densely shaded and consistently drop off, leaving a more uniform distribution of fluoride.

Fluoride enters the plant largely through the leaf, since this organ provides by far the greatest absorptive surface area. The outer bark of stems accumulates only minute amounts of fluoride directly, and no fluoride is translocated from the leaves to the stem (Brederman and Radeloff, 1954). Nor is fluoride translocated downward into the roots; rather, translocation is toward the leaves and away from roots and stems. Zimmerman (1956) found that tomato leaves contain about 25 times as much fluoride as the stems and over 100 times as much as the fruit. Leaves of alfalfa may contain 7 to 8 times as much fluoride as the stems. Studies by Brewer et al. (1959) showed that when leaves of fumigated orange trees contained 230 ppm, twigs contained only 28 ppm fluoride.

Fluoride accumulation by fruits is equally negligible. In an area where foliage levels generally exceeded 200 ppm, berries contained less than 3 ppm; when foliar fluoride levels in stone and pome fruits ranged between 100 to 300 ppm, the concentration in the fruit rarely exceeded 10 ppm.

The fluoride content of vegetation is affected first by the atmospheric concentration and second by the amount of vegetation produced per unit of land area. There is a tendency for the amount of fluoride deposited on, or absorbed by, the vegetation on a given area of land during a specific time period to be constant, regardless of the kind of vegetation; therefore, the greater the amount of vegetation present, the lower its fluoride content is likely to be.

The rate of fluoride absorption by large plants such as trees is probably different from that by forage plants, where the vegetation is concentrated nearer the ground. Also, a single plant standing alone comes in contact with more fluoride from the atmosphere than a plant surrounded by others in a dense stand where only its upper part is exposed. The degree of fluoride accumulation is subject to such additional variables as rate of growth, species capacity for absorption, and the leaching action of precipitation.

Further variation exists in the normal "background" amounts of fluoride accumulated by plants in rural areas far from major fluoride sources.

The total foliar fluoride levels found in nonindustrial areas are important to the evaluation of possible increases due to air pollution.

From the extremely high "background" levels of several hundred parts per million found in "accumulator" plants, the normal fluoride content ranges down to 1 ppm. Plants absorb sufficient fluoride from the trace amounts present in the atmosphere, and from the soil, so that measurable amounts are found in foliage wherever sampled.

Thousands of samples collected from nonindustrial areas revealed that the fluoride content of unwashed leaves of fruit crops generally ranged from 10 to 20 ppm. Fluoride content of pine needles collected in nonindustrial areas generally ranged between 1 and 5 ppm. Mixed pasture forage averaged from 5 ppm in early spring to 32 ppm in late winter. Pastures that have been trampled, flood-irrigated, or otherwise exposed to soil contamination contained nearly twice this; foliar concentrations of 1 to 15 ppm are found in almost any geographical area.

Normal fluoride concentrations in fruits are still lower. Stone, pome, and cane fruits rarely contained more than 2 to 5 ppm fluoride, and barley grain only 2 ppm.

Foliar fluoride concentrations in industrial areas are often much higher. Concentrations exceeding 100 ppm are not unusual, and concentrations over 1,000 ppm have occasionally been reported.

The amount of fluoride plants accumulate largely depends on the duration of exposure, concentration in the air, and the form of fluoride. Atmospheric concentrations are often given as total fluorides, but the gaseous fraction is most likely to be injurious. Fluoride gases pass readily through the stomates, while particulate forms are more apt to accumulate on the leaf surface where they are subject to removal by leaching.

The accumulation and possible toxicity of particulate fluorides was studied by McCune et al. (1965) at the Boyce Thompson Institute. Gladiolus, milo maize, corn, tomato, and alfalfa plants were fumigated with airborne particulate cryolite for nine to forty-nine days at concentrations of 1.2 to 15 $\mu gF/m^3$, raising the foliar fluoride level two- to fifty-fold depending upon the duration of exposure; but most of this could readily be washed off, indicating that limited amounts actually entered the plant. The resulting leaf injury was slight. Since fluorides that dissolve in moisture on leaves can be absorbed, their potential phytotoxicity may depend on their solubility and the availability of moisture.

Studies in Utah also demonstrated the physiological inactivity of particulate fluorides. Merrill Pack et al. (1959) found that exposure of the highly fluoride-sensitive Snow Princess gladiolus plants for four weeks to an average of 0.79 μgF/m^3 as hydrogen fluoride gas caused 15 percent leaf necrosis, while exposure to particulate fluoride collected from a sintering plant, used at concentrations averaging 1.9 μgF/m^3, failed to cause measurable amounts of necrosis.

Accumulation from the Soil

A few plant species have the normal capacity to accumulate substantial amounts of fluoride from the soil. The camellia family, of which tea is a member, is an outstanding example; leaves of field-grown tea plants in areas free from atmospheric fluoride may contain over 1,300 ppm, and the African plant gifblaar may contain several hundred parts per million.

The possibility of a general fluoride accumulation from soil contaminated by smoke from fluoride-emitting industries has posed a serious question for many years. As airborne pollutants settle out of the atmosphere, they accumulate in the soil. In the case of an essential element such as sulfur this could be viewed as desirable, but with one such as fluoride which is more toxic than essential to plants, its accumulation could be viewed with alarm. Despite the seemingly large quantities of fluoride emitted by certain industries, its dispersion is normally so great that accumulation in the soil is negligible. If 1 ton of elemental fluoride were released into the air each day, and half of this settled within a 10-mile radius, the average rate of fluoride accumulation in the surface 6 in. of soil would be 1 ppm per year. Decades would be required before this amount contributed significantly to the fluoride background levels of most soils.

Because of possible soil contamination, but due more to the naturally high fluoride content of certain soils, studies have been conducted to determine the ability of agricultural plants to take up and accumulate fluorides from the soil. In the 1940s, when fluoride pollution was first intensively studied in the United States, there was some doubt regarding the origin of the fluoride found in plants. Did fluorides emanate entirely from the atmosphere, or were they taken up from soils high in fluoride?

The surface foot of soil may easily contain 500 ppm or more fluoride, but most of this occurs in an insoluble form firmly fixed by lime and clay. But in acid soils, fluoride may occur in soluble forms such as sodium fluoride (NaF), potassium fluoride (KF), or hydrogen fluoride (HF). When fluorides are present in soluble forms, they are absorbed and accumulated by both roots and leaves. Nutrient studies have shown that in acid soils containing only a few parts per million of soluble fluoride, roots may accumulate 1,000 to over 6,000 ppm fluoride, while several hundred parts per million more are translocated to the leaves (Leone et al., 1948). Brewer et al. (1959) found the citrus trees also absorbed substantial quantities of fluoride from nutrient solutions, but that such absorption was unlikely under field conditions. MacIntire and his associates (1942) found that when superphosphates were used to improve soil fertility, their high fluoride content added materially to the total fluoride in the soil but not in the plant. Up to 2,300 lb of fluoride per acre was added to the soil as calcium silicate slag and calcium fluoride. Even such tremendous soil supplements failed to increase the fluoride content of above-ground tissues of the forage plants studied.

Vegetation Sampling

The fluoride content of vegetation often provides a major basis for diagnosing the symptoms of toxicity and can help reveal the extent of forage pollution and the possible hazard to livestock.

To obtain reliable fluoride data, however, several variables must be considered. Some of the greatest variation and error comes in collecting the plant samples. Species to sample, tissues, age of tissue, time of year, area of collection, and the number of samples and sites must all be considered.

The species and variety of plants to sample will depend largely on their availability and the purpose of the study. While different species vary in their capacity to accumulate fluorides, it is usually essential to sample only one or two since the concentrations present are representative of the area.

In an area where livestock is raised, forage species will be the most important to sample. Alfalfa has been intensively sampled and found to have remarkably uniform characteristics as to fluoride uptake even among different varieties.

In natural areas where evergreen species such as pine or fir are prevalent, these species are extremely useful to sample. However, care must be taken to sample needles of the same age. Current year's needles are preferred, since needles continue to collect fluoride as long as they remain on the tree. However, progressively less is absorbed each year as metabolic activity diminishes. Agricultural or ornamental plants are also suitable for sampling if the same species is grown throughout the area being studied.

The extent of the area over which to collect samples depends on the magnitude of the fluoride source in question, the topography, and patterns of air movement. Once the species in the area and their distribution are known, exploratory samples of these should be collected along representative transects. If the results of the preliminary samples indicate a need for more intensive studies, later sampling sites can be selected and samples collected closer together.

The next consideration in sampling is the portion and volume of plant material to collect. Leaves are best suited for sampling since they accumulate the greatest amount of fluoride.

When alfalfa or grasses are sampled, the upper 6 in. of the plant is collected. When deciduous trees are sampled, care must be taken to collect leaves of the same age. Leaves from the first cluster which emerged in the spring should be selected, since these will have accumulated fluoride during the entire growing season and will contain maximum, and relatively uniform, amounts of fluoride. Once collected, the leaves should be analyzed for fluorides as soon as possible, although they can be held in a dry condition for several days or weeks if necessary.

Chemical analysis of the samples is subject to further variation, and the results may be unreliable, unless the laboratory has had considerable experience in fluoride analysis. Even then, variation of 50 to 100 percent is not unlikely (Jacobsen et al., 1967).

16-4 EFFECTS OF FLUORIDE ON PLANTS

Symptomatology

LEAVES   When plants have been exposed to high atmospheric fluoride concentrations for a sufficient length of time, injury symptoms will develop on sensitive species. The intensity of symptoms varies greatly among different individuals, and even on leaves of a single plant, but the expression is definitive to the experienced observer.

Accumulation of toxic concentrations of fluorides in leaves of broad-leaved species causes characteristic injury symptoms consisting of necrosis, chlorosis, or both. Symptoms are most prominent at the

FIGURE 16.1 *Chlorotic mottle on Carolina poplar.*

FIGURE 16.2a *Fluoride-induced necrosis of apricot. Chinese apricot leaf fumigated with HF, showing chlorosis accompanying necrosis.*

leaf tip and margin, where fluoride accumulation is greatest. A chlorotic mottle characteristically develops toward the margins of citrus, poplar, and cherry leaves (Fig. 16.1), while the water-soaked, dull, gray-green discoloration of tissue along the leaf margin or tip is the earliest sign of injury appearing on such sensitive species as apricot, gladiolus, and grape (Figs. 16.2a and b). Apricot leaves exposed to low fluoride concentrations soon develop semicircular lesions along the margin and about ¼ to 1 in. wide. Affected tissue turns light to dark brown as soon as toxic amounts of fluoride accumulate. A sharply defined, narrow, often reddish-brown band separates the necrotic tissue from the adjacent healthy tissue. The necrotic tissue may soon separate along this band and drop out. Microscopic examination has

FIGURE 16.2b *Chinese apricot showing typical field expression.*

shown that this reddish-brown band consists of an abscission layer formed by excessive cell division. When fluoride concentrations are higher, or the duration of exposure prolonged, larger areas of tissue become affected. However, narrow dark reddish-brown bands produced by the deposition of resins and tannins in the peripheral cells characteristically delimit the necrosis. These narrow bands of darker, reddish-brown tissue form zones which delimit the necrosis formed by each successive fumigation, giving the necrotic lesions a wavy, zonate appearance. The necrotic tissue may form a continuous band around the leaf affecting up to 100 percent of the leaf, or be limited to trace amounts at the tip or along one edge. The necrotic tissue may drop out, leaving an irregular, ragged edge, but the leaf itself rarely drops, even when foliar fluoride levels exceed several hundred parts per million. Premature defoliation has, however, been reported to result from greenhouse fumigations at high fluoride concentrations, but this has not been observed in the field.

The most sensitive leaves on fruit trees, such as apricot and prune, are those on "sucker" shoots with their softer, rank growth. Brewer et al. (1960) reported that older leaves seem to develop some tolerance and that the youngest leaves have not had time to accumulate toxic quantities of fluoride when symptoms first appear. Leafbuds are highly resistant and have been observed to retain their capacity to produce new shoots even when over 50 percent leaf necrosis has been produced for each of ten successive years.

Symptoms of fluoride toxicity on monocotyledonous plants are essentially similar to those on broad-leaved species. On gladiolus, the reddish-brown waves denoting successive fumigations are striking and definitive (Fig. 16.3). The zonation is absent only when necrosis is caused by a single exposure. Necrosis appears first, and is usually most severe, at the leaf tip but tends to extend down one side of the leaf more than the other. When only a trace of necrosis develops, it generally appears at the leaf tip or an inch or so below.

Iris, tulip, narcissus, and related species, though more tolerant, develop similar symptoms. Necrosis on cereal crops tends to be lighter in color and sometimes almost bleached. These species are highly resistant to fluoride, and necrosis more frequently is due to hot, dry winds or other environmental stress.

Symptoms on monocots such as corn and sorghum consist of a chlorotic stipple or mottle (Fig. 16.4). Small, irregularly shaped chlorotic spots develop predominantly along the margins, where chlorosis

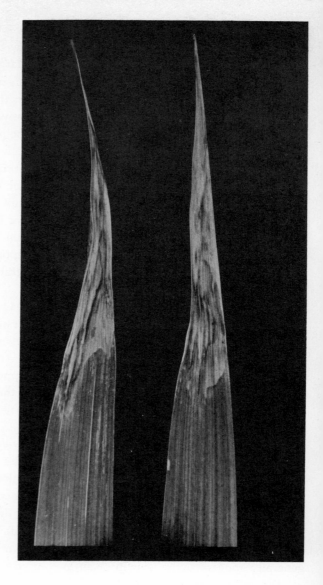

FIGURE 16.3 *Typical fluoride injury expression on gladiolus.*

merges into continuous bands. When more severe, most of the leaf may become chlorotic. Chlorotic areas are yellow to bleached in color and in the early stages of development stippled with minute, yet undamaged, green flecks. The green stippling diminishes as this tissue loses its chlorophyll. Chlorotic areas tend to be confined between the larger veins, but these may be bridged as chlorosis intensifies. Tissues along the larger veins tend to remain green. The chlorotic areas are sharply delimited, with little or no gradation in color between healthy and injured tissue. When injury is severe, the chlorotic areas become necrotic, particularly along the margins and at the leaf tip. Symptoms may appear on a few to all plants in a field, depending on the concentrations and duration of fumigation.

A more distinctive type of leaf injury is produced on lilac, on which a reddish-purple discoloration, developing first along the margin, progresses into a reddish mottle between the secondary veins.

Fluoride toxicity symptoms on needles of conifers consist of necrosis beginning at the tip of the current year's needles and progressing toward the base. Injured tissue first becomes chlorotic, then

FIGURE 16.4 *Chlorotic stipple characterizing fluoride injury on corn.*

FIGURE 16.5 *Tip necrosis of pine characterizing fluoride injury.*

buff, and finally reddish-brown (Fig. 16.5). Rarely, isolated bands of tissue become necrotic. Needles are most sensitive when they are elongating and emerging in the spring, and they become increasingly resistant as the season progresses. Needles formed during preceding years are highly resistant and are rarely, if ever, injured by subsequent fumigations.

Richard Solberg et al. (1955) conducted detailed histological studies of the effects of fluoride on ponderosa pine needles and observed that tissues collapsed only a few cells in advance of the necrotic areas. Epidermis, hypodermis, and xylem tissues were most resistant and parenchyma most sensitive. Phloem cells were first to be injured. Phloem and xylem parenchyma, together with transfusion parenchyma cells, enlarged greatly, becoming extended and distorted.

The protoplast then became granulated, vacuolated, and finally collapsed. Resin duct epithelial cells enlarged to an extent where duct canals were frequently occluded.

FLOWERS AND FRUIT Flowers are highly resistant to fluoride and rarely injured. One of the few observations of flower injury was cited by Thomas and Hendricks (1956), who listed petunia petals as "intermediate" in sensitivity. Fluoride injury was also reported on cyclamen petals even in the absence of leaf markings (Spiering, 1968). This is unusual, however.

Apricot, apple, cherry, and peach blossoms have been observed in areas where atmospheric fluoride levels often exceeded 1 ppb and leaves later in the season contained over one-hundred parts per million fluoride. No injury to apricot blossoms was observed, even when leaves became marked as soon as they emerged from the bud and later developed 70 to 90 percent necrosis. Gladiolus flowers are also resistant, despite the extreme sensitivity of leaves and bracts.

Fruits, on the other hand, may indirectly be more sensitive to fluoride than the leaves. The prime example of this fruit response is the *soft suture* disease of peach (Fig. 16.6). Soft suture, possibly

FIGURE 16.6 *Soft suture (suture red spot) of peach attributed to fluoride. Note prominence of symptoms at the basal end of the fruit.*

better called *suture red spot* since this denotes a specific type of "soft suture" associated with fluoride, is characterized by the premature reddening and ripening of a local, rather sharply delimited area along one or both sides of the suture line toward the lower third of the fruit. The red area contrasts with the pale yellow or yellow-green background color. In some instances, the premature ripening and swelling causes a local enlargement of the affected area. But more often the tissues along the suture ripen without undergoing excessive cell enlargement, so the affected portion is more typically soft and slightly sunken by the time the rest of the fruit is ripe. Premature reddening may appear two weeks to a month before harvest. The early, sharply defined coloring, even in shaded parts of the tree, is especially striking before the normal coloration begins. By the time the rest of the fruit is ripe, the suture area is overripe or even rotted. Another symptom is the separation of the flesh along the suture. This splitting beneath the skin may occur even when symptoms are extremely mild. Reddening of the affected flesh in a local region along the suture is another good diagnostic feature with yellow-fleshed varieties such as Hale and Elberta.

The disease was first described and given the name "fluoride suture" in the 1930s, when cryolite and sodium fluosilicate sprays were tested for control of insects and the bacterial spot disease (Anderson, 1931). The fruit symptoms described were identical with those later found associated with atmospheric fluorides. Subsequently, the disorder has been a problem in areas where atmospheric fluoride levels exceed roughly 1 ppb. When concentrations are about twice this, but still too low to cause necrosis of peach leaves, up to 90 percent of the fruit may be rendered unsalable if there is a light crop. Far less fruit is damaged when the trees have a full crop, although the reasons for this are not known.

Dr. Nels Benson (1959) at Washington State University has studied the disease intensively and developed control measures which have rendered suture red spot economically insignificant even closest to major fluoride sources. One calcium chloride spray (2 lb/100 gal water) applied to the trees at the pit-hardening stage of fruit maturity is sufficient to prevent the disease. If fluoride levels are especially high, a second spray, a month later, is recommended.

In considering control, though, it must be realized that fluoride is not the only pathogenic agent causing a "soft suture" type of symptom. Several other agents can cause the local, premature ripening characterizing such a syndrome, and the general name "soft suture" has been given this expression regardless of the cause. In all cases, the affected area colors and ripens early and is overripe by the time the fruit is ready for harvest.

A "physiological" type of soft suture, characterized mostly by the early ripening of one side of the fruit, was described and named *soft suture* by Dr. M. H. Dorsey at Illinois in 1944. It was associated with the delayed abortion of one of the two ovules in the fruit. The suture area of the best-developed half undergoes a greater final swell than the other half of the fruit, and this produces a suture "ridge" along one side. In contrast with the fluoride-induced "soft suture," the condition is associated with overloaded, weak trees rather than vigorous trees having a light crop.

Growth-regulating chemicals such as the phenoxyacetic acids (2,4-D; 2,4,5-T, etc.) can also cause a soft-suture expression when applied in excessive amounts to improve fruit size or to control weeds in orchards while fruits are developing. As little as 10 ppm 2,4,5-T applied before harvest can cause symptoms. In contrast with suture red spot, the premature ripening is usually most severe toward the stem end of the fruit.

Still another type of "soft suture," specifically named *red suture*, was described by Cation and Bennett in 1951. Fruits ripened several days prematurely, especially along the suture, which softened and was often ripe while the rest of the fruit was still green. A virus was found to be responsible.

Less striking, but basically similar, symptoms attributed to fluoride have been described on other fruits. Bolay and Bovay (1965) described and illustrated a local necrosis at the stylar end of apricot, cherry, and pear fruits. A round to irregular, sunken, dark-brown to black lesion developed at the basal end of the fruit as it ripened.

A similar condition called *shrivel tip*, has been observed to affect Orb cherry fruits (Treshow, unpublished). The stylar end of the fruit ripens and shrivels a few days prematurely. The disease may be caused by high fluoride alone, as demonstrated by Facteau (1968, personal communication), but is aggravated by water stress. Almost indistinguishable symptoms can be caused by water stress alone. A *black tip* condition of sour cherries and peach is also attributed largely if not wholly to water stress.

Metabolic Effects and Mechanism of Fluoride Action

Necrosis and chlorosis caused by high fluoride concentrations are obvious to any observer, but the effects this tissue injury may have on growth and production still are not well established. Even more uncertain are the possible biochemical effects of fluoride on plant metabolism and reproduction at "sublethal" concentrations too low to produce any clearly defined leaf markings. The visible lesions result from abnormally altered metabolic activity of the cell; but how much, and what kind of, alteration takes place before the cell is killed? Presumably, fluoride first affects plants at the molecular level of organization.

Fluoride has been postulated to affect most fundamentally the activity of enzymes essential to such plant processes as respiration, photosynthesis, carbohydrate metabolism, protein synthesis, cell wall formation, energy balance, and nucleotide and nucleic acid synthesis.

Fluoride has long been known to act as an enzyme inhibitor, and several studies have been conducted attempting to explain how fluorides affect basic enzyme reactions. Fluoride reacts with, and affects, so many different enzymes and other substrates in the cell that it is reasonable to assume that fluoride affects the cell in many ways. Certain enzymes of the glycolytic pathway of respiration, such as enolase and phosphoglucomutase, are among the most sensitive to fluoride, and their inhibition by fluoride may impair both carbohydrate metabolism and respiration.

Fluoride inhibition of both enolase and catalase was demonstrated by McCune et al. (1965). Inhibition was thought to be caused by the formation of a metal fluorophosphate complex tying up the iron or magnesium in certain enzymes and rendering them inactive. Fluoride also combines with the ferric ion in the heme groups of peroxidase, catalase, and cytochrome oxidase, inhibiting these enzymes (Hewitt and Nicholas, 1963). By its formation of metal fluoride complexes, fluoride may also alter the cells' nutrient status and exert a general metabolic effect, decreasing the activity of the metal-requiring enzymes and affecting physiological processes in a multitude of ways. One of the simplest, most direct effects on nutrition would be for fluoride to precipitate calcium as insoluble calcium fluoride, producing a calcium deficiency. Once the available calcium was precipitated, fluoride might act in the same way on magnesium at the expense of the chlorophyll molecule.

Some of the most intensive work concerning the inhibiting action of fluoride has dealt with its effect on respiration. The inhibiting action of higher fluoride concentrations on respiratory enzymes is well known; but at lower concentrations, respiration is stimulated rather than inhibited. McNulty and Newman (1957) reported that fluoride stimulated oxygen uptake in leaf tissue before any visible injury appeared. They later showed that this stimulus was not associated with death of the tissue and was readily reversible, that is, it was an effect of fluoride which took place at sublethal concentrations. Subsequently, fluoride-induced respiration stimulation has been demonstrated by a number of workers.

Applegate and Adams (1960) found that either subjecting bush bean to gaseous fluoride or growing plants in fluoride-containing solutions increased oxygen uptake. The average fluoride content in the fumigation chambers of 2.2 $\mu g/m^3$ was sufficiently low not to cause chlorosis. Fluoride-induced stimulation of respiration, as measured by oxygen uptake, has also been demonstrated in leaf discs of grape and apricot, in pea epicotyl sections, and in bush beans, gladiolus, and wheat (Lustinec et al., 1960; Pilet, 1963; and McNulty and Newman, 1956). Weinstein (1961) observed a 70 percent increase in the respiration rate of tomato and bean leaves fumigated eight days with 1.6 ppb HF. The way in which fluoride stimulates, or at higher concentrations inhibits, respiration is not completely understood; but Lords and McNulty (1965) suggested that stimulation may be due to a disruption of the balance between ADP (adenosine diphosphate) and ATP (adenosine triphosphate). ATP concentrations in control plants and plants fumigated with fluoride were measured using the firefly luminescent assay. This utilizes the fact that the intensity of light from firefly extract is proportional to the ATP concentration. The concentration of ATP and the rate of respiration were significantly and correspondingly higher in fluoride-treated bean leaf discs than in the controls. The enzyme adenosine triphosphatase, which breaks down ATP, has been reported to be inhibited by fluoride. Failure of ATP to be broken down could possibly account for the fluoride-induced ATP buildup. Inhibition of breakdown would alter the ADP/ATP ratio. If the respiration rate is regulated by a balance between inorganic phosphate ADP and ATP, then the correlation between ATP and respiration may be causally related, and a high ratio of ATP might increase oxygen consumption.

Effects of fluoride on photosynthesis have also been demonstrated.

Thomas and Hendricks (1956) reported the effects of fluoride fumigations at 2 to 10 ppb on photosynthesis of gladiolus. Photosynthetic activity of control and fumigated plants was practically identical until necrosis appeared. As necrosis developed, photosynthesis was reduced by an amount equal to the amount of necrosis. They concluded that the photosynthetic activity of the green area was not affected.

Brief fumigations with higher concentrations briefly inhibited photosynthesis. The photosynthetic activity of the fumigated plants fell off to a low level following fumigation and then recovered slowly. When a threshold of fluoride concentrations was exceeded, a temporary interference with the photosynthetic mechanisms occurred without necessarily causing any necrosis or chlorosis. Recovery of photosynthetic activity was slow, requiring one to three weeks.

Extensive photosynthetic studies on a number of plant species have been conducted by Dr. A. C. Hill and associates in Utah. In one study, gladiolus plants were subjected to average fluoride concentrations of 0.76 to 1.45 $\mu g/m^3$ for up to thirty-nine days. There was no reduction in photosynthesis until necrosis appeared; the reduced photosynthetic rate was then proportional to the amount of leaf necrosis. Fumigations of corn with an average of 7.2 $\mu g/m^3$ HF produced chlorotic mottling with a correspondingly reduced photosynthesis rate. Continuous fumigation of strawberry for eighteen days at fluoride levels from 1 to 9 $\mu g/m^3$ had no effect on photosynthesis, but on the nineteenth day the fluoride level was increased sharply to 37 $\mu g/m^3$ for twenty-four hours, reducing photosynthesis by 50 percent. Photosynthesis returned to about 75 percent of the previous rate within twenty-four hours and then recovered more slowly, until after three weeks it had returned to about 95 percent of the control plot. This was considered to represent complete recovery, since about 5 percent leaf necrosis had developed. Impairment of photosynthesis was demonstrated at high fluoride levels, but plants recovered fully when fluoride concentrations were reduced.

It appears, as proposed by Thomas (1958), that a threshold of fluoride concentration and exposure time exists, and that above it there is a reduction in photosynthesis, and presumably growth rate, in excess of that which can be accounted for by necrosis or chlorosis alone.

Impaired cell wall development was thought to be a basic reason for possible inhibition of growth by fluoride. But regardless of the effects of fluoride on enzyme activity, respiration, photosynthesis, and assimilation, the results fail to explain the influence that fluoride might have on growth and reproduction.

Plant Growth

Biochemical studies are essential to understanding the metabolic pathways and activity of fluoride, but fumigation studies with intact plants are necessary to determine if plant growth is affected. By its reactions with and inhibition of sensitive enzymes, fluoride may induce metabolic changes and affect growth in the absence of visible markings.

Curiously, the effects of fluoride on plant growth are not always negative. Aso (1906) concluded that sodium fluoride stimulated vegetative growth of barley and pea plants. More recently, Adams and Sulzback (1961) reported that bean plants fumigated with fluoride for ten to twenty days exhibited a longer initial internodal growth than control plants. In field studies with Douglas fir, Treshow et al. (1967) found that needles of trees exposed to slightly elevated fluoride concentrations were longer than those on trees grown in the absence of fluoride, although the radial growth diminished as fluoride levels increased. They also found that fluoride stimulated vegetative growth of bean plants fumigated with fluoride at sublethal concentrations in the range of 0.5 to 4 $\mu g/m^3$ throughout their life cycle. Higher fluoride concentrations inhibited growth; however, even when no chlorosis or necrosis was produced (Treshow and Harner, 1968). When fluoride concentrations in bean leaves exceeded 200 to 300 ppm fluoride, growth diminished; leaves containing 300 to 400 ppm fluoride had only half the weight of controls grown in filtered air.

Field studies have shown that the leaf size of woody plants such as aspen and Oregon grape exposed to fluorides was reduced as much as 30 percent (Anderson, 1966). Brewer (1960) conducted greenhouse studies in which orange trees were exposed to 1 to 5 ppb fluoride for about twenty-six months. The average leaf size of fumigated plants was reduced 25 to 35 percent. Tree height and crown volume were also reduced, and linear shoot growth of the HF-fumigated trees was 242 m, compared with 295 m for the controls. Total green weight of the tops of fumigated trees was also greatest in the controls.

Leonard and Graves (1966) found that leaf size decreased as fluoride levels increased. During a two-year period the relative leaf area was 50 percent less in outdoor checks compared with leaves in filtered air. Average leaf area was reduced 20 percent when trees received unfiltered air containing an average of 6.3 ppb fluoride.

Apparently, fluoride at sublethal concentrations may affect plant growth. Growth may be stimulated or inhibited depending on the fluoride concentrations, the sensitivity of the plant, and presumably environmental factors such as nutrient and other soil conditions.

Production and Reproduction

The possible effects of fluoride on production have been the basic concern of every grower raising crops near a fluoride source. Before the installation of fluoride control equipment became standard procedure, leaf necrosis of nearby sensitive species was common. The presence of this leaf burning led growers to suspect that fluoride was equally harmful to fruit development and caused reduced fruit size and production even when crops were not visibly marked. Later, when fluoride concentrations were reduced, and necrosis and chlorosis were no longer evident, growers continued to wonder if fluorides might be affecting production.

The relationship of fluoride to both necrosis and production was studied intensively by Robert Brewer and his associates at the University of California (1966). Growth, flowering, and corm production of gladiolus were studied under controlled greenhouse conditions. Plants on which fluoride had destroyed approximately 5, 15, and 25 percent of the total leaf area were compared with control plants grown in fluoride-free air. Flower size, and weight and number of florets, were reduced as the severity of injury increased. Corm size and weight were reduced in direct proportion to the extent of leaf injury. These data for two years are summarized in Table 16.1.

TABLE 16.1. *The effects of fluoride-induced necrosis on growth, flowering, and corm production of snow princess gladiolus.*

% Necrosis	0.6	5.0	15	26
Corm weight, g	37.3	34.7	29.3	23.5
Total plant weight, g	107.6	98.3	87.6	75.3
F content of leaves at flowering, ppm	5.0	19.0	52.0	88.0

The weight reductions, which could be considered analogous to yield loss, were attributed primarily to loss of photosynthetic area associated with leaf necrosis.

The possible effect of fluorides on citrus yields was studied by Leonard and Graves in Florida (1966). The experimental trees were located in an area exposed to relatively high atmospheric fluoride concentrations. Growth factors were compared in houses containing filtered versus ambient air. Dry calcium carbonate filters were used to remove fluoride from the air entering three greenhouses, while unfiltered, ambient air was used in the other three greenhouses. Three additional, uncovered trees were used as outdoor checks.

Greenhouse trees receiving unfiltered air produced 21 percent less fruit than trees receiving filtered air. A 27 percent decrease in average yield of fruit per tree was shown for each increase of 50 ppm fluoride in 5½-month-old spring flush leaves. These data are summarized in Table 16.2.

Yield reductions were partly attributable to the visible chlorosis which appeared when foliar fluoride levels exceeded 20 to 30 ppm. Also, the reduced average leaf size lowered the photosynthate comparably, reducing food production for support of growth and fruit development. The amount of yield loss (if any) attributable to excessive blossom or fruit drop was not reported.

Removal of the fluorides from the atmosphere around Valencia orange trees in plastic greenhouses in the field resulted in greater rates of photosynthesis, increased chlorophyll contents, and lower rates of respiration than those exposed to fluoride (Woltz and Leonard, 1964).

Merrill Pack at Washington State University (1966) described one of the few studies designed to reveal the possible effect of sublethal fluoride concentrations on fruit production. Tomato plants, which are relatively tolerant to fluorides, were divided into three groups. One served as a control while the other two were fumigated with hydrogen

TABLE 16.2. *Effect of airborne fluorides on yield of Valencia orange trees.*

	Foliar F concentrations, ppm	Mean air F concentrations, ppb	Fruit yield, lb	% yield
Filtered air	10	2.1	377	100
Unfiltered air	71	6.3	297	79
Outdoor checks	197	6.9	107	28

fluoride at concentrations of 2.9 and 6.4 $\mu g/m^3$, respectively, for twenty-two weeks. Calcium levels in the soil were also varied. Injury, consisting of marginal necrosis—especially of the younger leaves—developed only on plants exposed to the higher concentrations. Fluoride caused a significant reduction in the number of fruits when plants were grown at low (40 ppm) calcium levels, but in general the reduction in flowers or fruit per plant was not significant. The reduction was primarily due to the killing of the growing tips by HF and the production of only three flowering trusses per plant. Effects on fruit size were more striking. Regardless of calcium nutrition, fluoride caused highly significant reductions in both the weight of fruit per plant and the average weight per fruit. These data, summarized in Table 16.3, are condensed from Dr. Pack's paper.

Fluoride contents of the control plants were 3, 2, 1, and 0 ppm for the leaflets, petioles, stem, and fruit respectively. The corresponding tissues of fumigated plants contained 1252, 46, 10, and 10 ppm respectively.

This study showed that fluoride is not likely to affect fruiting unless continuously high fluoride levels are provided. The interrelation between fluoride and calcium nutrition was also shown. Either increasing the HF concentration or reducing the level of calcium nutrition had about the same effect on fruit size and seedlessness, and the effects were additive. It seemed that fluoride interfered with some function of calcium in fruit formation.

Brewbacker and Kwak (1963) showed that calcium is essential for pollen germination and pollen tube growth. If the calcium were precipitated by fluoride, insufficient calcium would be available for the germinating pollen, and fertilization would be prevented. This is supported by Dr. Pack's study, which showed that low levels of calcium, even in the absence of HF, produced many seedless or nearly

TABLE 16.3. *Effects of atmospheric fluorides, (6.4 $\mu g/m^3$) on tomato fruit production.*

Factor	Control chamber	HF chamber
Flowers per plant	68	51*
Fruits per plant	55	41*
% flowers producing fruit	81	80
Av. weight per fruit, g	46	24*

* Differences significant from controls.

seedless fruits. The smaller size of these fruits appeared to be related to the absence of seeds.

There seems to be only one other study treating the effects of fluoride on blossom set. This is presented in a master's thesis by Dayna Stocks at the University of Utah (1960). Again, high fluoride levels were required to produce an adverse effect. When bush beans were fumigated five to ten days with HF at concentrations of 10 to 30 ppb, the number and size of fruits were substantially reduced. Despite the high concentrations, no visible chlorosis appeared, but fumigated plants averaged only 5.1 fruits per plant compared with 7.0 for the control plants. The average weight of 0.097 g for the fumigated plants was significantly less than the 0.22 g of the controls.

Species Sensitivity

Different plant species, and even varieties or clones of a single species, vary tremendously in their sensitivity to fluorides. A highly sensitive species like gladiolus may be damaged when atmospheric fluoride concentrations are below 1 $\mu g/m^3$, while tolerant species such as cotton, celery, alfalfa, and hundreds of others are unaffected by many times this amount.

Differences in sensitivity may be partly due to environmental variables, as will be discussed later, but the main difference is inherent. Unfortunately, the genetic factors which render a minority of species highly sensitive to fluoride have not been studied and cannot be explained until the mechanism of toxicity is better understood. But extensive field observations and fumigation studies have provided excellent data on relative sensitivity. Relative sensitivity lists have been developed in different parts of the country and are not always the same; field observations which are valid in one region may not hold in another, where soil, climate, and other environmental conditions vary. The list of the relative sensitivity of some common plant species provided in Table 16.4 is based on fumigation studies and personal observations.

Relatively few species are sufficiently sensitive to fluoride to be injured significantly and cause any concern. In a most general way, injury to these may appear before the fluoride concentrations in the leaves have reached 50 ppm if the existing environmental conditions are condusive to predisposing the plants to injury. These species are listed in the "sensitive" column. A few other species are injured only

TABLE 16.4. *Relative sensitivity of plants to fluoride.*

Sensitive	Intermediate	Resistant
Gladiolus (some varieties)*	Walnut (English)	Linden (American)
Apricot (Chinese and Royal)	Apricot (Moorpark, Tilton)	Pyracantha
Oregon grape	Citrus (Lemon, tangerine)†	Ailanthus†
Peach (fruit)	Walnut (Black)	Elm (American)†
Corn†	Poplar (Lombardy, Carolina)†	Tomato
Plum (Bradshaw)	Grape (Concord)	Asparagus
Prune (Italian)	Aspen (Quaking)	Wheat
Grape (European var.)	Barley (young plants)	Birch†
Pine (Ponderosa)	Grapefruit†	Current
Larch (Western)	Cherry (Bing, Royal Ann)†	Mt. Ash (European)
Pine (Eastern white, Lodgepole, Scotch, Mugho)	Sumac	Elderberry
Fir (Douglas)	Orange†	Cherry (Flowering)
Spruce (Blue)	Lilac	Sunflower
Blueberry	Peach (leaves)	Pigweed
Tulip (some varieties)	Chokecherry	Squash
Box elder	Maple (Rocky Mt., hedge, silver)	Virginia creeper
	Serviceberry	Burdock
	Spruce (white)	Strawberry
	Arborvitae	Pear
	Chickweed	Bridal wreath
	Raspberry	Ash (Modesto)
	Rose	Willow (Laurel leaf)
	Yew	Juniper
	Apple (Delicious)	
	Aster	
	Ash (green)†	
	Mulberry†	
	Geranium	
	Paeonia	
	Linden (European)	
	Sorghum†	
	Lambs quarter	
	Goldenrod	
	Rhododendron	
	Yellow clover	

* Plants are listed in approximate order of increasing tolerance
† Predominant symptom chlorosis rather than necrosis

at somewhat higher fluoride concentrations; these are listed as "intermediate" in sensitivity. A third group consists of plants on which injury has been observed in the field, or in fumigations, only when leaf levels have exceeded a few hundred ppm. These are considered "resistant."

Influence of the Environment on Plant Responses

Every factor in the environment which influences the plant's vigor and thriftiness appears also to influence its sensitivity to fluorides.

Mineral nutrition has a marked effect on plant growth and might be expected to exert a corresponding influence on sensitivity. McCune et al. (1966) found that despite a limited effect on fluoride absorption, nutrient relations had a marked influence on necrosis severity. The rate at which necrosis developed was most rapid at optimal levels of nitrogen, phosphorus, or calcium. When potassium or phosphorus was deficient, tip burn increased. When calcium or nitrogen was deficient, tip burn decreased. Deficiencies of iron, magnesium, or manganese had no effect on necrosis.

Water relations have an important influence on the severity of leaf necrosis under field conditions, but data are conflicting. Zimmerman and Hitchcock (1956) found that turgid, succulent plants grown under optimum moisture and fertility conditions were most sensitive. Responses under greenhouse conditions, however, are not consistent with responses observed in the field. Extensive field observations consistently show that plants grown under neglected, unfavorable arid conditions are most sensitive and severely injured. Extreme examples have been observed in orchards where necrosis in neglected trees often exceeded 10 percent, while necrosis on irrigated, well-managed trees in adjacent orchards remained under 1 percent.

Ecological Considerations

Plants have evolved in harmony with their environment. As the environment is altered, the composition of the plant community will change correspondingly. When fluoride or other air pollutants are imposed, sensitive members of the community may be killed out entirely, while further from the source their metabolic activity may be impaired. Modification of growth and reproductive potential of species will reduce their ability to compete with more tolerant plants and thereby alter the population. The significance of this to the watershed, forest, or range will depend on the importance of the sensitive species to the stability of the plant community.

In one incident, Adams et al. (1952) reported that fluorides were responsible for needle burning and death of ponderosa pine over a 50 sq mile area near Spokane, Washington. Pine was eliminated as a dominant species in the immediate vicinity of the implicated aluminum-reduction plant. No mention was made of the understory composition at the time, but in 1967 observations showed that vegetation was limited to herbaceous species including largely cheatgrass, milkweed, *Linaria*, and *Anchusa*. Further from the operations, *Helianthus* sp., wild rose, and lupine prevailed.

Vegetation changes were described around an aluminum plant in Europe by Hajduk (1963). The most striking observation was the decrease in species numbers from twenty-six some distance away to two near the factory. Species of fescue, *Agropyron*, *Agrostis*, and *Rumex* were most resistant and comprised the major vegetation, replacing the hedge maple, European beech, Scots pine, plum, apple, hornbeam, and oaks which formerly occupied the sites.

One of the few studies to determine the effects of fluoride on plant populations was conducted by Frank Anderson at the University of Utah (1966). About 200 acres of Douglas fir trees, together with a few acres of quaking aspen and lodgepole pine, had been killed near the phosphate fertilizer plant around which the study was conducted (Fig. 16.7). Modifications to the plant populations were studied in the area of mortality and several miles into the forest where no visible injury was apparent. Study plots were selected at increasing distances from the operations and the population in each determined.

Certain species were found significantly more often in areas of high, as compared to low, fluoride concentrations, while others showed a reverse pattern. Oregon grape was among the first species killed and replaced. Frequency of lichens, moss, Douglas fir seedlings, and chickweed also declined significantly with increasing fluoride concentrations. Forbs, especially annuals, were significantly more abundant in areas of highest fluorides. Grasses, of which pine grass was most dominant, increased at high fluoride concentrations. The majority of plants were randomly distributed regardless of fluoride concentrations. While vegetation changes occurred in areas where trees were not killed, they were not as striking as where overstory species had

FIGURE 16.7 *Mortality of lodgepole pine and Douglas fir trees in the vicinity of a phosphate-reduction plant.*

been killed so that competition for light, minerals, and water was reduced.

16-5 PROBLEMS IN DIAGNOSIS

The foliar markings caused by fluoride and other air pollutants have proved to be highly sensitive criteria for revealing their presence. Accurate evaluation of symptoms can serve as a valuable and inexpensive tool for delineating an air pollution situation, but the value of this tool depends on accurate diagnosis.

Accurate diagnosis of fluoride injury may be complicated by the presence of pathogens causing similar symptoms. But the similarity of symptoms is generally limited to individual leaves; and if the total disease picture is evaluated, the observer should not be misled. He must, however, be familiar with the normal appearance and relative sensitivity of a wide range of species, also with an array of disorders caused by other pathogens. Knowledge of the fluoride concentrations of the leaves has often been mentioned as another aid to diagnosis. If fluoride levels are "normal," fluoride can be ruled out as the pathogen, but an elevated concentration of fluoride proves nothing since their presence does not mean they are causing any damage. Similarly, air concentrations provide only a general guide as to the possible threat of fluorides. These data are only of general value, and the diagnostician must rely mostly on visual observations.

Diagnosis of an air pollution situation consists first of locating and examining plants of "indicator" species. These highly sensitive plants are generally most severely affected. If fluoride-type symptoms are confined to resistant or tolerant species, some other causal agent should be sought.

Symptoms caused by moisture stress resemble fluoride-induced necrosis particularly closely (Treshow, 1965). Leaf burning from fluoride is most severe when soil moisture is inadequate, but even with a water supply which might appear ample, necrosis may arise from the water stress alone when temperatures are extreme or tissue sufficiently predisposed. Cool spring weather induces soft, rank growth which is especially sensitive to later moisture stress.

Necrosis caused by other pollutants, especially chloride, and occasionally sulfur dioxide, may also resemble fluoride toxicity symptoms. The relative sensitivity of different species must be considered.

Virus and genetic disorders such as prune leaf spot may also resemble fluoride toxicity, and other species should be examined for symptoms whenever possible. Pesticides, especially oil sprays, can also cause necrosis similar to that produced by fluoride.

Nutrient-deficiency symptoms most closely resemble fluoride-induced chlorosis. Manganese-deficiency symptoms on species such as peach and citrus are so similar to mild fluoride-toxicity symptoms that one must rely on the appearance of symptoms on other species in the area. Zinc-deficiency symptoms consisting of leaf chlorosis may also resemble symptoms caused by fluoride but the leaf distortions and rosetting associated with zinc deficiency are not associated with fluoride.

The fluoride-induced chlorosis of citrus has been likened to boron toxicity, and other symptoms associated with boron, such as gumming and dieback, must be sought.

Fluoride symptoms on corn are often similar to those produced by zinc or potassium deficiency, mite injury, genetic variation, and normal senescence. Generally, the genetic mottle is brighter yellow in color.

Growth and yield reduction sometimes ascribed to fluoride are especially difficult to establish in the field. In the first place, if growth or production losses occur, they are apt to be so slight as to require quantitative measurement and filtered chambers to detect them. Secondly, the multitude of well-established environmental factors causing growth and yield reduction are likely to be far more significant than any losses related to fluorides.

The considerations essential to accurate diagnosis of pollution injury in general were summarized by Treshow (1965) as follows:

1. The observer must be familiar with the relative susceptibility of a wide range of plant species to major air pollutants.

2. The overall syndrome on a number of affected plants of the same species when available must be studied, and the distribution and geographic relation of the marked plants to suspected sources of pollutants must be known.

3. The presence of possible sources of pollutants must be considered, even though injury may occur many miles from a suspected source.

4. Chemical analysis of leaf tissue may prove helpful to diagnosis of pollutants such as fluoride, but even in this case foliar levels often fail to reveal a quantitative correlation with injury.

5. Background information on the cultural, environmental, disease, and insect conditions is vital to correct evaluation, and the observer must be thoroughly familiar with symptoms associated with each of these pathogens.

6. Comparison of symptoms in "outside," pollutant-free areas with conditions in the polluted area is particularly valuable to diagnosis. Far superior, though, is quantitative knowledge of conditions in the given polluted area prior to the development of a real or alleged air pollution situation. Where the disease picture, including production-limiting factors, is understood prior to occurrence of an air pollution situation, a far more enlightened and accurate evaluation can be provided.

BIBLIOGRAPHY

Adams, D. F., D. J. Mayhew, R. M. Gnagy, E. P. Richey, R. K. Kappe, and I. W. Allen, 1952. Atmospheric pollution in the Ponderosa pine blight area. *Ind. Eng. Chem.* **44**:1356–1365.

———— and C. W. Sulzbach, 1961. Nitrogen deficiency and fluoride susceptibility of bean seedlings. *Science* **133**:1245–2426.

Anderson, F. K., 1966. Air pollution damage to vegetation in Georgetown Canyon, Idaho. M.S. thesis, Univ. of Utah, 102 pp.

Anderson, H. W., 1931. Problems on spray to the peach. *Trans. Ill. State Hort. Soc.* **65**:454–468.

Applegate, H. G., and D. F. Adams, 1960. Effect of atmospheric fluoride on respiration of bush beans. *Bot. Gaz.* **121**:223–227.

Aso, K., 1906. On a stimulating action of calcium fluoride on phaenerogams. *Bull. Coll. Agr. Imp. Univ. Tokyo* **7**:85–90.

Benedict, H. M., 1962. The physical and chemical fate of fluorides in plants. *Prog. Rep. S. Calif. Lab. Stanford Res. Inst.*, S. Pasadena, Calif.

———, J. M. Ross, and R. W. Wade, 1964. The disposition of atmospheric fluorides by vegetation. *Int. J. Air Water Pollut.* **8**:279–289.

Benson, N. R., 1959. Fluoride injury on soft suture and splitting of peaches. *Proc. Amer. Soc. Hort. Sci.* **74**:184–198.

Bolay, A., and E. Bovay, 1965. Observations sur les dégâts provoqués par les composés fluorés en Valais. *Agr. Romande* **4(A)**(6):43–46.

Bredermann, G., and H. Radeloff, 1954. On fluorine smoke injury absorption of fluorine by the bark of shoots and its effect. Translated from the German, Stanford Res. Inst., pp. 1–12.

Brewbaker, J. L., and B. H. Kwack, 1963. The essential role of calcium ions in pollen germination. *Amer. J. Bot.* **50**:856–859.

Brewer, R. F., H. D. Chapman, F. H. Sutherland, and R. B. McColloch, 1959. Effect of fluorine additions to substrate on navel orange trees grown in solution cultures. *Soil Sci.* **87**(4):183–188.

———, 1960. Effects of hydrogen fluoride gas on seven citrus varieties. *Proc. Amer. Soc. Hort. Sci.* **75**:236–243.

———, F. H. Sutherland, F. B. Guillemet, and R. K. Creveling, 1960. Some effects of hydrogen fluoride gas on bearing navel orange trees, *Proc. Amer. Soc. Hort. Sci.* **76**:208–214.

———, F. B. Guillemet, and F. H. Sutherland, 1966. The effect of atmospheric fluoride on gladiolus growth, flowering, and corm production. *Proc. Amer. Soc. Hort. Sci.* **88**:631–634.

Cation, D., and C. W. Bennett, 1951. Red suture, in "Virus diseases of stone fruits in North America." *U.S.D.A. Handb.* 10, pp. 11–13.

Chang, C. W., and C. R. Thompson, 1966. Effect of fluoride on nucleic acids and growth in germinating corn seedling roots. *Physiol. Plant.* **19**:911–918.

Dorsey, M. H., and R. L. McMunn, 1944. Tree-conditioning the peach crop. *Univ. Ill. Exp. Sta. Bull.* 507.

Hajduk, Juraj, 1963. Effect of fluorine exhalation production on the plant association and individual plants around the aluminum factories (English summary). Sympozio, Problematike Exhalator na Sloven sku, vol. 11, pp. 39–50.

Hewitt, E. J., and D. J. D. Nicholas, 1963. Cations and anions: Inhibitions and interactions in metabolism and in enzyme activity, in P. M. Hochster and J. H. Quastel (eds.), "Metabolic inhibitors," vol. 2, pp. 311–436.

Jacobson, J. S., L. H. Weinstein, D. C. McCune, and A. E. Hitchcock, 1967.

The accumulation of fluorine by plants. *J. Air Pollut. Contr. Assoc.* **16**(8):412–417.

Ledbetter, M. C., R. Mavrodineanu, and A. J. Weiss, 1960. Distribution studies of radioactive fluorine-18 and stable fluorine-19 in tomato plants. *Contrib. Boyce Thompson Inst.* **20**:331–348.

Leonard, C. D., and H. B. Graves, 1966. Effect of air-borne fluoride on Valencia orange yields. *Proc. Fla. State Hort. Soc.* **79**:79–86.

Leone, I. A., E. G. Brennan, R. H. Daines, and W. R. Robbins, 1948. Some effects of fluorine on peach, tomato and buckwheat when absorbed through the roots. *Soil Sci.* **66**(4):259–266.

———, E. Brennan, and R. H. Daines, 1956. Atmospheric fluoride: Its uptake and distribution in tomato and corn plants. *Plant Physiol.* **31**(5):329–333.

Lords, J. L., and I. B. McNulty, 1965. Estimation of ATP in leaf tissue employing the firefly luminescent reactions. *Utah Acad. Sci. Arts Lett.* **42**:163–164.

Lustinec, J., V. Krekule, and V. Pokorna, 1960. Respiratory pathways in gibberellin-treated wheat. The effect of fluoride on the respiration rate. *Biol. Plant Acad. Sci. Bohemoslov* **2**:223–226.

MacIntire, W. H., S. H. Winterberg, J. G. Thompson, and B. W. Hatcher, 1942. Fluorine content of plants. *Ind. Eng. Chem.* **34**:1469–1479.

McCune, D. C., A. E. Hitchcock, J. S. Jacobson, and L. H. Weinstein, 1965. Fluoride accumulation and growth of plants exposed to particulate cryolite in the atmosphere. *Contrib. Boyce Thompson Inst.* **23**(1):1–12.

McNulty, I. B., and D. W. Newman, 1956. The effects of a lime spray on the respiration rate and chlorophyll content of leaves exposed to a fluoride atmosphere. *Proc. Utah Acad. Sci. Arts Lett.* **33**:73–79.

——— and ———, 1957. Effects of atmospheric fluoride on the respiration rate of bush bean and gladiolus leaves. *Plant Physiol.* **32**:121–124.

——— and J. L. Lords, 1960. Possible explanation of fluoride-induced respiration in *Chlorella pyrenoidosa*. *Science* **132**(3439):1553–1554.

Pack, M. R., A. C. Hill, M. D. Thomas, and L. G. Transtrum, 1959. Determination of gaseous and particulate inorganic fluorides in the atmosphere. Symposium on air pollution control. *Spec. Tech. A.S.T.M. Publ.* no. 281, pp. 27–44.

———, 1966. Response of tomato fruiting to hydrogen fluoride as influenced by calcium nutrition. *J. Air Pollut. Contr. Assoc.* **16**(10): 541–544.

Pilet, P. E., 1963. Action du fluore et de l'acide B-indolylacetique sur la respiration de disques de feuilles. *Bull. Soc. Vaudoise Sci. Nat.* **68**(312):359–360.

Solberg, R. A., D. F. Adams, and H. A. Ferchau, 1955. Some effects of hydrogen fluoride on the internal structure of *Pinus ponderosa* needles. *Proc. 3d Nat. Air Pollut. Symp.*, pp. 164–176.

Spierings, F. H. G., 1968. Effect of air pollution on crop yield. *1st Int. Congr. Plant Pathol.*, London.

Stocks, D. L., 1960. The effects of fluoride on the growth and reproduction of bush bean plants, *Phaseolus vulgaris* L. M.S. Thesis. University of Utah, 48 pp.

Thomas, M. D., and R. H. Hendricks, 1956. Effect of air pollution on plants, in P. L. Magill, F. R. Holden, and C. Ackley (eds.), "Air pollution handbook," sec. 9. McGraw-Hill, New York.

————, 1958. Air pollution with relation to agronomic crops: I. General status of research on the effects of air pollution on plants. *Agron. J.* 50:545–550.

Treshow, M., 1965. Evaluation of vegetation injury as an air pollution criterion. *J. Air Pollut. Contr. Assoc.* 15(6):226–269.

————, F. K. Anderson, and Frances Harner, 1967. Responses of Douglas fir to elevated atmospheric fluorides. *Forest. Sci.* 13(2):114–120.

———— and F. M. Harner, 1968. Growth responses of Pinto bean and alfalfa to sub-lethal fluoride concentrations. *Can. J. Bot.* 46:1207–1210.

Weinstein, L. H., 1961. Effects of atmospheric fluoride on metabolic constituents of tomato and bean leaves. *Contrib. Boyce Thompson Inst.* 21(4):215–231.

Woltz, S. S., and C. D. Leonard, 1964. Effect of atmospheric fluorides upon certain metabolic processes in Valencia orange leaves. *Proc. Fla. State Hort. Soc.* 77:9–15.

Zimmerman, P. W., 1950. Impurities in the air and their influence on plant life. *Proc. 1st Nat. Air Pollut. Symp.*, pp. 135–141.

———— and A. E. Hitchcock, 1956. Susceptibility of plants to hydrofluoric acid and sulfur dioxide gases. *Contrib. Boyce Thompson Inst.* 18(6):263–279.

SELECTED REFERENCES

Adams, D. F., H. G. Applegate, and J. W. Hendrix, 1957. Relationship among exposure periods and foliar burn and fluorine content of plants exposed to hydrogen fluorine. *Agr. Food Chem.* 5(2):108–116.

———— and M. T. Emerson, 1961. Variations in starch and total polysaccharide content of *Pinus ponderosa* needles with fluoride fumigation. *Plant Physiol.* 36:261–265.

————, 1963. Recognition of the effects of fluorides on vegetation. *J. Air Pollut. Contr. Assoc.* 13(8):360–362.

Allmendinger, D. F., V. L. Miller, and F. Johnson, 1947. Fluorine analysis of Italian prune foliage affected by marginal scorch. *Phytopathol.* 38:30–37.

Brennan, E. G., I. A. Leone, and R. H. Daines, 1950. Fluorine toxicity in tomato as modified by alterations in the nitrogen, calcium, and phosphorus nutrition of the plant. *Plant Physiol.* 25:736–747.

Brewer, R. F., F. H. Sutherland, and F. B. Guillemet, 1960. Sorption of fluorine by citrus foliage from equivalent solutions of HF, NaF, NH₄F, and H_2SiF_6. *Proc. Amer. Soc. Hort. Sci.* 76:215–219.

Chang, C. W., and C. R. Thompson, 1965. Subcellular distribution of fluoride in navel orange leaves. *Int. J. Air Water Pollut.* **9**:685–691.

——— and ———, 1966. Site of fluoride accumulation in navel orange leaves. *Plant Physiol.* **42**(2):211–213.

———, 1967. Study of phytase and fluoride effects in germinating corn seeds. *Cereal Chem.* **44**(2):129–142.

Compton, O. C., and L. F. Remmert, 1960. Effect of air-borne fluorine on injury and fluorine content of gladiolus leaves. *Proc. Amer. Soc. Hort. Sci.* **75**:663–675.

———, 1960. Effects of leaf clipping upon the size of gladiolus corms. *Proc. Amer. Soc. Hort. Sci.* **75**:688–692.

Hansen, E. D., H. H. Wiebe, and W. Thorne, 1958. Air pollution with relation to agronomic crops. VIII. Fluoride uptake from soils. *Agron. J.* **50**:565–568.

Hildebrand, E. M., 1943. Peach-suture spot. *Phytopathol.* **33**:167–168.

MacIntire, W. H., 1957. Fate of air-borne fluorides and attendant effects upon soil reaction and fertility. *Agr. Chem.* **40**:958–73.

McCune, D. C., L. H. Weinstein, J. S. Jacobson, and A. E. Hitchcock, 1964. Some effects of atmospheric fluoride on plant metabolism. *J. Air Pollut. Contr. Assoc.* **14**(11):465–468.

McNulty, I. B., and D. W. Newman, 1961. Mechanism(s) of fluoride-induced chlorosis. *Plant Physiol.* **36**(4):385–388.

Thomas, M. D., 1961. Effects of air pollution on plants, in "Air pollution," pp. 233–278. *W.H.O. (Geneva) Monogr.* no. 46.

Woltz, S. S., 1964. Translocation and metabolic effects of fluorides in gladiolus leaves. *Proc. Fla. State Hort. Soc.* **77**:511–515.

CHAPTER SEVENTEEN

FROM "SMOG" TO PAN

Early in the 1940s, industrial production in the United States accelerated rapidly to meet the demands of a nation at war. Nowhere was expansion greater than in the sparsely settled and relatively unpolluted Los Angeles basin of southern California. The productive capacities of local industries were taxed to their limits; existing facilities were expanded, and new industries arose. In just ten years the area's population nearly doubled.

Air pollution intensified correspondingly, producing a musty, acrid, irritating pall of smoke which clung relentlessly over the basin. A disturbed public watched the pollutants thicken each year until the formerly picturesque vista of the mountains and sea disappeared to be replaced with a dirty brown haze which became known as "smog." Los Angeles became the "smog" capital of the world, but the same condition soon plagued every major city in the world.

There was nothing new about the term smog; it was originally used in 1905 in London to denote a combination of smoke and fog. In Los Angeles smog meant essentially the same, but its components, and the extent of damage it caused, were yet to be recognized. By 1944, a unique type of plant injury attributed to smog added to the nuisance of reduced visibility and eye irritation. Symptoms consisting of glazing, silvering, bronzing, and sometimes necrosis of the lower leaf surface appeared on sensitive species, causing serious crop losses. Highly sensitive crops could no longer be grown profitably. By 1949, in Los Angeles County alone, losses to eleven crops exceeded an estimated $480,000 exclusive of growth suppression and fruit losses which were not yet recognized (Middleton et al., 1950).

When Los Angeles smog first became a serious irritant to man and was discovered to cause serious losses to agricultural crops, sulfur dioxide was the only widely recognized air pollutant and was quickly blamed for the problem. As late as 1949, officials of the newly established Los Angeles Air Pollution Control District thought that the problem would be solved if sulfur compounds could be removed from the atmosphere. Stringent controls on industry nearly achieved this goal, but pollution continued to increase.

The nature of the atmosphere was studied intensively, and by 1951 contaminants were recognized to include "every state of matter": dusts, fumes and smoke, peroxides, aldehydes, organic acids, sulfur dioxide, oxides of nitrogen, and numerous hydrocarbons were listed as the principal ingredients.

While major industrial contributors of these pollutants were soon recognized, and their emissions largely brought under control, smog not only persisted but grew worse. Far more precise knowledge was needed to define the exact nature of the pollutants so that their sources could be traced and hopefully eliminated.

Intensive chemical research led to the identification of many specific components of smog. Middleton et al. (1950) provided evidence that the toxic components of smog which caused plant damage were oxidizing substances associated with automobile exhaust gases. Once the oxidizing nature of the toxic components of smog was recognized, "oxidant" injury became accepted as a preferred name for this type of pollution. But "oxidants" denoted simply an oxidizing atmosphere, and the specific ingredients responsible for plant injury still had not been discovered. One of the most abundant constituents was found to be ozone, with lesser amounts of nitrogen oxides and organic peroxides, but none of these was accepted as an important phyto-toxicant.

Dr. Aarie Haagen-Smit of the California Institute of Technology, who pioneered the study of the Los Angeles atmosphere, discovered in 1952 that hydrocarbon and oxides of nitrogen emitted from automobile exhausts react in the presence of sunlight to yield the toxic gases which caused plant damage, eye irritation, reduced visibility, and many other adverse effects. Nitrogen oxide was formed by any high-temperature combustion process which caused two constituents abundant in any normal atmosphere, nitrogen and oxygen, to combine. Light energy is absorbed by nitrogen dioxide, and, if the radiation is sufficient, it will split the molecule into nitric oxide and atomic oxygen. This atomic oxygen can either combine with molecular oxygen to form ozone or react with some of the unburned hydrocarbons to form various oxidation products.

The polluted atmosphere could now be produced in the laboratory and the response of plants, as well as man and other animals, studied under controlled conditions. Haagen-Smit et al. (1952) demonstrated that leaf injury identical with that found so widely on plants in the Los Angeles basin could be produced on test plants by exposing them to the reaction products of ozone and unsaturated hydrocarbons produced by mixing known amounts of ozone and hydrocarbon from an automobile engine in a reaction tube and passing the mixture into the fumigation chamber. This type of synthetic "smog," preferably called photochemical oxidant, was utilized extensively in research programs for about six years before the more exact nature of the phytotoxic substances was discovered.

In 1959, Darley et al. reported that the half life of the reacted ozone-olefin phytotoxicants was not more than a few minutes, indicating that a "transitory" ozone-olefin complex was formed. Dr. E. R. Stephens and others (1956) at the University of California at Riverside studied the smog reactions with mixtures of nitrogen oxides and organic compounds in oxygen using a new 250- to 500-meter-path infrared cell. They observed the formation of ozone along with the progressive disappearance of nitrogen dioxide. Reaction products from the ozonated olefins were found to include formaldehyde and higher aldehydes, formic acid, ozonide, carbonyl, ozone, organic nitrate, carbon monoxide, carbon dioxide, water, and an important but unknown substance initially designated as "Compound X." Subsequently, Stephens identified it to consist of a family of related compounds collectively named peroxyacyl nitrates. Three of the compounds, peroxyacetyl nitrate (PAN), peroxypropionyl nitrate (PPN), and peroxybutyryl nitrate (PBN), have been produced experimentally, isolated, and used in fumigation studies. PAN appeared to be the principal constituent causing the plant damage initially attributed to smog and subsequently to photochemical oxidants. The PAN compounds were detected in natural smog in 1957, and their importance as phytotoxic air pollutants was established shortly afterwards following extensive, controlled fumigation studies with the synthesized compounds.

In summary, the major classes of naturally occurring, phytotoxic photochemical oxidants consist of ozone and PAN-type oxidants. This latter group contains the many varied products of ozone-olefin reactions as well as PAN. The complex is the same as that known in the 1950s as smog, oxidants, and photochemical oxidants. The effects of these compounds, particularly PAN, will be discussed in this chapter.

17-1 CONCENTRATIONS IN THE ATMOSPHERE

The concentrations of pollutants in the atmosphere must be known in order to accurately evaluate their hazard to vegetation and man and to establish an acceptable air quality standard. Oxidant concentrations must be determined by air analysis and observations of their effects on sensitive vegetation; they cannot be determined by vegetation analysis, as was the case for fluorides. Most of the available information on the distribution and concentrations of PAN has been provided by vegetation surveys of damage; air sampling data are limited mostly to total oxidant concentrations, which consist largely of ozone.

Chemical analysis for PAN in California and Utah show that concentrations regularly range from 1 to 2 pphm on average days and up to 5 pphm on "smoggy" days (Darley et al., 1968).

Almost no data exist attempting to correlate ambient PAN concentrations with injury in the field, but controlled fumigations indicate the levels at which visible symptoms might be expected. Stephens and Scott (1962) reported lesions developing on sensitive petunia and tobacco varieties following five hours' fumigation with 10.0 pphm. Observations in Salt Lake City following "natural" fumigations indicate that injury can be produced by as little as 2.0 pphm for two to four hours. When PAN concentrations averaged 2.5 to 3 pphm for a few hours, damage to sensitive plants was observed the following day. Jaffe (1965) reports work of Taylor which revealed sublethal effects including a reduction in dry fresh weight at concentrations of 0.5 to 1.5 pphm for eight hours.

17-2 PLANT RESPONSES

Symptomatology

The glazing, silvering, and bronzing symptoms which were so prevalent on petunia, lettuce, spinach, endive, Swiss chard, and sugar-beet crops in southern California in the 1940s were at first regarded as a single syndrome (Figs. 17.1 to 17.3). The causal agent was recog-

FIGURE 17.1 *Characteristic "oxidant-type" glazing of petunia produced by Los Angeles smog (photo courtesy of Wilfred Noble).*

FIGURE 17.2 *Bronzing of Great Lakes lettuce following PAN fumigation.*

nized only as "smog." Gradually more specific components were identified, and individual phytotoxic chemicals were delineated. The first major symptom recognized to be distinct from the "oxidant" injury was that caused by ozone. Ozone is now known to be the major constituent of photochemical pollution and is treated in the next chapter.

A second type of symptom attributed to smog now seems to be caused primarily by PAN and its homologues, that is, the reaction product of photolyzed nitrogen oxides with unsaturated hydrocarbons. The symptoms produced include most of those once designated to be caused by smog or photochemical oxidants.

The precise nature of the symptoms depends as much on environmental factors, tissue maturity, concentrations, and duration of fumigation as on the toxicant. Many types of symptoms may arise. Consequently, it is desirable to designate one general category as "PAN-type" oxidant injury (Jaffe, 1966). The still broader term, oxidant, which is sometimes used, is inaccurate since it also denotes ozone.

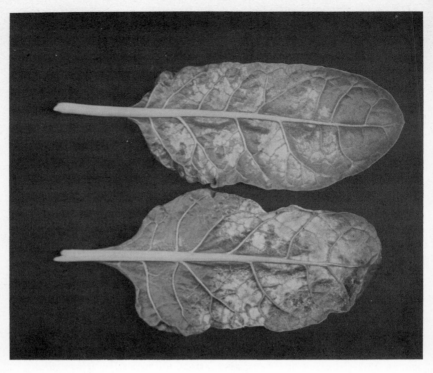

FIGURE 17.3 *Necrosis of Swiss chard attributed to ambient PAN at the University of Utah.*

Darley et al. (1963) distinguished the PAN-type symptoms from those caused by products of ozone-olefin reactions. The symptoms were essentially similar, but the ozone-olefin reaction products were most toxic to young, fully expanded leaves, causing uniform or indiscriminately distributed glazing over the lower leaf surface. The banding which appears on plants in the field, or when PAN is present, was lacking.

Symptoms of PAN injury cover a considerable range of expression depending on the concentrations, duration of exposure, and the nature and maturity of the leaf tissue exposed. High concentrations in the range of 50 to 100 pphm for thirty minutes may cause complete collapse of the tissue in a rather diffuse band across the leaf. Lower concentrations of 10 to 30 pphm for four to eight hours, still much higher than found in the field, may cause bronzing or glazing with little or no collapse visible on the upper leaf surface. Below 10 pphm, visible symptoms consist mostly of chlorosis.

Fumigations of sensitive species at the University of Utah (Anderson, unpublished data) with PAN at concentrations of 2 to 10 pphm for two to six hours provided a clear picture of symptomatology at realistic levels. Symptoms on Ranger alfalfa, one of the most sensitive species, first consisted of a very light yellow to white stippling appearing mainly on the upper surface but evident also on the lower (Fig. 17.4). Stippling characteristically developed between secondary veins, but distribution along the leaf varied with maturity. Bleaching was most prominent at the tip of terminal leaves and base of older leaves. Narrow bleached bands less frequently developed on leaves of intermediate age. Similar symptoms developed on sweet clover but at higher concentrations. When concentrations approached 10 pphm, local areas of tissue collapsed completely, producing bleached white necrotic lesions. Leaflet tips were most frequently affected, although basal portions or bands of necrosis appeared on other leaves. Marginal bleaching and curling also developed on some leaves.

Lima bean plants fumigated with PAN also occasionally developed a chlorotic stipple not unlike that produced by ozone. The pale green to white stipple was scattered over the entire upper leaf surface. More

FIGURE 17.4 *PAN injury of alfalfa showing fine, dense stipple pattern.*

typically, a silvery to lead-colored glaze developed in bands over the lower surface. Bronzing was often accompanied by a tan stippling.

Symptoms on corn consisted of chlorotic or necrotic streaks developing principally on the upper leaf surface. These streaks were largely interveinal, ½ to 2 in. long, and extended across the leaf at varying distances from the tip. Light gray to chlorotic stipple symptoms, developing in similar bands, appeared on other leaves.

The response of endive was especially interesting in view of the glazing and bronzing which develops so characteristically on plants in the field. When fumigated with PAN, light green, overall chlorosis with slight marginal necrosis and sometimes a yellowish stipple appeared; but lower-surface interveinal, silvery-to-bronzed glazing was most frequent and prominent.

In the field, the gross appearance of PAN injury may vary with each crop as well as its condition and maturity. Symptoms may be limited to a water-soaked appearance, as in dahlia; the lower-surface silvering or bronzing as in spinach, bean (Fig. 17.5), and petunia; browning or bronzing of either surface as in romaine lettuce, zinnia, and some chrysanthemum varieties; chlorotic or brownish mottling as

FIGURE 17.5 *Lower leaf surface (left) glazing of pinto bean attributed to ambient PAN at the University of Utah.*

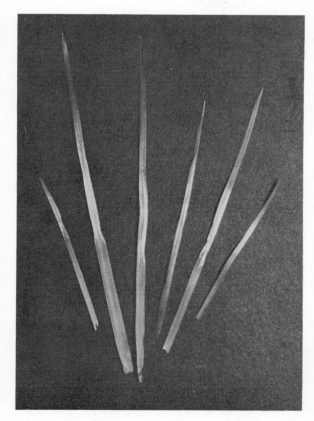

FIGURE 17.6 *Banding-type necrosis on* Poa annua *produced by Los Angeles smog (photo courtesy of Wilfred Noble).*

in tomato, pumpkin, and mallow; increase in anthocyanin and cork formation as in beets; tan banding as in barley and annual bluegrass (Fig. 17.6); or longitudinal streaking as in oats, corn, and many grasses.

Silvering varies from an almost glistening, silvery sheen on spinach to a bronzing on table beet or milky whiteness on snapdragon. Small plants and young leaves are most sensitive. Bronzing ranges from a glossy, bronze sheen to general necrosis.

The symptoms of smog injury on herbaceous plants in the field, as described in the Los Angeles area, generally consist of some type of glazing or bronzing distributed irregularly over the underside of the leaf. In 1952 it became apparent that the area of the leaf damaged was related to its maturity. When plants were exposed to a single fumigation, only a few leaves at a particular stage of maturity were marked. The youngest leaf would be injured at the tip, the next oldest about one-third of the way down from the tip, and the oldest near the base. This banding was especially prominent on leaves of petunia, ryegrass, and annual bluegrass (Fig. 17.7). Sensitive broad-leaved plants such as spinach were most frequently marked about one-third of the way down from the leaf tip. Banding was also observed in the field on beets, chard, *Mimulus,* chickweed, pigweed, dock, and others.

Banding is extremely helpful in identifying PAN-type oxidant injury. However, if exposure occurs on successive days, the newly differentiating cells may be injured each day, so that a general glazing appears over the entire leaf. This has been the more characteristic expression in the polluted areas of California where toxic PAN concentrations occur daily.

Pathological Histology

Photochemical pollutants, including PAN, enter the leaf through the mature, functional stomata. Once within the substomatal chambers, the pollutant attacks the mesophyll cells bordering the inter-

FIGURE 17.7 *PAN injury to petunia showing simultaneous injury to tip of youngest leaves and center of older leaves (photo courtesy of B. L. Richards).*

cellular spaces. Bobrov (1952; Glater, 1962) studied the morpho-logical and histological effects of ozonated olefins, and later of PAN, on representative sensitive species. She found that the earliest visible indication of injury was the oily, shiny, water-soaked appearance of sensitive tissues on the lower leaf surface. By the time this appeared, tissue alterations were already apparent. Tiny, raised blisters appeared which were formed by the swelling of guard cells and other cells nearest the stomata. These became engorged with water and increased in width, causing the stomata to enlarge further. If the guard cells become excessively distorted, they will be permanently injured and collapse, closing the stomata and preventing further entrance of PAN.

By the time the epidermal cells collapse, the entire leaf becomes turgid. Permeability may be damaged, so that excessive water enters the affected cells; these stretch as they engorge water, giving the underside of the leaf its shiny, water-soaked appearance. If the fumi-gation does not persist, or is not too severe, the turgid cells may recover after a few hours, leaving no trace of visible injury.

Cells of the spongy mesophyll nearest the intercellular spaces, or substomatal cavity, are affected first. In 1965, Thomson et al., using an electron microscope, found that small, electron-dense granules appeared in the chloroplast stroma soon after fumigation. Subse-quently "crystalline" arrays of granules appeared, and the shape of the chloroplast was altered. The granules seemed to fuse into rods and then into an organized system of plates which persisted and even seemed to continue to develop. Finally the integrity of the chloroplast was lost, and the membranes disrupted. As the chloroplasts broke down and became dispersed into the cytoplasm, the entire protoplast aggregated into a large mass which condensed in the interior of the cell, moving away from the cell walls and causing the cell to collapse. The plasmodesmatal connections with neighboring cells appeared to persist. While the lower surface was usually injured first due to the larger amount of intercellular space, the reverse may also be true. The tissue injured appears to depend to a great extent on its age. Palisade cells differentiate later than spongy parenchyma and may sometimes be in a less sensitive condition at the time of exposure. Smaller veins may be damaged together with the mesophyll, but the larger veins and midrib usually remain intact.

Internal, as well as external symptoms, vary with the structure of the plant. Broad-leaved herbaceous plants such as table beet, sugar-beet, lettuce, and spinach are most sensitive and injury is characterized by a silvering and sometimes bronzing, of the lower leaf surface.

Monocotyledonous plants, including oats, corn, barley, and grasses, appear dark green as if water were trapped beneath the epidermis. As cell damage progresses, the dark green water-soaked areas develop into yellow streaks which follow the zones where stomates are densest. The yellow soon turns to brown, and longitudinal necrotic streaks appear between the larger veins.

In all of these plants, damage seems to be partly proportional to the amount of internal air space existing within the leaf. Other factors which regulate sensitivity include its inherent susceptibility or immunity, the plant's metabolic activity, the age of the plant, the leaf, and leaf tissue, and many environmental factors such as temperature, soil moisture, and nutrition.

The different sensitivity of young, old, and middle-aged leaves and tissues has been apparent for many years, but the reasons were not clear. Also, leaves of different ages show damage in different positions, so that the tip of one young leaf and the base of a second leaf might be the only portions affected.

The relative smog sensitivity of different-aged tobacco leaves and cells was discussed in detail by Glater et al. (1962), who studied the developmental anatomy of tobacco leaf injury. Following a single fumigation, a band of injured tissue appeared across the central portion of most of the middle-aged leaves. The tip portion of younger leaves and the basal area of older leaves were also injured. The damage band moved from the tip through the middle and toward the base of the leaves from young to old. Damage areas were not absolutely restricted to a single band because all cells do not differentiate at exactly the same rate.

Apparently the cells which have just finished differentiation are most severely injured. These are most sensitive possibly due to a maximum rate of metabolic activity in such cells, but also because of the greater tolerance of other cells. Stomata in the young portion of the leaves are not fully formed; stomatal initials are still differentiating, and intercellular spaces are undeveloped. The resistance of older tissues was attributed to the normal development of a waxy layer around individual cells. Also, greater cuticle thickness, lignification, reduced stomatal activity, and possibly the reduced gas exchange rate and metabolic activity of older tissues may increase tolerance.

The same mechanism explains the transverse banding expression found in monocotyledonous plants. These plants possess a persistent basal meristem. Since new cells are formed at the leaf base, cells mature from the tip down, and sensitivity of cells varies with the distance from the meristem. The young basal cells and oldest tip cells are most tolerant. Cells a short distance from the meristem are most sensitive. Following fumigation, a single discrete band of cells will be injured. Following successive fumigations, a broader band, or even longitudinal streaks, will develop as the leaf matures.

Metabolic Effects

The histological changes and visible symptoms caused by air pollutants are basically caused by impaired metabolic processes. Basic processes such as enzyme activity, respiration, photosynthesis, ion absorption, and carbohydrate and protein synthesis all may be impaired by PAN concentrations far lower than necessary to produce any leaf chlorosis or necrosis.

To understand how PAN might affect metabolism, it is desirable first to know how the chemical is incorporated into the plant constituents. Stephens et al. (1961) synthesized PAN labeled with C^{14} so that its pathway in the plant might be followed.

When the cells were fractionated, much of the C^{14} in treated plants appeared in the chloroplasts, and it is in these organelles that one might first look for deleterious effects. Early damage to the chloroplasts was shown first by histological, microscopic studies, but later physiological studies confirmed the damaging effect of PAN on the chloroplasts. The mechanism of action is not clearly defined but may be partly caused by injury to the enzymes necessary to photophosphorylation. Photophosphorylation is essential to photosynthesis in providing the energy needed to split the water molecule. In 1963, W. M. Dugger and his associates at the University of California showed that PAN inhibited photosynthetic CO_2 fixation in pinto bean plants.

PAN may also be harmful in oxidizing critical sulfhydryl groups of certain enzymes. Mudd (1963) showed that enzymes which contained free sulfhydryl groups for catalytic action were especially sensitive to PAN. Dugger et al. (1966) have shown that a correlation exists between the sulfhydryl content of bean plants and their susceptibility to PAN. Damage may be prevented by adding sulfhydryl reagents, which protect the SH group, to enzyme reaction mixtures, chloroplasts, and mitochondrial suspensions, or spraying them on intact plants.

Oxidation of the -SH group would impair the activity of many enzymes. Inhibition of an enzyme such as phosphoglucomutase, which is vital to the synthesis of glucose, may be directly responsible not only for inhibition of glucose synthesis but synthesis of closely related carbohydrates including galactose, xylose, arabinose, and polymers that constitute the wall fractions, including cellulose. Inhibition of the metabolism of cellulose and cell wall noncellulosic polysaccharides by PAN would seriously limit cell wall synthesis, cell enlargement, and growth.

Continued normal growth requires not only the synthesis of new cell wall constituents but the continued breakdown of old cell wall constituents. The cell wall expansion phase of plant cell growth is dependent on growth hormone–induced changes in cell wall plasticity. PAN has been found to inhibit one of the enzymes, glucan hydrolase, which is necessary to this metabolic breakdown of the cell wall.

PAN-inhibition of enzymes essential to both synthesis and degradation of cell wall constituents may well account for reduced cell enlargement and growth associated with PAN damage.

Growth and Reproduction

A knowledge of the mechanisms by which an air pollutant acts on specific metabolic processes provides the background for understanding the effects air pollutants have on growth and reproduction in the field. This threat is the ultimate concern of anyone attempting to grow plants and may be vital to the continued well-being of natural plant populations. The multitude of variables affecting growth and production makes it extremely difficult to measure losses which might be attributed solely to air pollutants, and relatively few studies provide much of a clue as to the extent of such losses.

Studies of growth suppression have consisted largely of exposing plants to ambient, polluted atmospheres and filtering air in which comparable control plants were grown. Total oxidant concentrations in the field were known, but the specific composition of the polluted air was not.

While ambient-atmosphere studies fail to delimit the effects of PAN alone on growth and production, they do provide an excellent picture of the actual impact of ambient polluted atmosphere. The limited studies of the effects of Los Angeles smog on plant growth were summarized by Todd et al. in 1956. Kentia palms were especially sensitive. When grown in ambient air, plants were small and leaves stunted and chlorotic. Palms in carbon-filtered atmospheres were noticeably larger, averaging one leaf more than on plants in ambient air, and leaves were longer and dark green in color.

Avocado trees grown in ambient and filtered atmospheres responded similarly, with the stem diameters significantly greater in the filtered air after six months.

Comprehensive studies of the effect of ambient atmospheres on citrus production in the Los Angeles area have been undertaken by C. Ray Thompson and his associates at the University of California (1966, 1968). Trees exposed to the naturally polluted air have up to 30 percent more leaf drop, and the average yields are often only half of what they would be in clean air.

17-3 DIAGNOSIS

The glazing and bronzing of the lower leaf surface and the occasional chlorotic stipple are characteristic expressions to seek in diagnosing PAN injury. If the symptom occurs in bands caused by a single fumigation, the diagnosis is even more definitive. No other pathogen, with the possible exception of ozone, causes symptoms just like it. But symptoms resulting from a few other stresses may raise some questions.

Frost, for instance, sometimes causes a glazing in which the epidermis is damaged and separates from the mesophyll, but the injury is more prominent on the exposed upper leaf surfaces, and epidermal cells are damaged more than the mesophyll. Also, the more frost-sensitive species are not necessarily the same as those most sensitive to PAN.

Particularly at low concentrations below 5.0 pphm, PAN sometimes produces a general stipple symptom closely resembling ozone toxicity. (Fig. 17.8). Since the two chemicals often occur together, it is not always possible to be positive which was responsible for the injury. This is especially true when symptoms appear on plants such as petunias or alfalfa which are sensitive to both pollutants. In such cases, the symptoms are best regarded simply as "oxidant" injury.

Lower-surface whitening, and occasionally glazing, can also be caused by mite, leafhopper, or thrip infestations. Banding in these instances is unlikely, and some sign of the pest can usually be found. But since presence of the pest does not ensure that it (and not PAN)

FIGURE 17.8 *Upper row, ozone injury; lower row, PAN injury.*

was responsible for the symptoms, closer scrutiny may be necessary. Microscopic inspection of the tissue can then be valuable in locating feeding wounds caused when insects are present.

But, as with other pollutants, one of the best means of evaluating the cause of injury is to be familiar with the relative sensitivity of different species. Nobel (1965) gives the following list of relative sensitivity to PAN-type damage:

TABLE 17.1. *Relative sensitivity of plant species to PAN-type injury**

CROPS			
Sensitive		*Resistant*	
Spinach	Beet	Cabbage	Onion
Endive	Corn	Cauliflower	Corn
Oats	Celery	Rhubarb	Cucumber
Romaine lettuce	Pepper	Carrot	Strawberry
Swiss chard	Tobacco	Squash	
Alfalfa	Clover		
Beans			

TABLE 17.1. (*Continued*)

ORNAMENTALS			
Sensitive		Resistant	
Petunia	Sweet basil	Cactaceae	Orchids
Mimulus	Fuchsia	Anthurium	Coleus
Snapdragon	Impatiens	Bromiliad	Cyclamen
Primrose	Mint	Calendula	Ivy
Aster	Ranunculus	Camellia	Narcissus
		Carnation	Lily
		House plants	

WEEDS		
Sensitive		Resistant
Annual bluegrass	Mustard	
Pigweed	Jimson weed	
Chickweed	Dock	
Wild oat	Ground cherry	

* Adapted from data provided by Noble (1965).

BIBLIOGRAPHY

Bobrov, Ruth Ann, 1952. The effect of smog on the anatomy of oat leaves. *Phytopathol.* **42**:558–563.

Darley, E. F., E. R. Stephens, J. T. Middleton, and P. L. Hanst, 1959. Oxidant plant damage from ozone-olefin ractions. *Int. J. Air Pollut.* **1**:155–162.

———, W. M. Dugger, J. B. Mudd, L. Ordin, O. C. Taylor, and E. R. Stephens, 1963. Plant damage by pollution derived from automobiles. *Arch. Env. Health* **6**:761–770.

———, K. A. Kettner, and E. R. Stephens, 1963. Analysis of peroxyacetyl nitrates by gas chromatograph with electron capture detection. *Anal. Chem.* **35**:589–591.

———, C. W. Nichols, and J. T. Middleton, 1966. Identification of air pollution damage to agriculture crops. *Bull. Dept. Agr. Calif.* **55**(1):11–19.

———, O. C. Taylor, and J. T. Middleton, 1968. Photochemical and exhaust fume damage to plants. 1st Int. Congr. Plant Pathol., London.

Dugger, W. M., Jane Koukol, W. D. Reed, and R. L. Palmer, 1963. Effect of peroxyacetyl nitrate of $C^{14}O_2$ fixation by spinach chloroplasts and pinto bean plants. *Plant Physiol.* **38**:468–472.

———, ———, and R. L. Palmer, 1966. Physiological and biochemical effects of atmospheric oxidants on plants. *J. Air Pollut. Contr. Assoc.* **16**:467–471.

Glater, Ruth Bobrov, R. A. Solberg and Flora M. Scott, 1962. A developmental study of the leaves of *Nicotina glutinosa* as related to their smog-sensitivity. *Amer. J. Bot.* **49**:954–970.

Haagen-Smit, A. J., E. F. Darley, M. Zaitlin, H. Hull, and W. Noble, 1952. Investigation on injury to plants from air pollution in the Los Angeles area. *Plant Physiol.* **27**:18–34.

Jaffe, L. S., 1966. Effects of photochemical air pollution on vegetation. *59th Annu. Meet. Air Pollut. Contr. Assoc. Paper* 66-43, pp. 1–31.

Middleton, J. T., J. B. Kendrick, Jr., and H. W. Schwalm, 1950. Injury to herbaceous plants by smog or air pollution. *Plant Dis. Rep.* **34**:245–252.

Mudd, J. B., 1963. Enzyme inactivation by peroxyacetyl nitrate. *Arch. Biochem. Biophys.* **102**:59–65.

Noble, W., 1965. Smog damage to plants. *Lasca Leaves* **15**:24.

Stephens, E. R., P. L. Hanst, R. C. Doerr, and W. E. Scott, 1956. Reactions of NO₂ and organic compounds in air. *Ind. Eng. Chem.* **48**:1498–1504.

————, E. C. Darley, O. C. Taylor, and W. E. Scott, 1961. Photochemical reaction products in air pollution. *Int. J. Air Pollut.* **4**:79–100.

———— and W. E. Scott, 1962. Relative reactivity of various hydrocarbons in polluted atmospheres. *Proc. Amer. Petrol. Inst.* **42**:670–675.

Taylor, O. C., E. A. Cardiff, J. D. Mersereau, and J. T. Middleton, 1958. Effect of air-borne reaction products of O₃ and 1-N-hexene vapor on growth of avocado seedlings. *Proc. Amer. Soc. Hort. Sci.* **71**:320–325.

Thompson, E. R., and O. C. Taylor, 1966. Plastic-covered greenhouses supply controlled atmospheres to citrus trees. *Trans. A.S.A.E.* **9**:338–342.

Thompson, C. R., 1968. Effects of air pollutants on lemons and navel oranges. *Calif. Agr.* **22**:2–3.

Thomson, W. W., W. M. Dugger, Jr., and R. L. Palmer, 1965. Effects of peroxyacetyl nitrate on ultrastructure of chloroplasts. *Bot. Gaz.* **126**:66–72.

Todd, G. W., J. T. Middleton, and R. F. Brewer, 1956. Effects of air pollutants. *Calif. Agr.* **9**:7–8, 14.

SELECTED REFERENCES

Dugger, W. M., Jr., O. C. Taylor, E. Cardiff, and C. R. Thompson, 1961. Stomatal action in plants as related to damage from photochemical oxidants. *Plant Physiol.* **47**:487–491.

————, J. B. Mudd, and Jane Koukol, 1965. Effect of PAN on certain photosynthetic reactions. *Arch. Environ. Health* **10**:195–200.

Koukol, Jane, W. M. Dugger, Jr., and N. O. Belser, 1963. The inhibition of cyclic photophosphorylation by peroxyacetyl nitrate. *Plant Physiol.* **38**:12–14.

Leighton, P. A., 1961. "Photochemistry of air pollution." Academic, New York, 300 pp.

Middleton, J. T., J. B. Kendrick, and E. F. Darley, 1955. Air-borne oxidants as plant-damaging agents. *Proc. 3d Nat. Air Pollut. Symp.*, pp. 191–199.

———— and A. J. Haagen-Smit, 1961. The occurrence, distribution, and significance of photochemical air pollution in the United States, Canada, and Mexico. *J. Air Pollut. Contr. Assoc.* **11**:129–134.

————, 1961. Photochemical air pollution damage to plants. *Ann. Rev. Plant Physiol.* **12**:431–448.

Ordin, Lawrence, 1962. Effect of peroxyacetyl nitrate on growth and cell wall metabolism of avena coleoptile sections. *Plant Physiol.* **37**:603–608.

————, M. J. Garbor, B. P. Skoe, and G. Rolle, 1966. Role of auxin in growth of inhibitor-treated oat coleoptile tissue. *Physiol. Plant.* **19**:937–945.

Stephens, E. R., R. R. Burleson, and E. A. Cardiff, 1965. The production of pure peroxyacyl nitrates. *J. Air Pollut. Contr. Assoc.* **15**:87–89.

Taylor, O. C., E. A. Cardiff, J. D. Mersereau, and J. T. Middleton, 1957. Smog reduces seedling growth. *Calif. Agr.* **9**:9, 12.

————, W. M. Dugger, Jr., M. D. Thomas, and C. R. Thompson, 1961. Effect of atmospheric oxidants on apparent photosynthesis in citrus trees. *Plant Physiol. Suppl.* **36**:26.

————, 1962. Air pollution and its effect on plant growth. *Proc. 38th Int. Shade Tree Conf.*, pp. 54–62.

CHAPTER EIGHTEEN

OZONE AND PLANTS

Ozone is a natural but deadly component of the upper atmosphere, where it plays the vital role of absorbing and filtering out dangerous ultraviolet radiation. It is also produced in vast quantities in the lower atmosphere of every metropolitan center by the action of light energy on the waste products of combustion. Here, ozone concentrations are increasing at an alarming rate, threatening man's health and the productivity of his crops. Since the recognition of ozone as a major phytotoxicant in 1958 (Richards et al.), damage has been described on every type of ornamental and agronomic crop. Leafy vegetables; cereal, forage, and textile crops; shrubs; ornamental, fruit, and forest trees; all have been extensively damaged.

Minor amounts of ozone are added to the atmosphere by electrical discharges such as lightning flashes; more may be brought down from the upper atmosphere by vertical flux. But the truly significant quantities of ozone present in our immediate environment are formed chemically by the action of ultraviolet light on nitrogen dioxide.

Ozone formation is greatest in urban environments, where the necessary chemicals for reactions abound. A multitude of combustion processes and sources, particularly the inefficient internal combustion engines of automobiles, daily emit tons of waste hydrocarbons and nitrogen oxides into the atmosphere. Heat from any flame causes atmospheric nitrogen and oxygen to combine into nitrogen oxides. The hotter the flame, the greater the production of nitric oxides. The nitric oxide (NO) is oxidized to nitrogen dioxide (NO_2), utilizing the oxygen in the atmosphere. But energy from sunlight quickly splits the nitrogen dioxide (NO_2) back into nitric oxide (NO) and atomic oxygen, which combines with the molecular oxygen of the atmosphere to form ozone:

$$NO_2 \xrightarrow{\ hv\ } NO + O; \; O + O_2 \xrightarrow{\ m\ } O_3$$

where m = any inert molecule and hv = light energy

The net reaction is

$$NO_2 + O_2 \underset{}{\overset{hv}{\rightleftharpoons}} NO + O_3$$

The back reaction theoretically proceeds faster than the initial reaction, so that ozone should be removed from the atmosphere. But hydrocarbons present in urban atmospheres react with, and remove, the NO, stopping the back reaction so that ozone accumulates. In bright sunlight and in the absence of any competing or back reaction, NO_2 would have a half life of only one or two minutes. In the dark the net reaction is reversed, and nitrogen dioxide accumulates at the expense of ozone.

The wide dispersion of urban pollutants makes it difficult to find areas where reliable measurements of background levels, entirely free from pollution, can be obtained; but measurements in remote coastal and mountain areas have provided some data on "natural" ozone concentrations. Charles Berry at the Forest Experiment Station in Asheville measured day- and night-time ozone concentrations in a remote valley and mountain top in western North Carolina. Hourly average oxidant levels usually ranged between 1 and 2 pphm, with a high of 2.5 pphm. Measurements in the remote desert areas of the western United States also indicated 2 pphm as a normal background ozone concentration.

Higher, dangerous oxidant concentrations are broadly confined to air sheds in which urban centers are located, but they may extend a hundred or more miles beyond the metropolitan boundaries. The highest total oxidant concentrations have been reached in the Los Angeles basin of Southern California, where the daily maximum has reached 100 pphm, the highest ever recorded. Daily peaks commonly reach 15 to 38 pphm in the summer and fall, and 5 to 10 pphm in the winter.

While ozone reaches its highest concentrations in the Los Angeles area, concentrations in and around other cities almost daily exceed the roughly 5 pphm at which plant damage is expected. Ozone has reached 50 pphm in San Francisco, 22 pphm in Salt Lake City, 16 in Washington, D.C., and over 15 pphm in nearly every large city in which measurements have been taken.

Fortunately, ozone does not persist in the atmosphere, and the supply must be renewed daily. Ozone reacts rapidly with other chemicals in the atmosphere as well as structures, ground surfaces, and particularly plant surfaces, and is soon neutralized. Consequently, photochemically produced ozone is cleansed from the atmosphere shortly after it stops being produced.

18-2 ENTRANCE AND TOXICITY OF OZONE

When the leaves of plants are exposed to ozone, the guard cells of the epidermis quickly respond and lose their turgor, and the stomatal aperture closes. While this response may afford some protection by excluding the pollutant, it fails to prevent the entrance of ozone sufficiently to provide complete protection.

Metabolic Alterations and Sublethal Effects

By the time symptoms of ozone toxicity can be seen, a multitude of metabolic or physiological changes have already taken place within the plant cells. At least five different plant processes have been shown to be directly affected by ozone. After ozone has passed through the stomates, it contacts the cytoplasmic membranes and affects membrane structure and permeability. Ozone next attacks cellular enzymes and organelles, disrupting critical metabolic processes. Alterations may be expected in carbohydrate and amino acid metabolism, and other cellular components (including chloroplast or mitochondrial membranes) may be oxidized. Mitochondrial activity and photosynthesis may be inhibited and respiration stimulated.

STOMATAL MECHANISM AND PERMEABILITY   The effects ozone may have on photosynthesis, respiration, and other metabolic processes are intimately entwined with the initial effects of ozone on the stomatal mechanism and the penetration of ozone into the plant. Any impairment of stomatal opening limits gas exchange and directly affects any subsequent metabolic process which depends on the availability of carbon dioxide or oxygen.

Studies have shown (Hill, 1967; MacKnight, 1968) that the width of stomatal openings of many plant species is rapidly reduced in the presence of ozone. Stomatal closure was induced in oat leaves by 5 pphm ozone for four hours and in bean by 6 pphm. More tolerant species respond similarly at higher concentrations.

The basic question of why and how ozone influences the stomatal aperature or, more directly, the turgidity of the guard cells, has not been answered. However, ozone is known to affect cell permeability, and an increase in permeability might allow water to escape from the guard cells and cause the stomata to close.

Ozone may destroy the structural integrity of membranes and impair both membrane transport and ion uptake. Ozone, being an extremely strong oxidizing agent, would be expected to react at the first reducing site encountered and oxidize the component membrane molecules. McFarlane (1966), studying the effects of ozone on permeability, found that sublethal ozone concentrations significantly decreased ion uptake and permeability in both tobacco leaf and potato tuber tissues.

PHOTOSYNTHESIS AND ASSIMILATION   Every synthesis, assimilation, and growth process in the plant depends on the energy derived from sugars and other intermediate metabolites initially formed via the photosynthesis process. Todd and Propst (1963) showed that sublethal concentrations of ozone markedly reduced photosynthetic activity. Photosynthetic rates of coleus and tomato, as measured by CO_2

uptake, continued to drop as long as plants were exposed to ozone, but they recovered almost immediately when the fumigation stopped. Recovery of bean leaves, on the other hand, was appreciably slower. In this study, the extent of photosynthesis reduction was correlated with injury, since most of the concentrations were sufficiently high to cause tissue damage.

The effects of sublethal concentrations of ozone on photosynthesis and transpiration, as functions of stomatal closure, were studied by Hill (1968). Measurement of the apparent rate of CO_2 assimilation was used to evaluate the effects on growth. Carbon dioxide was added to recirculating, airtight growth chambers at the same rate as the plants removed it. Transpiration was determined by maintaining humidity and temperature within narrow limits and measuring the rate at which water condensed on the cold water coils.

At ozone concentrations averaging 57 pphm for fifty-five minutes, CO_2 assimilation was usually reduced to 40 to 70 percent of the control within thirty to ninety minutes. All fifteen species studied, which included cereals, tobacco, and vegetable crops, responded in a similar manner. After the ozone treatment was terminated, apparent photosynthesis frequently continued to decrease for a few minutes but usually began to increase within less than an hour, and occasionally it returned to the control rate within two to three hours. Recovery was generally complete by the next morning, so long as tissues were not killed. The same trend persisted for transpiration, which was rapidly reduced about 50 percent in the presence of ozone compared with a forty percent reduction for apparent photosynthesis. The adverse effects were more basically caused by stomatal closure, which correlated perfectly with the reduced carbon dioxide uptake and transpiration.

CARBOHYDRATE METABOLISM   The most immediate effect of ozone on carbohydrate metabolism lies in its influence on respiration. Tobacco leaf respiration may be inhibited initially by ozone, but as visible injury appears, respiration is stimulated. Dugger has shown, however, that in lemon leaves not visibly injured, respiration increases if low-level fumigations are continued over an extended period of time. This increase in respiration means an increase in the rate at which sugars are burned and the carbohydrate reserves depleted.

Studies by Dugger and his associates at the University of California (1966) have demonstrated that ozone treatment causes a significant reduction in total carbohydrate, mostly starch, but an increase in the amount of reducing sugars. The increase in reducing sugars, mostly hexose, would provide a substrate for respiration and may account for the increased respiration observed with ozone treatments. These changes are probably interrelated and indicate either an inhibition in starch synthesis from hexose or a stimulation in the breakdown of starch. J. Bennett (1968) found that ozone impaired both starch synthesis and hydrolysis, but inhibition of synthesis was greatest, and the net starch concentration diminished. It is possible that the increased rate of sugar decomposition by very low, sublethal ozone levels over an extended period of time starves the tissue and causes subsequent premature senescence, leaf abscission, and fruit drop. Abscission is most striking after a week or two of high temperatures, when photosynthesis and replenishment of the carbohydrate reserves is limited to a few cool hours in the morning and afternoon.

Any influence of ozone on carbohydrate metabolism might also affect the formation of cell walls by limiting the supply or utilization of the required glucose. At sublethal levels, ozone is thought to affect overall plant growth. Ordin (1965) showed that this might be due to ozone-retarding cell elongation by causing a marked inhibition of cellulose synthesis and subsequent cell wall formation. Inhibition was attributed primarily to the oxidizing effect of ozone of the sulfhydryl groups of enzymes critical to synthesis.

GROWTH AND PRODUCTION   All the sublethal effects of ozone on every plant process from ion uptake to photosynthesis suggest that growth and ultimately production must be affected to at least some degree. Regrettably, studies of growth responses are limited. Research using ambient atmospheres does not define the effects of ozone alone; these effects must be studied in prolonged, controlled atmospheres in which a multitude of extraneous pollutants are not allowed to confuse the results.

Symptomatology

The necrotic flecking developing on the upper leaf surface of sensitive plants exposed to ozone is widely recognized to be a characteristic symptom of ozone toxicity, but the expression may be misleading when other pathogens causing similar symptoms are present. Low temperatures, PAN, chlorine, and even normal senescence can cause plant markings which resemble those caused by ozone. In order to diagnose ozone toxicity, the observer must be familiar with the histological response of plants to ozone, the details of the variations in visible symptoms, and the relative sensitivity of species.

HISTOLOGICAL, MICROSCOPIC EFFECTS   The reasons why symptoms appear as they do, and to a large extent the mechanisms of toxicity, can be explained by a thorough understanding of the histological responses of plants to a pollutant.

Histological studies reported by Hill et al. (1961) revealed that chloroplasts were the first organelles to respond to ozone, becoming disrupted and lighter green in color than normal soon after exposure. The chloroplast membrane breaks down, chlorophyll becomes dispersed in the cytoplasm, the affected cells lose their green color, and the protoplast shrinks from the cell wall (Figs. 18.1 and 18.2). Cell

FIGURE 18.1 *Photomicrograph of incipient ozone-induced fleck on spinach, showing complete collapse of one outer palisade cell and various stages of chloroplast and cellular breakdown.*

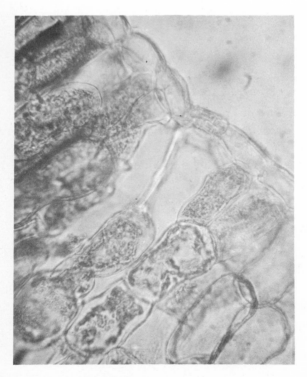

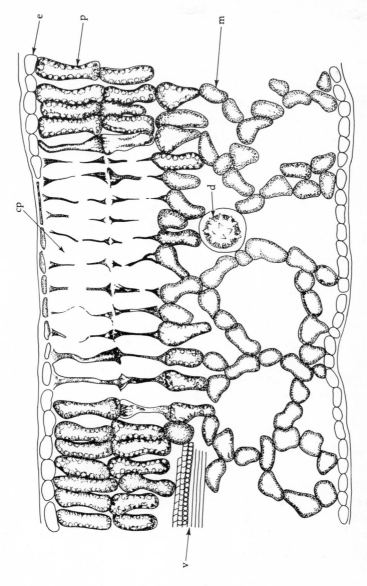

FIGURE 18.2 Cross section of spinach leaf showing collapsed palisade tissue (cp) overlain by collapsed epidermal tissue. Normal epidermis (e), palisade (p), and spongy mesophyll (m) cells, druse (d), and a sector of vascular tissue (v) are also shown.

wall and plasmodesmatal connections appear to be retained. When cell wall connections are lost, large intercellular areas are formed, giving the lesions a bleached appearance. When brown or reddish-brown pigments are formed as the cell dies, the lesions have a characteristic brownish color.

In 1966, Thomson et al., using electron microscopy to study the fine structures, also found the chloroplasts—more specifically the stroma of the chloroplast—to be the first organelles affected. The stroma was almost entirely granulated, with a considerable increase in electron opacity. In other cells, ordered arrays of granules and packets of plates of fibrils developed in the stroma. Nuclei of palisade cells within damaged areas were usually shrunken to spherical to irregularly ellipsoid in shape, and the nuclear membrane was clearly damaged. The earliest sign of a general cellular breakdown was the pulling away of the plasmalemma from the wall and a collapse of the vacuole, with a clumping of the cell organelles and cytoplasm toward the center of the cell. In the most severely damaged cells, the plasmalemma and tonoplast were almost completely degenerated; mitochondria were swollen, and there was an accumulation of electron-dense material in and near the mitochondria.

EXTERNAL INJURY SYMPTOMS While ozone may impair plant growth at concentrations far lower than those necessary to cause visible symptoms, these external symptoms are the only indication the observer has in the field that phytotoxic ozone concentrations are present.

The earliest macroscopic indication of ozone injury on broad-leaved, herbaceous species is the shiny, oily appearance of the upper leaf surface developing within one to two hours of exposure to toxic concentrations. The oily appearance disappears shortly after fumigation; tissues then either return to normal if the exposure was sublethal or become permanently damaged. Where the tissue is irreversibly damaged, affected areas become water-soaked and dull, gray-green in appearance. Gradually the water-soaked areas become chlorotic to bleached as the palisade cells collapse. Isolated groups of cells between the smallest veinlets are affected first, forming the characteristic punctate or flecked pattern on the upper leaf surface. This faint chlorotic or necrotic mottling or stippling is the first expression usually noted in the field (Fig. 18.3). Palisade tissue comprising the upper leaf mesophyll is most sensitive to ozone and the first injured; next the interior

FIGURE 18.3 *Ozone-stipple lesions on upper surface of avocado leaf (photo courtesy B. L. Richards).*

spongy mesophyll cells and finally the lower mesophyll cells are damaged. Epidermal cells, including the stomatal guard cells overlying the injured palisade cells, generally appear normal until a large area of palisade cells is destroyed. Vascular tissues, including phloem, xylem, and bundle sheath cells, are most tolerant of ozone and are damaged only when all the associated tissues are killed.

Two variations of injury have been observed. The more common is a dull white stipple in which affected areas collapse to form slight depressions. Less frequently, the surface has a glazed, pearly appearance (Fig. 18.4); injury is superficial, with little to no apparent collapse of the affected area. After a few days, the lesions frequently turn

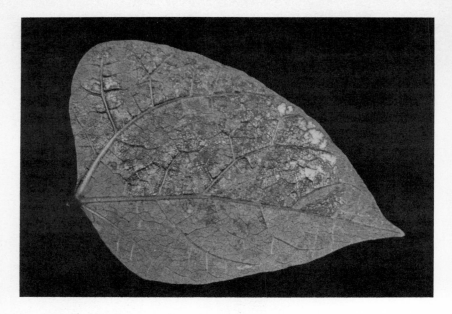

FIGURE 18.4 *Glazing of pinto bean following six hours of ozone fumigation at 30 pphm. This symptom is virtually identical to that produced by PAN.*

pale yellow to tan in color. However, in a few plants such as beet, lesions become reddish. Injury is usually most pronounced at the leaf tip and along the margins, but this varies somewhat with the age of the tissues.

Hill et al. (1961) found that ozone also caused hypertrophy in certain plants. Tumefaction developed on broccoli and tomato leaves subjected to ozone at a concentration of 25 pphm for two hours (Fig. 18.5).

Symptoms on legumes such as alfalfa may be so similar to those caused by PAN and SO_2 that they should be specifically described. Within a few hours of exposure to ozone, affected areas became finely stippled and dull gray-green to yellow-green (Fig. 18.6). Chlorosis tends to be most concentrated along the veins, but when severe it may extend over most of the interveinal tissue and be evident on both the lower and upper leaf surface. Minute islands of normal green tissue up to 1 mm in diameter and circular in outline may remain within the chlorotic areas. A second symptom which often appears is the silver-gray to bronzed necrotic bleaching of the upper leaf surface. Lesions are sometimes confined between the smaller veins but more commonly extend irregularly over much of the leaf surface. At higher ozone concentrations, the lesions extend through the leaf blade and overlap even the secondary veins. Affected tissue is completely desiccated, leaving dry, papery, necrotic lesions.

The pattern of injury on plants such as endive and grasses, which lack a palisade tissue, varies only slightly. The outermost mesophyll is again most readily injured, and symptoms develop randomly on both upper and lower leaf surfaces. The mildest symptom on such parallel-veined monocots as corn, barley, oats, wheat, rye, and annual bluegrass is a fine chlorotic stipple between the largest veins. When injury is more severe, the bleached flecks coalesce into elongated, chlorotic interveinal streaks which often extend through the leaf blade (Fig. 18.7). Lesions extend lengthwise, often overlapping smaller veins, and are characteristically pale tan to bleached in color.

Symptoms on woody deciduous ornamentals differ slightly from those on herbaceous species in more closely resembling premature senescence. Upon casual observation, affected leaves appear irregularly bronzed, but closer inspection shows that the bronzing results from a

FIGURE 18.5 *Blisters produced on broccoli leaves following two hours of ozone fumigation at 25 pphm.*

FIGURE 18.6 *Ozone stipple of alfalfa fumigated for 3½ hours at 25 pphm.*

fine, dull yellow to bronzed or brown stippling of the upper leaf surface (Figs. 18.8, 18.9). Fumigation studies by Harper (1967, unpublished) revealed that bronzing and premature senescence characterize the response of many woody species to ozone (Figs. 18.10, 18.11). Fumigations with ozone at 25, 40, 55, or 60 pphm for four hours daily for up to two weeks produced symptoms characterized primarily by bronzing and stippling of the upper leaf surface. At 25 pphm, dense upper-surface stippling developed on all but the youngest leaves of snowberry and sumac (Fig. 18.12). At the same concentration, close to 50 percent of the aspen and bridal wreath leaves showed upper leaf surface bronzing. Ozone injury on aspen was best characterized by irregular black, necrotic lesions (Fig. 18.13). Milder stipple and bronzing appeared on leaves of concord grapes, Gambel oak, Bing cherry, lodense privet, green ash, Hopa crab, honey locust, and English walnut. Different clones of Gambel oak varied greatly in sensitivity, and some of those fumigated were distinctly more tolerant than others. Distinct symptoms of marginal and leaf-tip necrosis developed on Chinese lilac and Elberta peach (Fig. 18.14).

Far more important than the visible symptoms which appeared were the subsequent untoward effects. Premature senescence and early

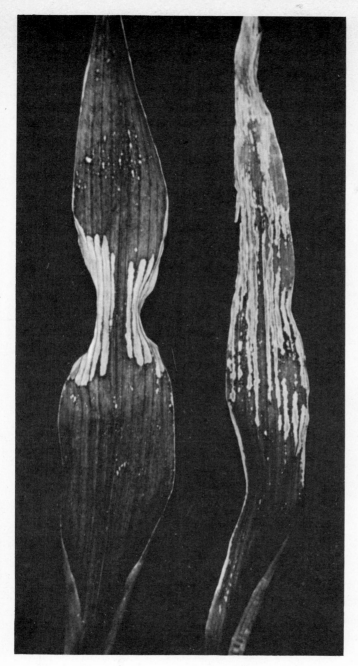

FIGURE **18.7** *Necrosis on corn following 2½ hours' exposure to ozone at 47 pphm.*

FIGURE 18.8 *Ozone stipple and bronzing of Hopa crab-apple fumigated for four hours at 55 pphm.*

defoliation were striking even when no visible symptoms developed. Hence, in instances where plants did not appear to be injured at 25 to 40 pphm, leaves became chlorotic a few weeks later and dropped, while control plants remained healthy. Early senescence has also been reported on bean leaves following prolonged exposure to 5 pphm ozone (Engle and Gabelman, 1967). In ambient, polluted air, leaves of such sensitive plants as the California sycamore become bronzed as early as June and drop soon afterwards, even though the stipple expression is poorly defined.

Chronic injury to flowers has also been reported (Feder and Campbell, 1968). Carnations grown at sublethal ozone concentrations of about 7 pphm for ten days had depressed bud formation. Tip burning developed only after exposure for about eight weeks at these levels.

Fruit damage has also been described (Miller and Rich, 1968), but only at very high concentrations. Apple fruits exposed continuously to ozone concentrations of 1.8 ppm for at least three days developed gray to brown pitted areas around the lenticels. The surface discoloration was often underlain with brown, corky, tissue.

Ozone-induced Disorders

The cause of many important widespread diseases now known to be caused by ozone remained a mystery until late in the 1950s, when elevated concentrations of ozone were discovered in urban atmospheres and fumigations with ozone revealed the range of symptoms the chemical produced. Grape stipple, onion blight, weather fleck of tobacco, and pine blight are among these.

Grape stipple, which seriously threatened the southern California

FIGURE 18.9 *Close-up (15X) of interveinal necrotic flecking on Hopa crabapple fumigated for four hours with ozone at 55 pphm.*

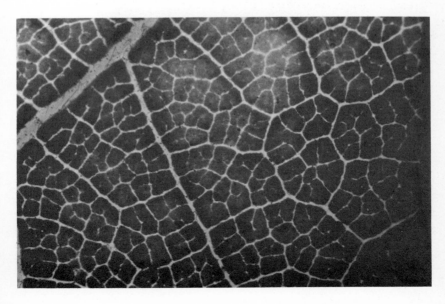

FIGURE 18.10 *Necrosis and bronzing of green ash following 4¾ hours of fumigation with ozone at 60 pphm. Bronzing most characteristic on lower leaflets.*

grape industry in the 1950s, was the first disease established to be caused by ozone. Symptoms of the disease were given as leaf bronzing, yellowing, premature senescence, and leaf fall (Richards et al., 1958). The early symptoms consist of small, discrete, brown to black groups of pigmented palisade cells forming lesions typically 0.1 to

0.5 mm across and bounded by the smallest veins (Fig. 18.15). As these lesions coalesce, larger flecks may develop. The lesions develop early in the growing season on the young, partially expanded leaves, but they are soon followed by bronzing of particularly the older leaves, premature senescence, and defoliation.

Other diseases now known to be caused by ozone can be traced back as far as 1903, when Whetzel (1904) first described *onion blight.* The characteristic burning of the leaf tips is preceded by flecking and tissue breakdown typical of ozone. The disease has occurred erratically over the years on highly sensitive varieties in many remote, nonindustrial areas of the country.

Weather fleck, a serious disease of tobacco in areas extending from Ontario, Canada, through the southeastern United States, has been observed since the 1930s and has been a serious problem since at least 1954 (Heggestad and Middleton, 1959). The necrotic flecking developing over the leaf severely mars its quality, rendering it worthless for cigar wrapping.

Fleck symptoms vary in detail depending on the variety, but are

FIGURE 18.11 *Bronzing of Gambel oak following four hours of ozone fumigation at 10 to 15 pphm.*

FIGURE 18.12 *Ozone-induced stipple predominantly on older leaves of snowberry following four hours of fumigation at 25 pphm.*

characterized by the appearance of initially light gray to tan flecks on the upper leaf surface (Fig. 18.16). Soon after exposure irregular, water-soaked lesions appear; by the next morning, these areas become bluish-black and finally brown. The irregularly shaped flecks are one to several millimeters in diameter and tend to be confined between the smaller veins or form a reticulate pattern following the small veinlets. The flecks are slightly sunken and bordered by narrow, diffuse, chlorotic tissue comprised of cells in varying stages of collapse.

Another disease, referred to as *white pine needle blight, emergence tip burn,* or *chlorotic dwarf,* was discovered in 1961 to be caused by ozone at concentrations so low that no urban source was needed. Outbreaks are likely to be widespread throughout the range of white pine distribution and are found in remote areas of Canada, throughout the American Northeast, and in the Appalachian mountain area, where needle blight has been known since at least 1908 (Clinton, 1908).

The cause of needle blight was established when Berry and Ripperton (1963) placed potted grafts from ozone-sensitive white pine trees in chambers in which only carbon-filtered air was admitted. Similar plants (ramets) from the same parent trees were placed in an

FIGURE 18.13 *Black, necrotic lesions on quaking aspen following four hours of ozone fumigation at 25 pphm.*

FIGURE 18.14 *Marginal necrosis of Elberta peach caused by six-hour ozone fumigation at 80 pphm.*

identical chamber but subjected to ambient, unfiltered air. A third group of plants was left outdoors. The day after a natural oxidant concentration of 7 pphm was recorded, severe tip burn developed on all plants exposed to unfiltered air. Subsequent fumigations with ozone produced identical symptoms of emergence tip burn.

Early in the summer, while the needles are emerging from the elongated new shoots, those of certain sensitive trees may develop pinkish spots each about the same distance back from the tips in a given fascicle. The spots coalesce into bands of dead tissue. The tips of the needles beyond this zone generally die, producing a leaf tip necrosis. Under milder conditions, or on more tolerant individuals, the symptoms may not progress more than to cause a chlorotic or necrotic banding, or general chlorosis and dwarfing of the needles in the absence of necrosis. The specific name chlorotic dwarf has been given this condition in Ohio, but ozone is still considered the causal agent (Dochinger et al., 1965).

While the above is regarded as the characteristic response of white pine to ozone, considerable variation exists among individual trees in

FIGURE 18.15 *Ozone-stipple lesions on Mission grape leaf (photo courtesy of B. L. Richards).*

FIGURE 18.16 *Ozone injury of Bel B tobacco following two hours' fumigation at 10 pphm. Lesions are typically larger than on other varieties.*

a population. Berry has found that necrosis, stipple, general chlorosis, and dwarfing can all be produced by ozone, depending on the genetics of the tree and its environmental predisposition; more variation exists in the response of different individuals to an air pollutant than of a single individual to different pollutants.

Another major disease caused by ozone, called "X disease," *chlorotic decline,* or *ozone needle mottle* of pine, appeared in southern California forests in the 1950s and was described by Parmeter et al. in

1962. The earliest expression of the disease is the appearance of minute, chlorotic flecks on the older needles (Fig. 18.17). Chlorosis gradually becomes more intense while affected leaves fade in color, becoming bronzed, then necrotic, and finally dropping prematurely, so that all but the current season's needles are lost as contrasted with the normal three to five years' retention. This loss of needles gives the branches a bare, tufted appearance (Fig. 18.18). As decline progresses, needle tip necrosis and shoot dieback develop, and shoot growth is markedly reduced; the length and number of needles is reduced until

FIGURE 18.17 *Chlorotic stipple of Ponderosa pine needles following six hours' ozone fumigation at 80 pphm.*

FIGURE 18.18 *Chlorotic decline of pine showing the difference in susceptibility among individuals and the bare, tufted appearance of sensitive trees. (Photo courtesy of Paul Miller)*

only a few dwarfed needles are found at the branch tips. In the final stages of decline, the lower, and gradually upper, branches die back over a period of three or more years.

Daily peak oxidant concentrations of 9 pphm were found in stands of declining ponderosa pine; but monthly maximum concentrations often reached 20 to 25 pphm in the same area.

Relative Sensitivity of Species

Despite the excellent studies which have been conducted on the physiological mechanisms of ozone toxicity, little information is available regarding the relative sensitivity of species or their response under field conditions. The few field observations of plant sensitivity which have been made in areas of intense oxidant pollution have had the obvious disadvantage that the precise nature of the pollutants was unknown. The plant response reported may have been due to ozone; but it may also have been caused by PAN, nitrogen oxides, ethylene, combinations of these, or pollutants still unkown.

Hill et al. (1961) reported the relative sensitivity of some thirty-two agronomic species to ozone in controlled fumigation studies. These data are summarized in Table 18.1. Preliminary fumigations in Utah (Harper, unpublished) have shown the relative sensitive of a few woody native and ornamental species (Table 18.2).

TABLE 18.1. *Relative sensitivity of herbaceous species to ozone**

Sensitive	Intermediate	Resistant
Spinach	Turnip	Beet
Tobacco	Parsley	Geranium
Alfalfa	Carrot	Annual bluegrass
Wheat	Petunia	Cucumber
Oat	Endive	Cotton
Barley	Parsnip	Lettuce
Rye		
Orchard grass		
Red clover		
Radish		
Bean		
Corn		
Tomato		
Broccoli		

* Adapted from A. C. Hill et al., 1961.

TABLE 18.2. *Sensitivity of woody plants to ozone*

Sensitive*	Intermediate	Resistant
Fragrant sumac	Chinese apricot	Siberian elm
English walnut	Pyracantha	European beech
Thornless honey	Thompson seedless	European white
locust	grape	birch
Chinese lilac	Blue-leaf honeysuckle	Bartlett pear
Bing cherry	Silverberry	Virginia creeper
Lodense privet		Norway maple
Concord grape		Viburnum
Quaking aspen		American linden
Gambel oak		Bur oak
Snowberry		
Hopa crab		
Green ash		
Bridal wreath		

* Sensitive category injured below 30 pphm for four hours; intermediate injured at 40 pphm for four hours; resistant damaged at 53–56 pphm for four hours.

Environmental Factors Influencing Sensitivity

Virtually every meteorological and edaphic factor in the environment acts on the welfare of the plant and influences its predisposition and sensitivity to ozone. Factors which regulate the degree of stomatal opening, namely soil moisture and light, exert the greatest influence.

Every environmental factor studied by Heck et al. (1966) was found to influence ozone sensitivity. Tobacco and bean plants provided with an eight-hour photoperiod were injured approximately three times as much as when provided with a sixteen-hour photoperiod. Plants were also more sensitive in the summer months than during the winter, when days were shortest and light intensity low. Plants grown in sandy, well-aerated soils were more sensitive to ozone than plants in heavier soils. Time of day influenced sensitivity, with the greatest injury being produced during midday.

The importance of soil moisture was demonstrated during the course of fumigations by Harper and Hill (unpublished) at the University of Utah. Plants provided with ample soil moisture were visibly injured by a two-hour ozone fumigation at 15 pphm, while plants provided with less water but never allowed to dry out were injured only when concentrations reached 45 to 60 pphm. It was speculated that adequate soil moisture might act in maintaining the turgidity of guard cells and limiting stomatal closure.

Similarly, light also helps regulate the degree of stomatal opening, and even slight shading tends to close the stomata and reduce the amount of ozone entering the leaf. Plants must be exposed to light before exposure to ozone in order for injury symptoms to begin development.

Leaf maturity also influences ozone sensitivity. The first leaves to mature are more sensitive and more readily marked than overmature or younger, rapidly expanding leaves. Ozone lesions develop rather uniformly over recently matured leaves, while only the tips of expanding leaves are damaged.

Menser et al. (1963*a, b*) speculated that the sensitivity of recently matured tissues may be related to the stomatal mechanism. Ozone is admitted to the leaf interior only through active stomata. Microscopic examination of the old, basal leaves showed stomata here were nearly always closed; some stomata were nonfunctional, with degenerative guard cells. Stomata of young leaves were more fully developed and functional at the tips than at the middle or base of the leaf. Stomata on the youngest leaf tissues, however, had not fully developed. Leaf development and stomatal behavior corresponded to the development of ozone symptoms.

A few studies have attempted to relate mineral nutrition to ozone sensitivity. Brewer et al. (1961) studied the influence of nitrogen, phosphorus, and potash nutrition of spinach and mangels. Nitrogen nutrition had a pronounced effect on the sensitivity of spinach and mangels to ozone. Injury increased significantly as the nitrogen level was increased. Addition of either phosphorus or potash without the other produced poor top growth that was relatively resistant to ozone. When both were added, top growth and injury both increased. The importance of these environmental factors lies in their role of predisposing plants to injury and modifying the threshold of injury. These and other factors not yet studied, such as the concentrations of nonessential as well as essential minerals in the soil, may greatly increase the sensitivity of plants to the point where damage may occur at ozone concentrations far lower than normally expected.

18-3 REDUCING OZONE INJURY

The principal approach to reducing crop losses from ozone has been to develop and use inherently resistant varieties. This has been largely successful in the tobacco industry, where many varieties are available, and relatively tolerant varieties have been selected in response to persistent damage of common commercial varieties.

There have also been many efforts to prevent injury by spraying sensitive plants with protective chemicals. Particulate minerals as diverse as clay and sulfur can protect plants against ozone damage. Dried particulate charcoal, diatomaceous earth, and powdered ferric oxide have all been found effective in decomposing the atmospheric ozone before it can enter the leaf (Jones, 1963).

Kendrick et al. (1962) tested a number of common fungicides, as well as antioxidants, as to their capacity to protect plants from subsequent ozone exposure. The most effective materials were dithiocarbamates and mercaptobenzothiozole and its derivatives. The degree of protection was directly related to the concentration of the chemical; the longevity of protection was related to the cumulative period of exposure to ozone. After ten hours of accumulated exposure, injury gradually increased. Action of the protectant was local rather than systemic, and the toxicant appeared to be deactivated at the leaf surface. Leaves remained free from injury only when the chemicals were applied to the lower leaf surface, and then only at the section that was actually covered with the chemical.

Rich and Taylor (1960) tested other antiozonants and found manganous 1,2-naphthoquinone-2-oxime to be especially effective in protecting tomato foliage in the field. The similar manganous and cobaltous chelates of 8-quinolinol were also highly effective.

18-4 INTERACTION OF OZONE AND OTHER POLLUTANTS

A single air pollutant is never present in the atmosphere alone; many different pollutants occur in combination with each other in any polluted atmosphere. Where photochemical pollutants are involved, they are constantly interacting, and the number of different chemicals present is exhaustive. Many are toxic, a few are harmless, and the hazards of most are simply unknown. Little is known about the possible mutual effects of two or more of these chemicals acting in combination.

Ozone, PAN, sulfur dioxide, nitrogen oxides, and other chemicals all occur together. Does the presence of one pollutant influence the toxicity of another? Do two or more pollutants interact on the plant so that together they are more toxic or less toxic than either alone?

Research at the U.S. Department of Agriculture Station at Beltsville, Maryland (Menser and Heggestad, 1966), suggests that "synergistic" action results when ozone and sulfur dioxide are present together. Tobacco plants were fumigated with ozone at concentrations which are normally noninjurious. When sublethal amounts of sulfur dioxide were added, however, typical ozone injury was produced. The synergistic metabolic effects of sublethal concentrations of various pollutants may be even more significant but have not yet been studied.

BIBLIOGRAPHY

Bennett, J., 1969. Effects of ozone on leaf metabolism. Ph.D. thesis. Univ. of Utah, 96 pp.

Berry, C. R., 1964. Differences in concentrations of surface oxidant between valley and mountaintop conditions in the southern Appalachians. *J. Air Pollut. Contr. Assoc.* **14**:238–239.

———— and L. A. Ripperton, 1963. Ozone a possible cause of emergence tipburn. *Phytopathol.* **53**:552–557.

Brewer, R. F., F. B. Guillemet, and R. K. Creveling, 1961. Influence of N-P-K fertilization on incidence and severity of oxidant injury to mangels and spinach. *Soil Sci.* **92**:298–301.

Clinton, G. R., 1908. Pine Blight. *Conn. Agr. Exp. Sta. Bull.*, pp. 353–355.

Dochinger, L. S., C. E. Seliskar, and F. W. Bender, 1965. Etiology of chlorotic dwarf of western white pine. *Phytopathol.* **55**:1055.

Dugger, W. M., Jr., Jane Koukol, and R. L. Palmer, 1966. Physiological and biochemical effects of atmospheric oxidants on plants. *J. Air Pollut. Contr. Assoc.* **16**:467–471.

Engle, R. L., and W. H. Gabelman, 1967. The effects of low levels of ozone on Pinto beans, *Phaseolus vulgaris* L. *Proc. Amer. Soc. Hort. Sci.* **91**:304–309.

Feder, W. A., and F. J. Campbell, 1968. Influence of low levels of ozone on flowering of carnations. *Phytopathol.* **58**:1038–1039.

Heck, W. W., J. A. Dunning, and I. J. Hindawi, 1966. Ozone: Nonlinear relation of dose and injury in plants. *Science* **151**:577–578.

Heggestad, H. E., and J. T. Middleton, 1959. Ozone in high concentrations as cause of tobacco leaf injury. *Science* **129**:208–210.

————, 1966. Ozone as a tobacco toxicant. *59th Annu. Meet. Air Pollut. Contr. Assoc. Paper* no. 66, p. 46.

Hill, A. C., M. R. Pack, M. Treshow, R. J. Downs, and L. G. Transtrum, 1961. Plant injury induced by ozone. *Phytopathol.* **51**:356–363.

————, and N. Littlefield, 1969. Ozone. Effect on apparent photosynthesis, rate of transpiration, and stomatal closure in plants. *Env. Sci. Tech.* **3**:52–56.

Jones, J. L., 1963. Ozone damage: protection for plants. *Science* **140**:1317–18.

Junge, Christian E., 1964. "Air chemistry and radioactivity," chap. 1, sec. 4, pp. 37–59, Ozone. Academic, New York.

Kendrick, J. B., Jr., E. F. Darley, and J. T. Middleton, 1962. Chemotherapy for oxidant and ozone induced plant damage. *Int. J. Air Water Pollut.* (Pergamon Press) 6:391–402.

MacKnight, Martha, 1968. Effects of ozone on stomatal activity of pinto bean. M.S. thesis, Univ. of Utah, 79 pp.

McFarlane, J. C., 1966. The effect of ozone on cell membrane permeability. M.S. thesis, Univ. of Utah, 59 pp.

Menser, H. A., H. E. Heggestad, O. E. Street, and R. N. Jeffrey, 1963a. Response of plants to air pollutants. I. Effects of ozone on tobacco plants preconditioned by light and temperature. *Plant Physiol.* 38:605–609.

———, ———, and ———, 1963b. Response of plants to air pollutants. II. Effects of ozone concentrations and leaf maturity on injury to *Nicotiana tabacum. Phytopathol.* 53:1304–1308.

——— and ———, 1966. Ozone and sulfur dioxide synergism: Injury to tobacco plants. *Science* 153(3734):424–425.

Middleton, J. T., and Arie J. Haagen-Smit, 1961. The occurrence, distribution, and significance of photochemical air pollution in the U.S., Canada and Mexico. *J. Air Pollut. Contr. Assoc.* 11:129–134.

Miller, P. M., and S. Rich, 1968. Ozone damage on apples. *Plant Dis. Rep.* 52:730–731.

Ordin, L., 1965. Effect of air pollutants on cell wall metabolism. *Arch. Environ. Health* 10:189–194.

Parmeter, J. R., R. V. Bega, and R. Neff, 1962. A chlorotic decline of ponderosa pine in Southern California. *Plant Dis. Rep.* 46:269–273.

Rich, S., and G. S. Taylor, 1960. Antiozonants to protect plants from ozone damage. *Science* 132:150–151.

Richards, B. L., J. T. Middleton, and W. B. Hewitt, 1958. Air pollution with relation to agronomic crops. V. Oxidant stipple of grape. *Agron. J.* 50:561.

Thomson, W. W., W. M. Dugger, Jr., and R. L. Palmer, 1966. Effects of ozone on the fine structure of the palisade parenchyma cells of bean leaves. *Can. J. Bot.* 44:1677–1682.

Todd, G. W., and Betty Propst, 1963. Changes in transpiration and photosynthesis rates of various leaves during treatment with ozonated hexene or ozone gas. *Physiol. Plant.* 16:57–65.

Whetzel, H. H., 1904. Onion blight. *Cornell Univ. Agr. Exp. Sta. Bull.* 218, pp. 135–163.

SELECTED REFERENCES

Darley, E. F., and J. T. Middleton, 1966. Problems of air pollution in plant pathology. *Ann. Rev. Plant Pathol.* 4:103–118.

———, E. W. Nichols, and J. T. Middleton, 1966. Identification of air pollution damage to agricultural crops. *Bull. Dept. Agr. Calif.* 55, pp. 11–19.

Hepting, G. H., and C. R. Berry, 1961. Differentiating needle blights of white pine in the interpretation of fume damage. *Int. J. Air Water Pollut.* **4**:101–105.

————, 1964. Damage to forests from air pollution. *J. Forest.* **62**(9):630–634.

Lee, T. T., 1965. Sugar content and stomatal width as related to ozone injury in tobacco leaves. *Can. J. Bot.* **43**:677–685.

Linzon, S. N., 1966. Damage to Eastern white pine by sulphur dioxide semimature tissue needle blight, and ozone. *J. Air Pollut. Contr. Assoc.* **16**:140–144.

MacDowall, F. D. H., E. I. Mukammoh, and A. F. W. Cole, 1966. Ozone dose and plant injury. *Science* **153**(3743):1552.

Menser, H. A., and O. E. Street, 1962. Effects of air pollution, nitrogen levels, supplemental irrigation, and plant spacing on weather fleck and leaf losses of Maryland tobacco. *Tob. Sci.* **6**:165–169.

————, 1964. Response of plants to air pollutants. III. A relation between ascorbic acid levels and ozone susceptibility of light-preconditioned tobacco plants. *Plant Physiol.* **39**(4):564–567.

————, H. E. Heggestad, and J. J. Grosso, 1966. Carbon filter prevents ozone fleck and premature senescence to tobacco leaves. *Phytopathol.* **56**:466–467.

Miller, P. R., J. R. Parmeter, Jr., O. C. Taylor, and E. A. Cardiff, 1963. Ozone injury to the foliage of *Pinus ponderosa. Phytopathol.* **53**:1072–1076.

Povilaitis, Bronium, 1962. A histological study of the effects of weather fleck on leaf tissues of flue-cured tobacco. *Can. J. Bot.* **40**:327–331.

Rich, S., 1964. Ozone damage to plants. *Ann. Rev. Phytopathol.* **2**:253–266.

Stephens, E. R., 1961. The photochemical olefin-nitrogen oxides reactions. Symposium on chemical reactions in the lower and upper atmosphere, pp. 51–69. Intersci. New York.

————, 1966. Reactions of oxygen atoms and ozone in air pollution. *Int. J. Air Water Pollut.* **10**:649–663.

Thompson, C. R., O. C. Taylor, M.D. Thomas, and J. O. Ivie, 1967. Effects of air pollutants on apparent photosynthesis and water use by citrus trees. *Environ. Sci. Tech.* **1**:644–650.

CHAPTER NINETEEN

POLLUTANTS
OF LESS RENOWN:

ETHYLENE, AMMONIA, NITRO-GEN DIOXIDE, CHLORINE, AND DUSTS

The pollutants previously discussed are responsible, on the whole, for over 95 percent of all plant damage attributed to air pollutants, but ethylene, ammonia, nitrogen dioxide, and chlorine can be just as harmful in local situations. These pollutants are not related to each other chemically nor in their distribution; they are treated in a single chapter only because of their limited significance and the paucity of relevant information concerning their phytotoxicity. Air pollution incidents involving these pollutants have been sufficiently sporadic and local in distribution that the demand for extensive studies has not been as great as for work on more widespread pollutants. Yet contamination of the environment by nitrogen oxides and ethylene is rapidly increasing; incidents of ammonia and chlorine pollution associated with industrial wastes and accidents are also becoming more frequent, and the associated symptoms must be recognized in order to make an accurate diagnosis and evaluation of an air pollution situation.

19-1 ETHYLENE

Ethylene, an unsaturated hydrocarbon ($CH_2^=CH_2$) of the olefin series, has been recognized as an air pollutant since at least 1871 (Kny; cited in Sorauer, 1914), but the potential threat of this gas could not be foreseen. At that time the only recognized sources of ethylene were coal and illuminating gases escaping from utility systems. Manufactured illuminating gas generally contains about 3 percent ethylene gas (Crocker and Knight, 1908)—more than enough to be toxic. Gas leaking from the underground pipes often damaged roots of nearby trees, while the escape of gas in such confined quarters as greenhouses led to accumulation of toxic quantities which damaged or killed greenhouse plants.

Illuminating gas provided a local source of ethylene, but the true potential of ethylene as a general pollutant was recognized only after its presence was discovered in automobile exhausts (Rohrbauch, 1943). Automobile engines provide the major source of ethylene in metropolitan atmospheres, but additional ethylene is contributed by combustion of natural gas, coal, or wood, or as a blow-off gas from the cracking of natural gas in petrochemical plants. As a waste product from the incomplete combustion of almost every organic substance, ethylene is released whenever organic substances are burned. Agricultural burning, home incinerators, and other refuse fires all contribute. Even smoke from cigars and cigarettes has been reported to release sufficient ethylene to damage plants (Knight, 1913). Ethylene is a natural plant product, and trace amounts emanate from growing plants. But the ethylene contributed by plants is likely to be significant only in closed fruit storage.

Measurements of ethylene concentrations in ambient atmospheres are relatively scarce, since analysis requires special equipment which is not readily available. But by using gas chromatographic methods, ethylene concentrations as low at 1 ppb can be detected (Bellar et al., 1962). Using this method, ambient concentrations have been determined in a few areas. Stephens and Burleson (1967) reported concentrations of 3 to 15 pphm during air pollution episodes in Riverside, California. Data from the San Francisco Bay area are especially noteworthy because of the extensive greenhouse crop losses in this region. Here, ethylene content of samples collected over a ten-month period averaged 7 pphm. Over a four-month period in Los Angeles ethylene averaged 20 pphm, with peaks reaching 142 pphm.

It is surprising that a natural plant product like ethylene is so highly toxic to plants—so toxic, in fact, that it may well be responsible for many of the crop losses near metropolitan areas. The ambient levels of ethylene may be 100 times those necessary to injure sensitive species. Yet even at far lower concentrations ethylene is important, even essential, to normal plant development, playing a role in regulating normal maturation and abscission of leaves and fruits. Some of the most important processes regulated by ethylene include (1) shoot elongation, (2) fruit ripening and coloration, (3) leaf and flower senescence and abscission, (4) root induction, and (5) lateral growth. Acting as a growth regulator, ethylene appears to be related to the normal processes influencing dormancy. Excessive ethylene too early in plant development hastens these processes and accelerates respiration, causing premature senescence and abscission. These responses of plants to ethylene suggest that it is not directly caustic to tissue but acts in the role of a growth regulator, upsetting the normal growth and aging patterns.

The deleterious effects of prolonged exposures to low ethylene concentrations, well under 1 ppm, were recognized following tremendous economic losses to the commercial orchid growers of Los Angeles. The orchid industry in that area was once a multimillion dollar enterprise, but during the 1940s a strange group of symptoms appeared which rendered the affected flowers unmarketable. The first indication of the disease was that the showy, petaloid sepals of the flower became chlorotic and dried from the tips down just as they were emerging from the bud. When growers later became more familiar with the disease, they found the earliest expression was the slight yellowing of the bud. Often the bud dropped before opening, but if it remained, the sepal tips were typically translucent to necrotic. Even when necrosis was not perceptible, the flower remained fresh for only a few days rather than the ten days to two weeks required in the orchid trade. Davidson (1949) found that the *dry sepal disease* as it became known, could be produced by as little as a twenty-four-hour exposure to 0.2 pphm ethylene or 5 pphm for six hours. Even 0.2 pphm (2 ppb) destroyed the commercial value of the bloom, dropping the price 25 percent.

The costly loss of orchids caused many growers to relocate beyond the urban limits of Los Angeles and San Francisco, but those who remained suffered losses estimated to exceed $100,000 a year in the San Francisco area alone (James, 1964). Additional greenhouse crops, and many field-grown floricultural species, sustained equal or even greater losses. The demise of the cut flower industry in Los Angeles and San Francisco was due at least in part to ethylene. Carnations, snapdragons, roses, camellias, and chrysanthemums were among the most critically damaged.

The *"sleepiness"* disease of carnation caused an estimated $700,000 loss to growers in 1963. This ethylene-caused disease has been known since 1908 (Crocker, 1908) but has never been more destructive than since automobile pollution became intense. Petals turn yellow and wither, buds remain partly or wholly closed, and flowers open slowly if at all.

Studies concerning the air pollution potential of ethylene received a substantial stimulus following severe losses to cotton and other economic crops near a Texas Gulf Coast polyethylene manufacturing plant. Ethylene concentrations in the field ranged from 4 to 300 pphm depending on distance and direction from the source. The cotton crop within a mile of the facility was completely destroyed. Less severe field symptoms included leaf abscission, scattered seedling death, vine-like growth habit, and fruit abscission. Vegetative and reproductive organs were malformed—symptoms much like those produced by 2,4-D (Hall et al., 1957). Young cotton plants were severely defoliated in fields close to the ethylene source. Further distant, plants exhibited "leaf puckering," reddening, and chlorosis; apical dominance was lost, and axillary buds were forced. Flowering was stimulated, but most fruits abscised.

In 1962, Heck and Pires fumigated ninety-three different species with ethylene at concentrations of 2.5 to 10 ppm. Symptoms on sensitive species consisted of yellowing and occasionally necrosis of the lower leaves, chlorosis of the flower buds, and inhibition of terminal growth, with an increase in the number of nodes and young leaves. These plants recovered rapidly when fumigation was stopped, but leaves formed during treatment never expanded normally.

A still milder type of expression was characteristic on grasses and other more tolerant species. Plants developed no apparent injury, but leaf elongation was permanently and measurably inhibited. Abscission in the absence of chlorosis occurred on some plants. The oldest small branchlets of juniper and arborvitae dropped readily, and light shaking after fumigation caused more than 50 percent of the older branchlets to fall. More than half of the arborvitae cones dropped on touch. Floral injury appeared on all twenty-two species which were in bloom during fumigation. Death and/or abscission were noted in eight species. Such an effect could result in the complete loss of economically valuable crops.

The profound effects ethylene exerts on the normal development of plants are due to its hormonal nature in being absorbed, transported, and assimilated readily by the vegetative organs. Abscission, for instance, is normally caused by the balance between auxin and ethylene in the petiole (Hall, 1952). An artificial increase in ethylene through pollution would upset this balance and cause early defoliation.

19-2 NITROGEN OXIDES

When internal combustion engines with improved efficiency and reduced hydrocarbon emissions are manufactured, some types of air pollutants will diminish—but not the oxides of nitrogen. Nitric oxide (NO), nitrogen dioxide (NO_2), and nitrogen tetroxide (N_2O_4) will all remain as significant contributors to air pollution. Any combustion process which produces high temperatures in the presence of nitrogen and oxygen will yield nitrogen oxides as combustion products. Over 70 percent of the NO_2 in the atmosphere originates from automobile exhausts; but significant contributions are also added by combustion of fuels used in heat and power generating, and in domestic operations. Range-top burners, ovens, furnaces, etc. all produce approximately 3 to 9 lb of NO_2 for every 100 million Btu produced (Faith, 1956). While the individual emissions seem negligible, the large number of sources in a city gives it increased significance.

Ambient air measurements (Schuck et al., 1966) have shown the average monthly concentration in Los Angeles to be 20 to 30 pphm during the winter and about 10 pphm in the summer. Concentrations on smoggy days may be much higher; on some occasions, nitrogen oxide concentrations in the Los Angeles area have exceeded the "alert" level of 3.0 ppm (Bush et al., 1962). Katz (1956) contrasted the "normal" nitrogen oxide level of about 8 pphm on clear days to the 40 pphm occurring on days of reduced visibility. Highest levels occurred at night since daytime concentrations were reduced by participation of nitrogen oxides in photochemical reactions. Also, with leaf stomates open during the day, substantial amounts are absorbed and removed by plants.

Long before nitrogen oxides were recognized as urban air pollutants, plant damage from nitric acid fumes was described. Leaf tips of grains and conifer needles turned bright yellow in color; margins of broad-leaved species became brown, and dark spots appeared on affected leaves.

Nitrogen dioxide injury in the field has been reported (Janone, 1954) in Italy to cause necrotic stem lesions, defoliation, dieback, and death of peach and black locust trees. Symptoms appearing on peach and cherry leaves consisted of small necrotic lesions between the veins. Field injury has never been reported in the United States, but laboratory studies have provided some preliminary data concerning the effects of nitrogen oxides on plants.

MacLean et al. (1967) described some of the drastic effects of exceedingly high NO_2 concentrations. At 10 to 250 ppm for a duration of ten minutes to eight hours, NO_2 caused rapid tissue collapse, necrosis, and up to 100 percent defoliation. The early "waterlogging" symptoms appearing first on the upper leaf surface indicated that palisade cells were most readily injured. Intercostal areas of tissue, limited to the upper surface, also collapsed and became necrotic.

Benedict and Breen (1955) found that nitrogen oxides caused two main types of injury. The primary symptoms were the collapsed, white to tan, irregular-shaped, small, necrotic lesions appearing between the large secondary veins near the leaf margin. A second expression consisted of a waxy, shiny green "coating" on the leaves of certain species. The glossy sheen developed on both upper and lower leaf surfaces of pigweed, the upper surface of mustard, and the lower surface of Kentucky bluegrass. The high concentrations of 20 to 50 ppm nitric oxide used are not likely to be encountered in the field; but the less striking symptoms produced by lower concentrations may be far more important in the long run.

Taylor and Eaton (1966) attempted to learn the response of plants to concentrations more likely to be encountered in ambient atmospheres. At concentrations of about 2 ppm for eight hours, tissue collapse and bleaching, especially of the leaf tips and margin areas, developed on tobacco plants. Exposures of four hours failed to cause visible damage. Pinto bean plants were not visibly damaged at concentrations below about 10.0 ppm for four hours. The intercostal necrotic markings resembled lesions caused by high SO_2 or ozone concentrations. Continuous exposures to concentrations of 30 to 50 pphm caused a gradual change in the appearance of the plants without any necrosis. The most striking change was the downward cupping of leaves and darker green leaf color of treated plants. Growth was noticeably suppressed within a week. The reduced plant weight and leaf size was associated with an increase in total chlorophyll content accounting for the deeper green leaf color. Tomato plants responded in the same way.

Tingey (1967), in studying the foliar absorption of NO_2 by plants, found that separate three-hour fumigations with concentrations of 6.0 and 4.0 ppm NO_2 were sufficient to visibly injure oat and alfalfa leaves during the day, causing bleached interveinal lesions (Figs. 19.1,

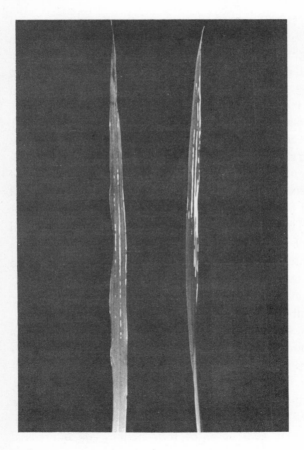

FIGURE 19.1 *Bleaching of oats two days after three hours of fumigation with 6 pphm. NO_2.*

FIGURE 19.2 *Marginal and intercostal bleaching of Ranger alfalfa following three hours of fumigation with 6 ppm NO₂.*

19.2). In the dark, 8.0 ppm were required to injure oats, but only 2.0 ppm injured alfalfa. Studies of stomatal responses revealed that oat stomata quickly closed in the dark, but alfalfa stomata remained open when NO_2 was present. Even in daylight, the stomatal aperture was 50 percent larger in the presence of NO_2 than in filtered air. Measurements of NO_2 uptake also showed a continued absorption by alfalfa in the dark.

Tingey also found that NO_2 may have a cumulative effect. Six-hour fumigations for each of three successive days at 16.0 pphm failed to cause injury, but on the fourth day alfalfa leaves showed characteristic necrotic markings. Basal leaves were flaccid and the older leaves often dehydrated, brittle, and brown. Pitting sometimes developed on one or two of the basal leaves. Cells of the lower epidermis were dead, and many of the spongy mesophyll cells were partially collapsed. No correlation was found between damaged mesophyll cells and the stomata; rather, the damage was generalized throughout the spongy mesophyll area. No damage to palisade mesophyll cells was observed.

Bush et al. (1962) have shown that nitrogen oxide concentrations are noticeably higher in afterburner exhaust than in exhausts where no afterburner is used. If more afterburners are used to control exhaust emissions, increased NO and NO_2 concentrations—together with the "atypical" smog injury to plants consisting of desiccation and necrosis of lower leaves—are likely to occur. Bush believes these changes are already taking place in the Los Angeles area.

19-3 AMMONIA

Ammonia has been implicated as an air pollutant at least since 1893 (cited in Sorauer, 1914), but not in the usual sense. The damaging ammonia has generally escaped from refrigerator precooling systems of cold storage rooms. Less frequently, anhydrous ammonia used as a fertilizer has escaped from tanks, or escaped during manufacture of anhydrous ammonia or nitric acid, and damaged nearby vegetation.

Ammonia is also found in the atmosphere above metropolitan areas, having originated from various combustion processes including domestic incineration and automobile engines. Ambient concentrations from these sources as high as 20 pphm have been reported (Cholak, 1952).

The concentrations of ammonia in the atmosphere following industrial incidents have not been measured but would be expected to approach 100 percent near the source and diminish with distance. The best idea of the concentration necessary to cause injury has been obtained from fumigation studies which have demonstrated the type and extent of injury produced at varying concentrations.

Sorauer summarized the knowledge regarding ammonia phytotoxicity up to 1914. Symptoms consisted of dark spots or the complete blackening of leaves; leaves of barley were bleached white, while rye and wheat developed rusty spots, especially along the margin. Azalea and chestnut leaves developed dark brown lesions between the veins. These turned black the next day, and the leaves later dried up. Red azalea flowers developed white, wedge-shaped spots; white-flowered varieties developed small brown spots. Old spruce needles became black, while young needles became reddish-yellow.

Slight amounts of the gas produced color changes in the pigments of fruit and vegetable skins. Apples, pears, peaches, plums, and onions have most frequently been reported injured. The gas readily penetrates the stomates or enters through cracks in the epidermis. The alkaline reactions cause a breakdown in the pigments and subsequent discoloration (Ramsey, 1953).

Fruit blemishes may also develop, consisting of brownish to black lesions and general browning and softening of peach, banana, and onion; prominent, corky, darkened lenticels of apple and peas may occur. Brennan et al. (1962) found that concentrations of 15 percent ammonia for five hours were sufficient to injure peach fruits, while slightly higher concentrations damaged apples. On plums, brownish or blackish blotches were localized. The dry, outer scales of red onions became greenish or black, while scales of yellow or brown onions turned dark brown.

Thornton and Setterstorm (1940) were among the first to fumigate whole plants with ammonia. They found that 40 ppm for one hour could markedly injure tomato, sunflower, buckwheat, and coleus; concentrations down to 8.3 ppm for five hours produced slight injury. Mildest symptoms consisted of marginal chlorosis, with the remainder of the leaf remaining normal.

Far lower concentrations were used by Benedict and Breen (1955) in studying the response of common weed species. Mustard and sunflower, the most sensitive species, developed necrotic lesions when exposed to 3 ppm ammonia. Fumigation studies in Utah (unpublished) showed that typical symptoms consisted of light tan to dark brown bronzing of the upper epidermis and dark brown, irregular-shaped lesions sometimes extending through the leaf. Lesions were densest at the leaf tip and along the margins. Symptoms varied among the several species studied. Interveinal bronzing and chlorosis typified ammonia injury on cocklebur (Fig. 19.3), while symptoms on corn consisted of interveinal bleaching (Fig. 19.4). Prickly lettuce leaves showed a red to brownish discoloration of the upper epidermis, particularly at or near the margins, with the larger veins usually remaining green. Dandelion, aster, and knotweed leaves exhibited a rather uniform reddening of the upper surface. Tip and margin necrosis characterized symptoms on Canada thistle, perennial ryegrass, milkweed, and white sweet clover. Irregular necrotic lesions developed on apricot which sometimes overlapped the small veins. The earliest symptom to develop on alfalfa was a bright green, glossy, water-soaked appearance followed by a mild interveinal chlorosis. More severe injury consisted of tip and marginal bleaching to light tan necrosis which extended inward between the larger veins (Fig. 19.5).

FIGURE 19.3 *Interveinal chlorosis of cockelbur following ammonia fumigation.*

19-4 CHLORINE

Chlorine escaping into the atmosphere from industrial accidents or manufacturing processes, or from leaking valves of chlorine cylinders used in water or sewage treatment and disposal processes, has both sporadically and chronically damaged plants and man for many decades. Additional chlorine is emitted in manufacturing glass and the processing of various products containing it. The few reports of plant damage and the paucity of details regarding such incidents have belied their frequency.

Little is known of the ambient-air chlorine concentrations following spillages, and measurements of levels near manufacturing plants releasing chlorides are not generally available. Concentrations at the source can be presumed to be close to 100 percent, but the chemical is rapidly diluted with distance. A general idea of the ambient concentrations can be obtained by knowing the amount of chlorine required to injure sensitive plants and relating this to the injury in question. A few fumigation studies have indicated the chlorine concentrations which are harmful (Thornton and Setterstrom, 1940; Benedict and

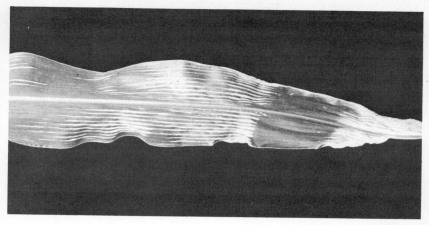

FIGURE 19.4 *Interveinal bleaching of corn following two hours of ammonia fumigation at 40 ppm.*

Breen, 1955; Brennan et al., 1965). The threshold of injury for the more sensitive species such as radish and alfalfa is about 10 pphm for two hours. Slightly less sensitive species, including tobacco, zinnia, onion, corn, mustard, and sunflower, were injured when exposed to 10 pphm for four hours (Brennan et al., 1965). The majority of species studied were damaged by 50 to 80 pphm chlorine for four hours.

Studies at the University of Utah (unpublished) showed that the symptoms of chlorine injury were of four basic types: chlorosis, stipple, necrosis, and reddening.

FIGURE 19.5 *Chlorosis and necrosis on alfalfa following two hours of fumigation with 40 ppm ammonia.*

The mildest symptom on broad-leaved species, consisting solely of chlorosis, is indistinct and has been observed on relatively few species, including only holly, hibiscus, and barberry. Chlorosis was mostly marginal but sometimes extended interveinally. More often, chlorosis was accompanied by stipple or progressed into a general bronzing and necrosis symptom, as on oak (Fig. 19.6).

Stipple reminiscent of ozone injury was by far the most characteristic expression of chlorine toxicity (Fig. 19.7). Stippling consisted of tiny islands of bleached, dead cells appearing between the smallest veinlets or tracheids. The stipples were most concentrated along the leaf margins and leaf tips and along the veins, although the veins themselves remained green in sharp contrast to the bleached background.

Necrosis often accompanied stippling on broad-leaved species but occasionally appeared as the sole symptom. Where both symptoms developed, necrosis was generally marginal, with the stipple appearing in a zone between necrotic and healthy tissue (Fig. 19.8). Where necrosis alone developed, it was mostly along the margin but tended to extend inward between the larger veins; sometimes it consisted of intercostal lesions tan to bleached in color or was scattered over the

FIGURE 19.6 *Chlorosis and necrotic fleck (left) of white oak following chlorine fumigation. Leaves on right show progressive general chlorosis and bronzing.*

FIGURE 19.7 *Chlorotic stipple of petunia following chlorine exposure.*

leaf (Fig. 19.9). The bleaching characterizing injury on grasses and cereals was most severe at the leaf tip and extended down the margins sometimes the full length of the leaf.

Leaf reddening was pronounced only on *Polygonum,* Johnson grass, and dandelion. Injury was characterized by the reddish discoloration of irregular, diffuse areas of the upper leaf surface, particularly toward the leaf tip and margin.

Symptoms were most pronounced on the middle-aged and older leaves and usually appeared on both leaf surfaces. Where symptoms were limited to one surface, it was always the one possessing stomates. There was a strong tendency for leaves to drop soon after exposure to chlorine.

Necrosis was the most prominent expression of injury on pines, fir, and spruce. Striking, severe injury consisted of bright reddish-brown necrosis extending halfway down the needle or further (Fig. 19.10). The coloration alone was not distinctive from that caused by other air pollutants, but the presence of a narrow band of stipple about ¼ in. wide subtending the necrotic tissue was definitive for

FIGURE 19.8 *Marginal necrosis and chlorotic fleck of Italian prune following chlorine fumigation.*

diagnosis. Lower concentrations of chlorine favored development of more chlorotic stipple and less necrosis.

Histological studies of damaged leaves of broad-leaved species revealed several stages of tissue and cellular degeneration. The mildest symptom, observed in cells adjacent to normal-appearing cells, was the more amorphous shape of the chloroplasts. This was followed by a breakdown of the chloroplast membrane, diffusion of chlorophyll throughout the cytoplasm, plasmolysis with the protoplast separating from the cell wall, and finally complete cellular collapse.

Efforts have been made to correlate the accumulation of chlorine in the leaf with injury. Such a correlation exists when chlorine is absorbed through the roots, but attempts to relate atmospheric chlorides with injury have been inconsistent.

The limited data available concerning relative sensitivity of species to chlorine indicate the following to be among the more sensitive: catalpa, buckeye, alfalfa, radish, dandelion, petunia, chrysanthemum, lodgepole and Scots pine, and blue spruce.

19-5 DUSTS

No clear line of demarcation exists between gaseous and particulate air pollutants. Particles as tiny as 0.01 μ in diameter may be classified as particulates or aerosols, and so may particles several microns in diameter. However, the larger particles soon settle to the ground, and their persistence in the atmosphere is brief. The most prevalent and persistent suspended solid and liquid aerosols fall within the size range from about 0.01 to 100 μ. In general terms, aerosols are also called "dust" or "smoke." The term dust is usually applied only to solid particles, while smoke refers to any product of combustion.

Particulate matter of all types pollutes the atmosphere. Partic-

FIGURE 19.9 *Chlorine-induced necrosis of Russian olive.*

FIGURE 19.10 *Bright, reddish-brown needle-tip necrosis of pine following chlorine exposure.*

ulates may consist of such natural contaminants as fungus and algal spores, pollen, and ordinary soil dust. Man's activities add to the atmospheric dust loading. The pollutants include smoke from industrial combustion and manufacturing processes, oxides from refining operations, and dusts from construction. McCrone Associates (1967) divided particulate materials into three broad groups: wind erosion particulates, industrial dusts, and combustion products. The first of these, consisting mostly of biological substances, soil, rock, and minerals, is not likely to be harmful to plants. The second group, consisting of products of metal refining, foundry operation effluents, and cement dust may be phytotoxic in some instances, but these industrial dusts soon settle out by gravity and present largely local problems. Combustion products, except for fly ash, are largely gaseous and have been discussed previously. As gases, they readily enter the plant and are highly toxic, but when the gases condense, or agglomerate, to form aerosols, their phytotoxicity diminishes rapidly.

Before discussing the toxicity of dusts, it is helpful to understand their concentrations and composition in the atmosphere. The traditional measurement of dust fall—tons deposited per square mile—fails to give the dust concentration at any one time and has generally been replaced by micrograms per cubic meter, which provides a preferred measure of suspended matter. The particulate loading may range from less than 30 $\mu g/m^3$ in completely rural areas to well over 200 $\mu g/m^3$ in large urban regions. In an era when coal was widely used in households and little effort was made to control industrial smoke, dust fall over cities in the United States often exceeded 1,000 $\mu g/m^3$, or, in earlier terms, ranged between 100 to 350 tons/mi²/mo.

The amount of dust is not nearly so important to the plant's welfare as its composition. Unfortunately, early reports of smoke and dust damage were concerned more with the tons of dust settling out than the composition. For this reason, reports of the tolerance of plants to city smoke are misleading. Smoke and soot are general terms which denote a multitude of undefined components. Where their origin is coal, as is so often the case, the principal effluent is sulfur dioxide, but this is accompanied by various other sulfides, by metallic elements such as lead, copper, and iron, and also by arsenic, nitrogen oxides, polycyclic hydrocarbons, and various other organic pollutants. These combine in various agglomerates and are dispersed in particles usually under 1 μ in size.

With a few exceptions, these particles themselves are relatively inert and harmless. Their main threat exists when they are present in concentrations sufficient to plug the stomata or smother the leaf, partially or completely preventing gas exchange. Plants with a pubescent surface most likely to catch the soot would be expected to be most sensitive. Arthur Ruston (1921) doubted that conifers could survive in areas where the annual soot deposit exceeded 50 tons/mi². One of the principal undesirable effects was premature defoliation. Leaves of ash trees dropped six to eight weeks early in areas where 539 tons/mi² of dust were deposited. Ruston attributed this effect to the choking of the stomatal openings with a tarry deposit. Over 75 percent of the stomata were more or less choked this way on conifers grown in polluted areas. Chemical analysis of plant tissues and the atmosphere in this region revealed that plants had been exposed to arsenic and sulfur dioxide as well as the "smoke." Hence it is impossible to distinguish the true cause of the smoke injury.

In 1934, O. E. Jennings of the University of Pittsburgh discussed smoke injury to shade trees and the relative sensitivity of a few common varieties. The smoke contained a number of different materials, including sulfur oxides. Consequently, this widely used sensitivity list may denote SO_2 tolerance more than smoke tolerance. Jennings mentioned SO_2 as an important constituent and reported that the levels "as low as 3.0 to 4.0 ppm" which occurred might be damaging. Unfortunately, this misleading list is still widely used by horticulturists for smoke tolerance even where oxidants are the more damaging pollutant.

Another particulate pollutant which has received some attention is cement dust. The dusts emanating from cement kilns contain varying quantities of different components, particularly oxides of calcium, potassium, and sodium, with lesser amounts of silica, aluminum, iron, manganese, magnesium, and sulfur. The principal constituent, however, is calcium oxide, which may comprise over 30 percent of the total effluent.

Cement dust was reported to be harmful to vegetation in California as early as 1909 (Pierce, 1909, 1910). A thick crust of dust was demonstrated to interfere with light absorption and subsequent starch formation. Darley (1966) observed that alfalfa plants subjected to moderate amounts of cement dust were stunted and had fewer leaves than plants in another portion of the field receiving no dust. Darley also reviewed work by Lacrenies and Piquer which showed that yields of bean and tomato plants subjected to cement dust were reduced 40 and 13 percent respectively in the absence of leaf injury. The yield loss was attributed to the shading effect of the dust. In Germany, Bohne (1963) reported a marked reduction in the growth of poplar trees following a doubling of production by a cement plant a mile away. Pines near the factory had been killed.

Darley (1966) applied cement dust in concentrations of 0.6 to 3.8 g/m^3 to bean leaves for eight- to ten-hour periods for two to three days. He found that the average CO_2 exchange following dusting was reduced over 30 percent. In addition to interfering with gas exchange, two of the three different types of dusts tested caused wilting and leaf necrosis. The greater toxicity of one dust was attributed to the concentrated amount of potassium chloride in this dust. Work in Germany indicated that the calcium content, expressed as calcium oxides (CaO), exerted the greatest influence on toxicity. Dusts containing over twenty-four percent CaO were more injurious than those having less than 24 percent.

Appreciable quantities of particulate pollutants are also contributed by field burning of agricultural brush, debris, and refuse. The burning of agricultural wastes to rid orchards and vineyards of dead trees and prunings, to remove stubble and straw from grain fields, and to clear range lands of brush is a standard procedure. This open burning contributes a notable burden of dust and gas to the surrounding atmosphere. Hydrocarbons of every description, together with every conceivable material found in plants, are evolved. Some of these, particularly the nitrogen oxides and hydrocarbons, enter into photochemical reactions and threaten plant life in the area. The remaining aerosol wastes are relatively inactive and exert their principal influence on visibility rather than on vegetation.

BIBLIOGRAPHY

(Ethylene)

Bellar, T., J. E. Sigsby, C. A. Clemons, and A. P. Altshuller, 1962. Direct application of gas chromatography to atmospheric pollutants. *Anal. Chem.* **34**:768.

Crocker, W., and L. I. Knight, 1908. Effect of illuminating gas and ethylene upon flowering carnations. *Bot. Gaz.* **46**:259–276.

Davidson, O. W., 1949. Effects of ethylene on orchid flowers. *Proc. Amer. Soc. Hort. Sci.* **53**:440–446.

Hall, W. C., 1952. Evidence of the auxin-ethylene balance hypothesis of foliar abscission. *Bot. Gaz.* **113**:310–322.

————, G. B. Truchelut, C. L. Leinweber, and F. A. Herrero, 1957. Ethylene production by cotton plants and its effects under experimental and field conditions. *Physiol. Plant.* **10**:306–317.

Heck, W. W., and E. G. Pires, 1962. Effect of ethylene on horticultural and agronomic plants. *Tex. Agr. Exp. Sta.* MP-613, pp. 3–11.

Herrero, F. A., and W. C. Hall, 1960. General effects of ethylene on enzyme systems in the cotton leaf. *Physiol. Plant.* **13**:736–750.

James, H. A., 1964. Commercial crop losses in the Bay area attributed to air pollution. Bay Area Air Pollut. Contr. Dist. mimeogr. release, pp. 1–11.

Knight, L. E., and W. Crocker, 1913. Toxicity of smoke. *Science* **37**:380.

Rohrbauch, P. W., 1943. Measurement of small concentrations of ethylene and automobile exhaust gases and their relation to lemon storage. *Plant Physiol.* **18**:79–89.

Sorauer, Paul, 1914. "Non-parasitic disease," pp. 730–732. Record Press, Wilkes-Barre, Pa.

Stephens, E. R., and F. R. Burleson, 1967. Analysis of the atmosphere for light hydrocarbons. *J. Air Pollut. Contr. Assoc.* **17**:147–153.

(Nitrogen Oxides)

Benedict, H. M., and W. H. Breen, 1955. The use of weeds as a means of evaluating vegetation damage caused by air pollution. *Proc. 3d Nat. Air Pollut. Symp.*, pp. 177–190.

Bush, A. F., R. A. Glater, J. Dyer, and G. Richards, 1962. The effects of engine exhaust on the atmosphere when automobiles are equipped with afterburners. *Univ. Calif. Dept. Eng. Rep.* no. 62–63.

Faith, W. L., 1956. Nitrogen oxides. *Chem. Eng. Prog.* **52**:342–344.

Janone, G., 1954. Agricoltura e industria in Liguria con particolare rigriardo a un case di fitotossicita da ipoazatite (Agriculture and industry in Liguria with special reference to a case of NO₂ injury). *Humus* **10**:17–19.

Katz, M., 1956. City planning, industrial-plant location and air pollution, in P. L. Magill, F. R. Holden, and C. Ackley (eds.) "Air pollution handbook." McGraw-Hill, New York.

MacLean, D. C., L. H. Weinstein, and R. H. Mandl, 1967. Effects of acute hydrogen fluoride and nitrogen dioxide exposures on citrus and ornamental plants of central Florida. *60th Annu. Meet. Air Pollut. Contr. Assoc. Paper* no. 67, p. 158.

Schuck, E. A., J. N. Pitts, Jr., and J. K. S. Wan, 1966. Relationships between certain meteorological factors and photochemical smog. *Int. J. Air Water Pollut.* **10**:689–711.

Taylor, O. C., and F. M. Eaton, 1966. Suppression of plant growth by nitrogen dioxide. *Plant Physiol.* **41**:132–135.

Tingey, D., 1967. Foliar absorption of nitrogen dioxide. M.S. thesis, Univ. of Utah, 46 pp.

(Ammonia)

Benedict, H. M., and W. H. Breen, 1955. The use of weeds as a means of evaluating vegetation damage caused by air pollution. *Proc. 3d Nat. Air Pollut. Symp.*, pp. 177–190.

Brennan, E., I. Leone, and R. H. Daines, 1962. Ammonia injury to apples and peaches in storage. *Plant Dis. Rep.* **46**:792–795.

Cholak, J., 1952. The nature of the atmosphere in a number of industrial communities. *Proc. 2d Nat. Air Pollut. Symp.*, Stanford Res. Inst. Los Angeles, Calif.

Ramsey, G. B., 1953. Mechanical and chemical injuries, in "Plant disease," U.S.D.A. Yearbook, pp. 835–837.

Thornton, N. C., and C. Setterstrom, 1940. Toxicity of ammonia, chlorine, hydrogen sulfide and sulfur dioxide gases on green plants. *Contrib. Boyce Thompson Inst.* **11**:343–356.

(Chlorine)

Benedict, H. M., and W. H. Breen, 1955. The use of weeds as a means of evaluating vegetation damage caused by air pollution. *Proc. 3d. Nat. Air Pollut. Symp.*, pp. 177–190.

Brennan, E., I. A. Leone, and R. H. Daines, 1965. Chlorine as a phytotoxic air pollutant. *Int. J. Air Water Pollut.* **9**:791–797.

Thornton, N. C., and C. Setterstrom, 1940. Toxicity of ammonia, chlorine, hydrogen sulfide and sulfur dioxide gases on green plants. *Contrib. Boyce Thompson Inst.* **11**:343–356.

(Dust)

Bohne, H., 1963. Schädlichkeit von Staub aus Zementwerken fur Wald-bestande. *Allg. Forstz.* **18**:7, 107–111.

Darley, E. F., 1966. Studies on the effect of cement-kiln dust on vegetation. *J. Air Pollut. Contr. Assoc.* **16**:145–150.

Jennings, O. E., 1934. Smoke injury to shade trees. *Proc. 10th Nat. Shade Tree Conf.*, pp. 44–48.

McCrone, W. C., R. G. Draftz, and J. G. Delly, 1967. "The particle atlas." Ann Arbor Sci. Publ. Inc., Ann Arbor, Mich., 406 pp.

Pierce, G. J., 1909. The possible effect of cement dust on plants. *Science* **30**:775, 652–654.

————, 1910. An effect of cement dust on orange trees. *Plant World* **13**:12, 283–288.

Ruston, A. G., 1921. The plant as an index of smoke pollution. *Ann. Biol.* **7**:390–403.

SELECTED REFERENCES

(Ethylene)

Abeles, F. B., and R. E. Holm, 1966. Stimulation of RNA synthesis, protein synthesis and abscission by ethylene. *Plant Physiol.* **41**:1337–1342.

Akamine, E. K., and H. I. Sakamota, 1951. Brominated charcoal to prevent fading of Vanda orchid flowers. *Amer. Orchid Soc. Bull.* 20(3), pp. 149–152.

Darley, E. F., W. M. Dugger, J. B. Mudd, L. Ordin, O. C. Taylor, and E. R. Stephens, 1963. Plant damage by pollution derived from automobiles. *Arch. Environ. Health* **6**:761–770.

Hall, W. C., and H. C. Lane, 1952. Compositional and physiological changes associated with the chemical defoliation of cotton. *Plant Physiol.* **27**:754–768.

Heck, W. W., E. Gerald Pires, and W. C. Hall, 1961. The effects of a low ethylene concentration on the growth of cotton. *J. Air Pollut. Contr. Assoc.* **11**:549–556.

Stone, G. F., 1913. Effects of illuminating gas on vegetation. *Mass. Exp. Sta. Rep.*, pp. 45–60.

Thornberry, H. H., 1964. Cattleya blossom brown necrotic streak. *Plant Dis. Rep.* **48**:936–940.

Warner, H. L., and A. C. Leopold, 1967. Plant growth regulation by stimulation of ethylene production. *Bioscience* **17**:722.

(Nitrogen Oxides)

Blewett, Stephen E., 1954. Smog. *Fortnight*, **17**:11–38.

(Chlorine)

Brennan, Eileen, Ida A. Leone, and R. H. Daines, 1966. Response of pine trees to chlorine in the atmosphere. *Forest. Sci.* **12**:386–390.

Miller, E. J., and F. C. Strong, 1940. A case of chlorine gas injury to shrubs, vines, grass, and weeds. *Arborists' News* 5, p. 73.

(Dust)

Czaja, A. T., 1966. The effect of dust, especially cement dust, upon plants. *Angew. Bot.* **40**:106–120.

Steinhubel, G., 1963. Resistance of evergreens to various dusts. *Acta Bot. Acad. Sci. Hung. (Budapest)* **9**:433–435.

CHAPTER TWENTY

PESTICIDES AS PLANT POISONS

Pesticides for controlling unwanted plants, diseases, and insect pests are an integral, vital part of modern agriculture. Every minute of every day and night, insects are chewing, sucking, biting, and boring away at our crops, livestock, timber, and gardens, reducing yields, lowering quality, and increasing production costs. Similarly, fungal and bacterial pests are living on, and deriving their nourishment from, the same plant products intended for man's use, and weeds are competing for space, light, nutrients, and water with the crop species intended for man. Man's success in combating and controlling the many competitive weeds and agricultural pests has been one of the outstanding developments of the twentieth century.

Yet this success has not been achieved without an appreciable threat to man and his environment. The hazards and detrimental effects of many agricultural chemicals might well outweigh the benefits derived were they not used with discrimination and sagacity. While biological activity degrades the majority into harmless end products and incorporates them into normal constituents of the environment, others persist for years or decades, accumulating in dangerous concentrations and providing a continued threat to normal growth.

Sufficient research has now been completed for biologists to be able to apply the chemicals intelligently and derive the vast benefits of their use with a minimum hazard to man and his environment. While the total impact is not completely understood, much work has been done to provide a picture of the adverse effects which might be expected. The many new chemicals released on the market each year have varying effects, but the general mechanisms of toxicity and plant responses remain much the same. The overall effects that a few of the more important, widely used, and best-studied pesticides—including herbicides, insecticides, miticides, and fungicides—may have on plants are discussed in this chapter.

20-1 HERBICIDES

There are many types of weed killers. All are intended to kill plants and can be expected to harm desirable as well as undesirable species when used at higher concentrations. Sometimes the line between weed-killing concentrations and those that kill or damage desirable plants is dangerously thin. Thus extreme caution must be exercised when applying herbicides near desirable species. Herbicides provide a valuable tool in eliminating weeds in nursery and seed beds, around young trees, in lawns, and wherever else it is desirable to reduce competition for nutrients, water, and space. They are valuable in destroying unwanted vegetation wherever it is found. They are used to eliminate undesirable plants in forests and on the range, to facilitate growth of desirable browse and forage plants, to reduce fire hazards in woods and fields, and to prevent regrowth on fire breaks, wilderness roads, fence lines, ditches, drainage canals, and railroad embankments.

The obvious hazard of herbicides lies in getting too much on desirable plants, but the possible persistence and accumulation of toxic residues in soils can be more significant in the long run. Crops planted a year or more after soil treatment may be seriously damaged, and roots of plants which eventually penetrate the treated soil are equally in danger.

Phenoxyacetic Acids

2,4-dichlorophenoxyacetic acid (2,4-D) was the first in a long line of organic herbicides which revolutionized the pesticide industry. Work in the 1940s showed that 2,4-D was translocated in the plant and toxic to it whether taken up from the soil by the roots or absorbed through the leaves. Unfortunately, under semiarid conditions much of the chemical is retained in the soil after treatment and may injure sub-sequent crops, unless special measures are taken to supplement leaching and accelerate biological degradation. In areas having little precipitation, residues tend to persist in the surface layers of the soil, acting as soil sterilants (Crafts, 1949). The persistence of 2,4-D also depends on temperature conditions, soil reactions, and texture. Breakdown is slowest under dry, cold conditions and in alkaline or neutral soils. In areas where leaching must be practiced to remove chemicals, sufficient water must be used to carry them beyond the zone of root growth.

2,4-D is a hormone, or, more specifically, an auxin-type herbicide, acting in the same manner as the natural plant growth regulators. Concentrations of naturally occurring auxins rise and fall to regulate normal plant development. The phenoxyacetic acids (2,4-D, 2,4,5-T, MCPA, etc.) saturate the cells, preventing their orderly normal growth and differentiation; meristematic cells containing surplus auxin stop dividing and elongating but continue to grow broader. Mature plant parts are also affected; cells swell and begin to divide, roots lose their ability to absorb water and salts, photosynthesis is reduced, phloem transport is inhibited, and respiration is accelerated to an extent where the sugar reserves are soon depleted. Mature cytoplasm seems to revert to the immature stage, and the immature cytoplasm is prevented from maturing. The number of ribosomes increases, and the extra RNA they produce causes extra, anomalous growth, 2,4-D is inactivated by combining with proteins, and it is likely that the more tolerant plants contain proteins which bind the chemical, thus deactivating it.

The visible symptoms of injury by 2,4-D and related compounds are easy to recognize. Leaves become thickened, roughened, and leathery in texture (Fig. 20.1). Edges are often crinkled and roll downward, giving a cupped effect. Veins become pronounced, and cell enlargement may be much reduced. Crowded, clearer veins with accentuated dentations often develop. A dense white flecking expression may also appear.

FIGURE 20.1 *2,4-D injury of apricot showing twisted, crinkled, flecked foliage. Normal leaves on right.*

Box elder, tomato, cotton, and grape, are especially sensitive and may develop symptoms even when exposed to 2,4-D drifting in from fields several miles away. Cotton has been injured at least 6 miles from where grain fields were treated. Even where leaves remain free from visible symptoms, 2,4-D has been found to cause an increase in the sugar content of branches and a decrease in sugar stored in the roots (Woodbridge, 1962).

Fruits are equally or more sensitive. Stone fruits such as peaches may be severely damaged when phenoxyacetic acid derivatives are used near orchards. Affected peach fruits ripen prematurely along the suture, and the suture area becomes swollen from abnormal increases in cell size (Fig. 20.2). Swelling is followed by cracking and local ripening and rotting of the suture area. The symptoms can easily be mistaken for suture red spot caused by fluoride, except that where herbicides are responsible, the swollen area is predominantly toward the stem end rather than the tip of the fruit.

The effects of 2,4-D are not confined to the leaf or fruit. At least one bark disorder has been described (Hamilton and Holtzmann, 1958) and shown to be associated with 2,4-D (Fukunaga, 1963). The symptoms consisted of "bleeding" and splitting of macadamia trunks

FIGURE 20.2 *Soft suture of peach caused by 2,4,5-T spray. Swelling, reddening, and rotting primarily at the stem end. Flesh reddening is revealed in the cutaway sections.*

and deposition of dried, resinous exudate. The point of bleeding was slightly raised or consisted of a short, longitudinal crack. The exudate was first amber and resin-like, later becoming black. In severe cases, the cracking and gumming was accompanied by dieback of the terminal shoots. In no case, though, was there any evidence of the leaf curling caused by higher concentrations of 2,4-D. Fukunaga reproduced the disorder by spraying weak solutions of 2,4-D on exposed roots or painting them on the trunk. The disorder was very similar to early stages of fungus and bacterial trunk infections, causing one to wonder how often such symptoms on the other fruit and ornamental trees are associated with herbicides.

Triazines

The trazine compounds include such widely used herbicides as simazine, atrazine, propazine, ipazine, and prometrine. All have a simple ring structure upon which alkoxy, methoxy, or other groups may be substituted. Their low water solubility and long residual effect make the triazines valuable for long-term weed control. Triazines are absorbed primarily through the roots and are applied to soils in concentrations of $\frac{1}{2}$ to 8 lb/acre, depending on the crop, the degree of control desired, and climatic conditions. But safety is a major consideration, and even the lower concentrations can be harmful.

The symptoms of phytotoxicity are a general chlorosis accompanied by shoot and leaf dieback. In red pine seedlings severe chlorosis, needle dieback, seedling distortion, and depressed growth precede mortality. Low concentrations which produce no foliar markings may suppress growth of desired species as much as 50 percent. Soil-incorporated atrazine was particularly toxic to red pine seedlings (Kozlowski et al., 1967), causing nearly 100 percent mortality even when applied at only 2 lb/acre. Other triazine herbicides were less toxic but substantially reduced seedling survival, even when germination was not impaired.

Under field conditions, triazine is injurious to most fruit trees at rates as low as 2 lb/acre. Triazines are reported to reduce sugar synthesis by interfering with photosynthesis. More specifically, simazine inhibits the Hill reaction with as little as 1 ppm, almost completely blocking CO_2 fixation (Ashton et al., 1960).

Triazines are broken down or degraded in the soil principally by microbiological decomposition. High temperatures may hasten degradation, but they also enhance toxicity (Kozlowski et al., 1967). For example, herbicides are usually far more effective in midsummer than in the spring or autumn, since plants are more readily killed at 25 to 30°C than at lower temperatures.

Substituted Ureas

The substituted ureas, principally monuron and diuron, also known as CMU, telvar, karmex and marmex, are widely used as pre-emergence herbicides and soil sterilants. Too often the chemicals persist, or roots of desirable plants grow into treated areas and are damaged. Substituted ureas have the particular advantage of extreme persistence in the soil, lasting through over twenty-five croppings, but this can also prove detrimental.

The mode of action is much like that of the triazines. The chemicals enter the plant through the root and are translocated upward in the transpiration stream. Once in the leaves, they block photosynthesis and cause a variety of effects leading to the death or functional impairment of the tissues. The principal mechanism appears to be in inhibiting the photolysis of water (Hill reaction). Monuron can do this in concentrations so low that one molecule prevents photosynthetic activity of over 125 chlorophyll molecules (Audus, 1964). Monuron appears to enter into association with a molecule essential to photosynthesis, blocking its ability to react normally.

Light energy normally knocks electrons from the chlorophyll molecules which are trapped by electron acceptors such as molecular oxygen. This leaves "holes" in the chlorophyll which must be filled by other electrons. Normally electrons are provided by the cleaving of the water molecule and the passing of electrons through a complicated system of carriers back to the chlorophyll. When a substituted urea is present, this backflow of electrons is blocked. Eventually the chlorophyll molecules become so shot full of holes that they are irreversibly damaged, or oxidized. More specifically, monuron is thought to form a complex molecule with one of the carriers which should return electrons to the chlorophyll. Oxidation of chlorophyll inhibits photosynthesis and limits food production, causing starvation. Hence, growth and development may be suppressed even when no visible symptoms appear.

In sublethal doses, monuron causes loss of apical dominance, retards mitosis in meristems, deforms influorescences, produces aberrant forms of leaves, and reduces shoot and root growth. The residual effect of the chemicals can prevent or impair growth of plants on previously treated land. In lethal doses, monuron causes loss of turgor, chlorosis, progressive dieback of leaves, collapse of young leaves, disorganization of palisade tissue and epidermis, reduced vascular differentiation, and nuclear breakdown. The most clearly defined visible symptom of injury is the extreme chlorosis, almost bleaching, of tissues around the margin of the affected leaves (Fig. 20.3). Chlorosis extends inward rather uniformly about ¼ in. from the margin. A few chlorotic spots may develop just inside this chlorotic border between the veins. When symptoms are more severe, the outer edge of the chlorotic tissue becomes necrotic.

Aliphatic Acids

Aliphatic acids denote open-chain molecules with a carboxyl, or acid, end group. Dalapon, the sodium salt of dichloropropionic acid, is the principal herbicide in this group. It acts as a protein precipitant,

FIGURE 20.3 *Marginal and interveinal chlorosis of cherry caused by systemic herbicides.*

degrading proteins to amino acids. Specifically, dalapon disrupts normal metabolism by competing with the substrates from which pantothenic acid is synthesized. Pantothenic acid, one of the B vitamins, is essential to growth.

Dalapon in the soil may inhibit seed germination and plant growth for thirty days or more, depending on the dosage, soil type, temperature, and rainfall. Dalapon is particularly effective against grass seedlings, but is phytotoxic to a wide variety of plants. Desirable plants may also be injured if the chemical is allowed to contact foliage or roots.

General

Many other herbicides and soil sterilants, including dinitro chemicals, oils, borates, chlorates, arsenicals, carbonates, and picinylic acids, are available and present a hazard to plants when used improperly. New chemicals are constantly being sought and developed. Ideally these should be highly selective and toxic only to certain undesirable species, but this ideal is rarely achieved.

20-2 INSECTICIDES

Insecticides are designed to control insect pests and are not necessarily injurious to plants. However, plant and insect cells share many basic metabolic pathways; chemicals which block a pathway, or are otherwise toxic to an insect cell, may also damage plant cells.

Despite the threat of insecticides to cells of plants, animals, and man, the chemical approach to insect control remains the most practical and effective means of controlling most noxious species. Alternative methods (particularly biological control) are more effective against some pests, and their value will continue to increase in the future, but they are not likely to fully replace chemical methods for many decades, if ever, and the need to recognize plant damage caused by insecticides will persist.

Inorganic Insecticides

These have been largely replaced by organic chemicals, which are more efficient and far less harmful to the environment. But the impact of older, more toxic inorganic chemicals is still being felt. Arsenicals in particular, which are lost from the soil extremely slowly, have been widely used for several decades and provide the classic example of how a persistent chemical can damage plants for many years. Continued use, e.g. against lawn weeds, will cause an accumulation in the soil which will damage plants for years to come. Lawns treated too frequently with arsenicals become pale green to yellow and may develop characteristic necrotic or chlorotic leaf spotting.

Acute arsenic poisoning causes leaves of affected herbaceous and woody plants to turn black, collapse, and die. Chronic poisoning causes the leaves to turn pale green and yellow over a period of weeks. Purple, reddish, or brown spots of varying size develop, especially on the older leaves, together with chlorosis (Fig. 20.4). In severe cases the necrotic areas drop out, leaving a shot-hole appearance. Leaves of peach and apricot trees drop prematurely, so that the tree is open and fruit exposed to sunburning by midsummer. These symptoms may develop on sensitive plant species such as peaches and apples planted twenty years or more after arsenicals stopped being used. Fruit russeting may appear in the absence of leaf markings, and fruit maturation may be abnormally accelerated. Fruits may be small and misshapen, in much the same way as when zinc is deficient.

Arsenic toxicity may be expected when the concentration in the soil exceeds about 75 ppm; over 1,000 ppm may accumulate from a single year's spraying.

Even when no visible effects appear, arsenic may impair such

FIGURE 20.4 *Necrotic and purple lesions associated with chronic arsenic toxicity of apricot.*

metabolic processes as respiration. Studies of arsenic toxicity in the low-bush blueberries are considered indicative of plant responses at sublethal concentrations (Anastasia and Kender, 1966). Significant growth reductions occurred which were correlated with foliar arsenic levels when the leaf concentration exceeded 7 ppm. Veinal necrosis began to appear on youngest leaves of plants treated with 60 ppm arsenic. Seven weekly treatments at this concentration killed many plants.

Sensitivity to arsenic is influenced by the general vigor of plants. Thus a tree bearing a full crop may be more sensitive than a non-bearing tree. This explains why trees may not express injury until they are mature and productive.

Sulfur insecticides and petroleum oils have also been widely used in the past and occasionally are still responsible for leaf chlorosis, burning, and twig killing (Fig. 20.5). But these chemicals are no longer widely used, and chances of plant injury are remote. The inorganic insecticides of the pre-1940s have rapidly given way to organic chemicals which are far more effective, less hazardous, and less likely to be phytotoxic.

Organic Insecticides

CHLORINATED COMPOUNDS   Chlorinated hydrocarbons, made up of carbon, hydrogen, chlorine, and occasionally oxygen, have provided man with his greatest weapon against insects and the diseases and famine associated with them. Epidemics of typhoid and other insect-transmitted diseases of man have been halted by timely use of these chemicals, and both livestock and crops often have been saved from the ravages of insects.

But this group of chemicals, which includes DDT, chlordane, methoxychlor, aldrin, benzine hexachloride (BHC), and dieldrin, to name a few, is persistent and slow to decompose, and they accumulate in the environment over the years.

DDT and its analogs accumulate in the soil following applications as sprays or when used in the soil against soil-borne pests. DDT has been applied to apple and other fruit trees for control of codling moth and other pests since 1944 and has accumulated in soils in much the same manner as arsenic over the preceding twenty years. Gripp and Ryugo (1966) calculated that some 100 to 300 lb/acre of actual DDT has been applied to California apple and pear orchards in this way, and residue was detected in twenty-two of the twenty-three orchards sampled.

FIGURE 20.5 *Lime-sulfur injury on Newtown apple leaves (photo courtesy of D. L. Coyier).*

To determine the persistence of DDT, Clore et al. (1961) in Washington analyzed soil samples ten years after DDT was incorporated. The decrease in the remaining DDT varied only slightly with the initial dosage. Decreases of 89, 79, and 73 percent were recorded when the initial application was 24, 119, and 238 lb/acre of technical DDT. Most of this decrease occurred in the first two years, after which the breakdown was extremely slow. Studies have shown that DDT is broken down at the rate of only 5 percent per year.

The residual effects on plant growth were also studied. Rye, snap beans, alfalfa, and strawberries were grown in soils previously treated with DDT or related chemicals. Straw and grain yields of rye were affected adversely by only 15 lb/acre of technical BHC and were observable as long as five years after application. Bean yields were drastically reduced when over 24 lb/acre of technical DDT was applied. Plant height was also affected adversely, but this result was not as striking as the stand reduction and delayed maturity. Plant growth was markedly reduced by 119 and 238 lb of technical DDT, although no leaf markings developed. There was a characteristic thickening of the main and secondary roots and a loss of the ability to form fibrous roots; yields of strawberry plants were reduced, plants were stunted, and runner development was limited.

Lima beans are also highly sensitive to DDT, and as little as 25 to 50 lb/acre can adversely affect their growth (Chapman and Allen, 1948). A buildup of 200 lb has reduced yields by one-half. Notable depressions in growth of spinach, beet, and tomato have been observed at 25 lb/acre. When DDT was used in combination with aldrin, application at rates of 14 lb/acre inhibited root development and reduced size of tomato, cauliflower, and Chinese cabbage plants (Hagley, 1965).

Dieldrin, a hydrocarbon widely used against soil insects, is considered relatively safe, yet it is more phytotoxic to most crops than DDT or chlordane. While dieldrin tends to persist in the soil, it is generally considered harmless to natural soil microflora at recommended dosages. In hot weather, though, its toxicity increases, and it may cause general chlorosis of lawns or other plants on treated soils.

Dr. L. D. Anderson (1967) of the University of California at Riverside has been evaluating the long-term effects of pesticides on crops including carrot, lettuce, and lima bean. The chemicals studied, including DDT and toxaphene at 20 lb/acre, dieldrin and endrin at 5 lb/acre, and lindane at 1 lb/acre, were applied annually for five years. At the end of this period, plants grown in treated soils were far smaller than the controls, and yields were reduced over 80 percent. After five years of treatment with DDT, Dr. Anderson found it nearly impossible to grow watermelons on treated soils. Accumulations of 40 to 60 ppm of lindane made it impossible to grow many crops.

Chlordane in combinations with other chemicals may act indirectly by damaging the soil microflora, particularly mycorrhizal fungi (Persidsky and Wilde, 1960; Wilde and Persidsky, 1956). Insecticides may kill the mycorrhizal-forming organisms and depress their growth-promoting capacity.

DDT is spread over the earth by wind and water and has thoroughly permeated our environment from pole to pole. It is a particular hazard to animal communities because of the way it accumulates to catastrophic levels in the food chain. Carnivorous and scavenging birds have been hardest hit, but populations of animals feeding on native plants have also been drastically reduced in many areas (Woodwell, 1967).

ORGANIC PHOSPHORUS COMPOUNDS Organophosphates include a large and important group of chemicals vital to successful agriculture. The more common include parathion, tetraethylpyrophosphate (TEPP), malathion, diazinon, guthion, trithion, ethion, demeton, disyston, and thimet. All are capable of damaging plants when applied in excess amounts, when mixed with an incompatible material, when used at a sensitive stage of plant development, or when temperatures are too high. Many of these chemicals enter the plant cells in low concentrations and possess varying degrees of systemic action on the plant cells as well as the insect or mite pests.

Organic phosphates disappear from the environment far more rapidly than the chlorinated hydrocarbon chemicals. Experiments conducted in Wisconsin (Lichtenstein, 1965) showed that 50 percent of the applied doses of parathion and guthion disappeared after twenty days, and 97 percent of the malathion residue was lost after eight days.

Chemicals used for pest control are selected almost exclusively on the basis of their efficiency in killing pests, with some attention to their obvious phytotoxicity and little concern for their physiological effect on growth or production. Yet all of these processes can be affected.

When parathion is applied as a soil treatment, germination is markedly depressed. When applied as a foliar spray or dust when temperatures are over 35°C, leaves of sensitive species such as beans may be severely burned. The typical response is a general necrosis of the most exposed interveinal tissue (Fig. 20.6). Applications of parathion as prebloom and postbloom sprays have caused noticeable yield reductions in citrus (Gressman, 1959).

Heinicke and Foott (1966) studied the effect of diazinon, guthion, and ethion on photosynthesis. Diazinon and ethion markedly reduced the assimilation rate for several days following application. Photosynthesis reduction in diazinon-treated leaves was most marked immediately after treatment, but the effect persisted, and fourteen days after treatment it was only 75 percent that of the control. When applications must be made more frequently than this to achieve adequate pest control, the plant is never able to attain its photosynthetic potential. Leaf chlorosis and fruit russeting may also develop (Fig. 20.7).

Related effects of pesticides have also been noted. Parathion and systox increased the sugar content of young pear leaves (Woodbridge, 1962). Guthion at 0.25 lb/acre was found to cause cotton plants to produce more flowers than normal, but if DDT was added, or DDT used with dieldrin, blossom set was reduced (Hacskaylo and Scales, 1959).

FIGURE 20.6 *Parathion injury of spinach.*

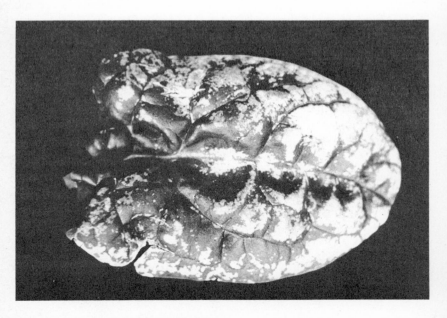

FIGURE 20.7 *Russeting of Golden Delicious apple caused by applications of diazinon during hot weather. Identical symptoms can also be produced by Bordeaux, karathane, and other pesticides.*

Disyston may also be phytotoxic, especially when used with other pesticides. Growth and fruiting of cotton was markedly altered when disyston was used together with phorate (Leigh, 1963). Mild phytotoxicity was evident as "shot-holing" and marginal necrosis of the young leaves. Leaves were larger and more blue-green due to an increase in chlorophyll stimulated by the phosphate pesticides. Plants were taller by as much as 10 to 12 in. in some treatments. Flowering was reduced, though, and maturity hastened.

20-3 FUNGICIDES

Inorganic Compounds

Elemental sulfur is relatively safe and rarely responsible for acute plant damage when used carefully. But in hotter regions where the daily summer high temperature often exceeds 35°C, it may be hazardous. Where used on apples at high temperatures, lesions may develop on the sunny side of the fruit, or leaves and fruit may drop prematurely. Sulfur dusts on apple blossoms may also inhibit pollen germination and thus reduce fertilization, seed formation, and fruit set.

The fungicidal properties of copper have been recognized since

1807. The potential of copper as a protective spray was suggested and tested in 1861, but so much foliage damage resulted that the idea was dropped. Twenty years later, however, copper sulfate was combined with hydrated lime, and the world's leading fungicide, which came to be known as Bordeaux was born. Lime rendered the copper substantially less phytotoxic but did not entirely eliminate the hazard. The most characteristic visible expression of Bordeaux injury usually consists of brown or purple flecks on the leaves of fruit and purpling of the leaf veins and veinlets. The initial purplish-brown leaf spots may enlarge and even drop out, giving a shot-hole effect. On cherries, some ornamentals, and vegetable crops, the vein purpling is followed by chlorosis and early leaf drop. Fruits, particularly apples, develop small brownish flecks which may coalesce to become extensive enough to cause a characteristic roughening or russeting of the surface. In more severe cases fruit may be cracked and malformed.

When visible symptoms are lacking, more insidious responses may be evoked. In general, Bordeaux-sprayed plants tend to be dwarfed, and leaves appear darker green due to the stimulation of chlorophyll production.

Bordeaux has been reported to dwarf cherry fruits and citrus. Since some blossom and fruit drop results from early sprays, however, there would be fewer fruits whose size would develop sufficiently to offset the stunting effects of the spray. In regions where soils are deficient in copper, copper sprays have been reported to have a beneficial effect and cause better growth. Copper sprays have also been blamed for hardening, thickening, and scorching of leaf tissues, stunted, abnormal growth, and reduced yields.

Fungicidal properties of mercury have also been recognized for many decades, but the extreme toxicity of mercury has precluded it as a foliar protectant and limited its use primarily to soil and seed treatments. Even when used to protect bulbs, mercury compounds may be harmful (Gould et al., 1961). Blooms developing from treated bulbs were paler than normal, narrower, and smaller; leaves were occasionally stunted and distorted. Bulbs may develop interveinal necrosis, and flower and foliage primordia may break down.

Organic Compounds

The increasing popularity of organic fungicides is due largely to their low degree of phytotoxicity. While not completely harmless, most are safe when used at the prescribed concentrations. New, more effective materials are continually being developed, supplementing and replacing the older chemicals, but much of the information available concerning fungicide toxicity is limited to the older formulations.

The era of organic fungicides was ushered in by the discovery of dithiocarbamates. These are derivatives of dithiocarbamic acid linked to sulfur, ethylene, or various metals, particularly including iron, zinc, or manganese, to produce fungicides of slightly different properties.

Dithiocarbamates are most likely to be injurious when used in combination with more toxic pesticides. Cristoferi et al. (1966) found that pollen viability and fruit set of apricot and pear were adversely affected by 0.2 percent ziram or 0.15 percent TMTD. Zineb used with iodine reduced pollen viability and fruit set in apples under controlled laboratory conditions, but no differences could be detected in the field. This is characteristic of responses to other pesticides under field conditions, where the influence of other environmental factors is more critical.

Winter (1962) found that yields of young apple trees were significantly reduced by phenylmercury acetate (PMA) sprays. Occasional reductions were caused by other fungicides, including sulfur and dichlone, glyodin, wettable sulfur, and iodine. Yield reductions were associated with a reduction of the number of fruits retained.

Fungicides may also adversely affect such basic processes as DNA, RNA, and chlorophyll synthesis (Shishiyama et al., 1965). Cycloheximide inhibited synthesis of both chlorophyll-*a* and DNA. At higher levels, cycloheximide may cause slight leaf wilting and chlorosis, with yellow lesions developing on the older leaves. Shishiyama suggested that cycloheximide, an antibiotic, may act as a general inhibitor, having a similar effect on living cells of both higher and lower plants.

Phytotoxicity may be influenced by several environmental factors. Daines et al. (1957) showed that the phytotoxicity of captan increased with temperature, especially above 30 to 35°C. Further experiments showed that shading plants immediately before or after captan treatments increased foliage injury over that of unshaded plants. This indicated that phytotoxicity was influenced by internal as well as external factors. The increased injury produced at low light intensity may have been due to the thin cuticle and more open cell structure more readily penetrated by the fungicides.

Ross and Longley (1963) studied the effects of captan, dodine, dichlone, and phenylmercury acetate-captan on the performance of mature McIntosh apple trees. No adverse effects were reported. Tree growth, bloom, and fruit color were unimpaired, yields were substantially increased, and fruit had superior quality.

Karathane, widely used for powdery-mildew control, may be phytotoxic and cause leaf chlorosis and necrosis when applied at higher temperatures. Karathane toxicity may be further accentuated when used in combination with other pesticides.

20-4 PESTICIDES IN COMBINATION

As more and more chemicals are being developed and used, the likelihood of applying them in harmful combinations increases. Two or three chemicals, desirable in themselves, may exert a destructive influence when used in combination. Some cropped soils receive a succession of treatments, starting with a preemergence herbicide, followed by fungicide, nematocide, systemic insecticide, postemergence herbicide, one or more insecticides or fungicides, and possibly a defoliant, all in a single season. Chemicals must be used with the realization that they can, under some conditions, impair production.

Such deleterious effects are illustrated by the use of certain fungicides such as dodine or glycodin with the miticide kelthane (Szkolnik, 1963). Necrotic spotting of apple fruit was observed following the last cover spray of the season. The initial injury was a superficial brown fleck or spot beneath the spray residue. Lesions were dark brown and about 3 to 5 mm in diameter. Lesions enlarged and often involved the underlying flesh, which became brown and desiccated, rendering affected fruits unsalable.

Captan was most toxic when used in combination with phorate or disyston in soil treatments. Seedlings from treated plots showed abnormal root growth, with necrotic lesions.

The hazards of pesticide combinations are especially acute when one of the members is a herbicide. When cotton seeds are treated with systemic insecticides such as phorate or disyston and planted in monuron- or diuron-treated soils, losses may be severe. Deleterious interaction of such chemicals may cause extensive seedling mortality. Hacskaylo et al. (1964) found that while herbicides exhibited some phytotoxicity in themselves, addition of disyston markedly increased phytotoxicity. Growth reduction was accompanied by increased mortality and a progressive increase in leaf necrosis of surviving seedlings. The decided reduction in margin of safety of chemicals could only be diminished by switching to other combinations or reducing the dosage. Combinations of phosphate insecticides with herbicides may be equally hazardous.

The almost limitless possible combinations of pesticides, while largely not deleterious, should always cause the observer to consider the possibility of phytotoxicity when diagnosing a problem involving treated plants.

BIBLIOGRAPHY

Anatasia, F. B., and W. J. Kender, 1966. Arsenic toxicity in the lowbush blueberry. *Hort. Sci.* 1:26–27.

Anderson, L. D., 1967. Symposium on: Damage to plants from soil and water pollution: Effects of pesticides in the soil. Unpublished paper, Annual Meeting American Phytopathology Society, Washington, D. C.

Ashton, F. M., G. Zweig, and G. W. Mason, 1960. The effect of certain triazines on $C^{14}O_2$ fixation in red kidney beans. *Weeds* 8:448–451.

Audus, L. J., 1964. "The physiology and biochemistry of herbicides." Academic, New York, 913 pp.

Chapman, R. K., and T. C. Allen, 1948. Stimulation and suppression of some vegetable plants by DDT. *J. Econ. Entomol.* 41:616–623.

Clore, W. J., W. E. Westlake, K. C. Walker, and V. R. Boswell, 1961. Residual effects of soil insecticides on crop plants. *Wash. Agr. Exp. Sta. Bull.* 627, pp. 1–9.

Crafts, A. S., 1949. Toxicity of 2,4-D in California soils. *Hilgardia* 19:141–169.

Cristoferi, G., 1966. Effetti dei trattamenti antiparassitori eseqiute in fiore su alauni fruittiferi (The effects of fungicidal treatments during flowering on some fruit trees). *Riv. Ortoflor.* 50:225–250 (*Hort. Abstr.* 37:746.).

Daines, R. H., R. L. Lukens, E. Brennan, and I. A. Leone, 1957. Phytotoxicity of captan as influenced by formulation, environment, and plant factors. *Phytopathol.* 47:576–572.

Fukunaga, E. T., 1963. Macadamia bark disorder caused by 2,4-D. *Hawaii Farm Sci.* 12:6–7.

Gould, C. J., W. D. McClellan, and V. L. Miller, 1961. Injury to narcissus from treatment of bulbs with certain mercury compounds. *Plant Dis. Rep.* 45:508–511.

Gressman, A. W., 1959. Effect of timing of parathion and malathion sprays on orange production. U.S.D.A. ARS-33-53.

Gripp, R. H., and K. Ryugo, 1966. DDT soil residues in mature pear orchards: A Lake County survey. *Calif. Agr.* 20(6):10–11 .

Hacskaylo, J., J. K. Walker, Jr., and E. G. Pires, 1964. Response of cotton seedlings to combinations of pre-emergence herbicides and systemic insecticides. *Weeds* 12:288–291.

Hagley, E. A. C., 1965. Effect of insecticides on growth of vegetable seedlings. *J. Econ. Entomol.* **58**:777–778.

Hamilton, R. A., and O. V. Holtzmann, 1958. Macadamia bark disorder discovered. *Hawaii Farm Sci.* **7**:3–4.

Heinicke, D. R., and J. W. Foott, 1966. The effect of several phosphate insecticides on photosynthesis of Red Delicious apple leaves. *Can. J. Plant Sci.* **46**:589–591.

Kozlowski, T. T., S. Sasaki, and J. H. Torrie, 1967. Influence of temperature on phytotoxicity of triazine herbicides to pine seedlings. *Amer. J. Bot.* **54**(6):790–796.

Leigh, T. F., 1963. The influence of two systemic organophosphates on growth, fruiting and yield of cotton in California. *J. Econ. Entomol.* **56**:517–522.

Lichtenstein, E. P., 1965. Problems associated with insecticidal residues in soils, in C. O. Chichester (ed.), "Research in Pesticides." Academic, New York, 380 pp.

Persidsky, D. J., and S. A. Wilde, 1960. The effect of biocides on the survival of mycorrhizal fungi. *J. Forest.* **58**:522–524.

Ross, R. G., and R. P. Longley, 1962. Effect of fungicides on McIntosh apple trees. *Can. J. Plant Sci.* **43**:497–502.

Shishiyama, J., M. Fukutomi, and S. Akai, 1965. Effects of some fungicides on the synthesis of chlorophylls, deoxyribonucleic acid, and ribonucleic acid in onion leaves. *Phytopathol.* **55**:844–847.

Szkolnik, M., 1963. Necrotic spotting of apple fruit from spray combinations of certain fungicides with kelthane. *Plant Dis. Rep.* **47**:79–80.

Wilde, S. A., and D. J. Persidsky, 1956. Effect of biocides on the development of ectotrophic mycorrhizae in Monterey pine seedlings. *Soil Sci. Amer. Proc.* **20**:107–110.

Winter, H. F., 1962. The comparative effects of various fungicide programs on fruit numbers and yields of apple trees. *Plant Dis. Rep.* **45**:560–564.

Woodbridge, C. G., 1962. The effects of some insecticides and 2,4-D on the sugar content of Bartlett pear tissues. *Proc. Amer. Soc. Hort. Sci.* **81**:123–128.

Woodwell, G. M., 1967. Toxic substances and ecological cycles. *Sci. Amer.* **216**(3):24–31.

SELECTED REFERENCES

Ashton, F. M., E. G. Uribe, and G. Zweig, 1961. Effect of monuron on $C^{14}O_2$ fixation in red kidney beans. *Weeds* **8**:448–451.

Crafts, A. S., 1961. "The chemistry and mode of action of herbicides." Interscience Publ., New York, 269 pp.

——— and W. W. Robbins, 1962. "Weed control." McGraw-Hill, New York, 660 pp.

Egler, F. E., 1964. Pesticides in our ecosystem. *Amer. Sci.* **52**:110–136.

Groves, A. B., and H. A. Rollins, Jr., 1966. A study of the comparative influence of captan and dodine fungicides on fruit set, fruit size, return bloom and cropping of the Stayman apple. *Hort. Sci.* **1**:11–12.

Hilton, J. L., L. L. Jansen, and H. N. Hull, 1963. Mechanisms of herbicide action. *Ann. Rev. Plant Physiol.* **14**:353–384.

Kozlowski, T. T., and J. E. Kuntz, 1963. Effects of simazine, atrazine, propazine and eptam on growth and development of pine seedlings. *Soil Sci.* **95**:164–174.

———— and J. H. Torrie, 1965. Effect of soil incorporation of herbicides on seed germination and growth of pine seedlings. *Soil Sci.* **100**:139–146.

Middleton, J. T., 1965. The presence, persistence, and removal of pesticides in air, in C. O. Chichester (ed.) "Research in pesticides," pp. 191–197. Academic, New York.

Miller, C. W., 1966. Dieldrin persistence in cranberry bogs. *J. Econ. Entomol.* **59**:905–906.

Myburgh, A. C., A. J. Heyns, and D. K. Strydom, 1966. The inducement of russetting by pesticides on Golden Delicious apples. *Fruit Grower* **16**:314–318.

Nash, R. G., 1966. Phytotoxic pesticide interactions in soil. *Agron. J.* **59**:227–230.

Ranney, C. D., 1964. A deleterious interaction between a fungicide and systemic insecticides on cotton. *Plant Dis. Rep.* **48**:241–245.

Richardson, L. T., 1960. Effect of insecticide-fungicide combinations on emergence of peas and growth of damping-off fungi. *Plant Dis. Rep.* **44**:104–108.

Roberts, J. E., R. D. Chesholm, and L. Koblitsky, 1962. Persistence of insecticides in soil and their effects on cotton in Georgia. *J. Econ. Entomol.* **55**:153–155.

Schweizer, E. E. and C. D. Ranney, 1965. Interaction of herbicides, a fungicide, and a systemic insecticide on cotton. *Miss. Agr. Exp. Sta. Bull.*, p. 877.

Walker, J. K., Jr., J. Hacskaylo, and E. C. Pires, 1963. Some effects of joint applications of pre-emergence herbicides and systemic insecticides on seedling cotton in the greenhouse. *Tex. Agr. Exp. Sta. Bull.*, p. 2284.

APPENDIX

SCIENTIFIC NAMES OF PLANTS

Acacia	*Acacia* Willd.
Ailanthus	*Ailanthus altissima* (Mill.) Swingle
Alfalfa	*Medicago sativa* L.
Almond	*Prunus amygdalus* Stokes
Apple	*Malus sylvestris* Mill.
Apricot, Chinese	*Prunus armeniaca* L. var. Chinese
Aquilegia	*Aquilegia* sp. L.
Arborvitae	*Thuja* sp. L.
oriental	*T. orientalis* L.
Arnica	*Arnica* L.
Ash	
black	*Fraxinus nigra* Marsh.
European	*F. excelsior* L.
European mountain	*Sorbus aucuparia* L.

German	*Fraxinus holotricha* Kaehne
green	*F. pennsylvanica* Marsh. var. *lanceolata* Sang.
modesto	*F. veluntina* Torr.
red	*F. pennsylvanica* Marsh.
Aspen	*Populus* sp. L.
quaking	*P. tremuloides* Michx.
Aster	*Aster* sp. L.
Avocado	*Persea americana* Mill.
Azalea	*Rhododendron* sp. L.
Banana	*Musa* L.
Barberry	*Berberis* sp. L.
Juliana	*Berberis juliana* Schneid.
Sargent's	*B. sargenti* Schneid.
Basil, sweet	*Ocimum basilicum* L.
Barley	*Hordeum vulgare* L.
Bean	*Phaseolus* sp. L.
bush	*P. vulgaris* L. var. *humilis* Alef.
lima	*P. limensis* Macf.
pole	*P. vulgaris* L.
red kidney	*P. vulgaris* L.
snap	*P. vulgaris* L.
Beech	*Fagus* L.
European	*Fagus sylvatica* L.
Beet	
red	*Beta* L.
sugar	*Beta vulgaris* L.
Begonia	*Begonia* L.
Birch	*Betula* sp. L.
European white	*Betula pendula* Roth
paper	*B. papyrifera* Marsh.
river	*B. nigra* L.
white or gray	*B. populifolia* Marsh.
yellow	*B. lutea* Michx.
Black medic	*Medicago lupulia* L.
Blue-leaf honeysuckle	*Lonicera zabeli* Rehd.
Box elder	*Acer negundo* L.
Boysenberry	*Rubus* L.
Bridalwreath	*Spiraea prunifolia* Sieb. & Zucc.
Broccoli	*Brassica oleracea* L. var. *italica* Plench.
Brussels sprouts	*B. oleracea* L. var. *gemmifera* Zenher Plench.
Buckeye	
California	*Aesculus californica* Nutt.
Ohio	*A. glabra* Willd.
Buckwheat	*Fagapyrum esculentum* Maench.
Burdock	*Anctium* L.

Cabbage	*Brassica oleracea* L. var. *capitata* L.
Chinese	*B. pekinensis* Rupr.
Cacao	*Theobroma cacao* L.
Calendula	*Calendula* sp. L.
Camellia	*Camellia* sp. L.
Cantaloupe	*Cucumis melo* L. var. *cantalupensis* Naud.
Carnation	*Dianthus* sp. L.
Carrot	*Daucus carota* L.
Catalpa	*Catalpa bignoniodies* Walt.
Cattail	*Typha latifolia* L.
Cattleya	*Cattleya* sp. Lindl.
Cauliflower	*Brassica botrytis* L.
Cedar, red	*Juniperus virginiana* L.
Celery	*Apium graveolens* L.
Chard, Swiss	*Beta vulgaris* L. var. *cicla* L.
Cheeseweed	*Malva neglecta* Wallr.
Cherry, Bing	*Prunus avium* L. var. Bing
Chestnut	*Castanea* Mill.
Chickweed	*Cerastium* L.
Chrysanthemum	*Chrysanthemum* L.
Citrus	*Citrus* sp. L.
Clematis	*Clematis orientalis* L.
Clover	
ladino	*Trifolium repens* L.
red	*T. pratense* L.
strawberry	*T. fragiferum* L.
white sweet	*T. repens* L.
yellow sweet	*Melilotus officinalis* Lam.
Cocklebur	*Xanthium pennsylvanicum* Wallr.
Coleus	*Coleus* sp. Lour.
Corn	*Zea mays* L.
Cosmos	*Cosmos* Cav.
Cotoneaster, rock	*Cotoneaster horizontalis* Decne.
Cotton	*Gossypium* sp. L.
Cottonwood	*Populus* sp. L.
Cowpea	*Vigna sinensis* Savi.
Crabapple	
Hopa	*Malus adstringens* Zabel.
flowering	*M. floribunda* Sieb.
Cranberry	*Vaccinum oxycoccus* L.
Creosote bush	*Covillea glutinosa* Rydb.
Cucumber	*Cucumis sativus* L.
Curly dock	*Rumex crispus* L.
Currant, black	*Ribes petiolare* Dougl.
Cyclamen	*Cyclamen* sp. L.
Cypress, Lawson	*Chamaecyparis lawsoniana* Parl.

Dandelion	*Taraxacum officinale* Web.
Dill	*Antheum graveolens* L.
Dock	*Rumex* sp. L.
curly-leaf	*R. crispus* L.
Eggplant	*Solanum melongena* L. var. *esculentum* Nees.
Elderberry	*Sambucus* L.
Elm, American	*Ulmus americana* L.
Siberian	*U. pumila* L.
Endive	*Cichorium endiva* L.
Euonymus	*Euonymus* sp. L.
Fescue	*Festuca* L.
Fig	*Ficus carica* L.
Fir, balsam	*Abies balsamea* Mill.
Douglas	*Pseudotsuga taxifolia* Britt.
Spanish	*Abies pinsapo* Boiss.
white	*Abies concolor* Hoopes
Flax	*Linum* sp. L.
Forsythia	*Forsythia* sp. Vahl.
Fragrant sumac	*Rhus aromatica* Ait.
Fuchsia	*Fuchsia* L.
Geranium	*Geranium* sp. L.
Gerbera	*Gerbera* sp. Cass.
Gifblaar	*Dishapetalum cymosum*
Ginkgo	*Ginkgo biloba* L.
Gladiolus	*Gladiolus* sp. L.
Golden chain	*Laburnum anagyroides* Med.
Goldenrain tree	*Koelreuteria paniculata* Laxm.
Goldenrod	*Solidago* sp. L.
Gooseberry	*Ribes grossularia* L.
Goosefoot, nettle-leaf	*Chenopodium* L.
Grape	*Vitis* sp. L.
Grapefruit	*Citrus paradisi* Macf.
Grass	
blue	*Poa pratensis* L.
buffalo	*Buchloe dactyloides* Englm.
cheat	*Bromus tectorum* L.
crab	*Digitaria ischanemum* Schreb.
Johnson	*Sorghum halepense* Pers.
June	*Koeleria cristata* (L.) Pers.
orchard	*Dactylis glomerata* L.
Rhodes	*Chloris gayana* Kunth.
rye	*Lolium* sp. L.
salt	*Distichlis stricta* Rydb.

sudan	*Sorghum vulgare* Pers, var. sudanesis, Hitchc.
switch	*Panicum virgatum* L.
wheat	*Agropyron* sp. Beauv.
Guava	*Psidium guajava* L.
Gumweed	*Grindelia squarrosa* Dunal.
Hawthorn	*Crataegus* L.
Hemlock, Eastern	*Tsuga canadensis* Carr.
Hibiscus	*Hibiscus* sp. L.
Hickory	*Carya* Nutt.
Holly	
American	*Ilex opaca* Ait.
Mahonia	*Mahonia aquifolium* Nutt.
Honeysuckle, tartar	*Lonicera tartarica* L.
Hops	*Humulus* L.
Hornbeam	*Carpinus caroliniana* Walt
Impatiens	*Impatiens* sp. L.
Iris	*Iris* sp. L.
Ivy, English	*Hedera helix* L.
Japanese box	*Buxus microphylla* Sieb. & Zucc.
Juniper, alligator	*Juniperus pachyphloea* Torr.
Chinese	*J. chinensis* L.
mountain	*J. communis* L.
Rocky Mountain	*J. scopulorum* Sarg.
tam	*J. sabina* L. var. *tameriscifolia*
Kale	*Brassica oleracea* L. var. *acephala.* DC
Kochia	*Kochia* Roth
Lamb's quarter	*Chenopodium album* L.
Larch	*Larix* Mill.
Laurel, sweet bay	*Laurus nobilis* L.
Lemon	*Citrus lemon* Burm.
Lettuce, head	*Lactuca capitata* L.
prickly	*L. serriola* L.
Lilac	*Syringa vulgaris* L.
Lily, Easter	*Lilium eximum* Baker
Lily-of-the-valley	*Convallaria* L.
Linden, American	*Tilia americana* L.
Liquidambar	*Liquidambar styraciflua* L.
Locust, black	*Robinia pseudoacacia* L.
honey	*Gleditsia triacanthos* L.
Loganberry	*Rubus ursinus* Cham. & Schlect. var. *loganbaccus* Bailey

Macadamia	*Macadamia ternifolia* F. Muell.
Maize	*Zea mays* L.
Maple, hedge	*Acer capestre* L.
Norway	*A. platanoides* L.
red	*A. rubrum* L.
silver or white	*A. saccharinum* L.
sugar or rock	*A. saccharum* Marsh.
Marigold	*Tagetes* L.
Marsh elder	*Iva frutescens* L.
Milkweed	*Asclepias syriaca* L.
Mimulus	*Mimulus* sp. L.
Mint	*Mentha* sp. L.
Morning-glory	*Ipomoea purpurea* Lam.
Mulberry	*Morus rubra* L.
Muskmelon	*Cucumis melo* L.
Mustard	*Brassica* sp. L.
Myrtle	*Myrtus* sp. L.
Narcissus	*Narcissus* L.
Nightshade	*Solanum* L.
Ninebark	*Physocarpus opulifolium* Maxim.
Oak, bur	*Quercus macrocarpa* Michx.
Gambel	*Q. gambelii* Nutt.
pin	*Q. palustris* Muenchh.
red	*Q. rubra* L.
white	*Q. alba* L.
Oats	*Avena sativa* L.
Oceanspray	*Holodiscus* Maxim.
Olive, Russian	*Eleagnus angustifolia* L.
Onion	*Allium* L.
Orange, mock	*Philadelphus* sp.
Pagoda, Japanese	*Sophora japonica* L.
Pansy	*Viola tricolor* L. var. *hortensis* DC.
Parsley	*Petroselinum crispum* Nym.
Parsnip	*Pastinaca sativa* L.
Peach	*Prunus persica* Batsch.
Peanut	*Archis hypogaea* L.
Pear, Bartlett	*Pyrus communis* L. var. Bartlett
Peas	*Pisum sativum* L.
Pecan	*Carya illinoensis* Koch.
Pepper	*Piper nigrum* L.
Peppergrass	*Lepidium* sp. L.
Persimmon	*Diospyros virginiana* L.
Petunia	*Petunia* Juss.
Pigweed	*Amaranthus retroflexus* L.

Pine	
Austrian	*Pinus nigra* Arnold.
black	*P. jeffreyi* A. Murr. (and *P. nigra*)
Chinese	*P. tabulaeformis* Curr.
digger	*P. sabiniana* Dougl.
fox	*P. aristata* Engelm.
jack	*P. banksiana* Lamb.
Jeffrey	*P. jeffreyi* A. Murr.
limber	*P. flexilis* James
lodgepole	*P. contorta* Loud.
Monterey	*P. radiata* D. Don
pinyon	*P. edulis* Engelm.
pitch	*P. rigida* Mill.
ponderosa	*P. ponderosa* Laws.
red	*P. resinosa* Ait.
Scots, Scotch	*P. sylvestris* L.
white	*P. strobus* L.
Plane tree	*Platanus* L.
Plaintain	*Musa paradisiaca* L.
Plum, Red or purple leaf	*Prunus cerasifera* Ehah, var. *pissadi* Koehne.
Plumbago or lead wort	*Plumbago* sp. L.
Poplar, Carolina	*Populus deltoides* Michx.
White	*P. alba* L.
Yellow	*Liriodendron tulipifera* L.
Primrose	*Primula* sp. L.
Privet, Lodense	*Ligustrum vulgare* v. pyramidale
Prune tree	*Prunus domestica* L.
Purslane	*Portulaca* L.
Quince	*Cydonia* Mill.
Radish	*Raphanus sativus* L.
Ragweed	*Ambrosia* sp. L.
Ranunculus	*Ranunculus* sp. L.
Raspberry	*Rubus idaeus* L.
Redbud	*Cercis canadensis* L.
Redwood	*Sequoia sempervirens* Endl.
Dawn	*Metasequoia glyptostroboides,* Hu & Cheng.
Rhododendron	*Rhododendron* L.
Rhubarb	*Rheum rhaponticum* L.
Rice	*Oryza sativa* L.
Rockspirea, creambush	*Holodiscus discolor* Maxim.
Romaine	*Lactuca longifolia* Lam.
Rose, tea	*Rosa odorata* Sweet
Rye	*Secale cereale* L.

Salt cedar	*Tamarix pentandra* Pall.
Serviceberry	*Amelanchier alnifolia* Nutt.
Shepherd's purse	*Capsella bursa-pastoris* Medic.
Silverberry	*Eleagnus argentea* Pursh.
Silver lace vine	*Polygonum auberti* L.
Smoke tree	*Cotinus americanus* Nutt.
Snapdragon	*Antirrhinum majus* L.
Snowball, Japanese	*Viburnum tomentosum sterile* K. Koch.
Snowberry	*Symphoricarpos* sp. Blake
Sorghum	*Sorghum vulgare* Pers.
Soybean	*Glycine* max. Merr.
Spinach	*Spinacia* L.
Spruce	*Picea* A. Dietr.
black	*P. mariana* BSP.
Colorado blue	*P. pungens* Engelm.
Norway	*P. abies* Karst.
red	*P. rubens* Sarg.
western white	*P. glauca* Voss.
white	*P. abies* Karst.
Squash, summer	*Cucurbita pepo* L. var. *melopepo* Alef.
Strawberry	*Fragaria* sp. L.
Sumac	*Rhus* sp. L.
Sunflower	*Helianthus* sp. L.
Sweetgum	*Liquidambar styraciflua* L.
Sycamore	*Platanus occidentalis* L.
Tamarack or Eastern larch	*Larix laricina* Koch.
Tea plants	*Thea sinensis* L.
Thistle	*Cersium* sp. L.
Tobacco	*Nicotiana tabacum* L.
Tomato	*Lycopersicon esculentum* Mill.
Tulip	*Tulipa gesneriana* L.
Tuliptree	*Liriodendron tulipifera* L.
Tung tree	*Aleurities fordii* Hemsl.
Turnip	*Brassica rapa* L.
Typha	*Typha* L.
Veratrum	*Veratrum californicum* Dur.
Viburnum, Burkwood	*Viburnum burkwoodii* Busk.
Vicia	*Vicia* sp. L.
Violet	*Viola* sp. L.
African	*Saintpaulia ionantha* Wendl.
Virginia creeper	*Parthenocissus quinquefolia* Planch.
Walnut, English	*Juglans regia* L.
Wheat	*Triticum aestivum* L.
buck	*Fagapyrum esculentum* Moench.

Willow	*Salix* L.
black	*Salix nigra* Marsh.
Yew, Japanese	*Taxus cuspidata* Sieb. & Zucc.
Zinnia	*Zinnia* sp. L.
Zucchini	*Cucurbita* L. var. *medullosa* Alef.

INDEX

INDEX